Revisiting Race in a Genomic Age

Revisiting Race in a Genomic Age

EDITED BY
BARBARA A. KOENIG,
SANDRA SOO-JIN LEE,
AND SARAH S. RICHARDSON

RUTGERS UNIVERSITY PRESS

NEW BRUNSWICK, NEW JERSEY, AND LONDON

LIBRARY OF CONGRESS CATALOGING-IN-PUBLICATION DATA

Revisiting race in a genomic age / edited by Barbara A. Koenig, Sandra Soo-Jin
Lee, and Sarah S. Richardson.
    p.   cm. -- (Studies in medical anthropology)
Includes bibliographical references and index.
ISBN 978-0-8135-4323-9 (hardcover : alk. paper) -- ISBN 978-0-8135-4324-6
(pbk. : alk. paper)
    I. Race.   2. Genomics--Social aspects.   3. Genomics--Moral and ethical aspects.
I. Koenig, Barbara A.   II. Lee, Sandra Soo-Jin, 1966–   III. Richardson, Sarah S.,
1980–
    GN269.R48   2008
    305.8--dc22   2007039073
                                                                            CIP

A British Cataloging-in-Publication record for this book is available
from the British Library.

Visit our Web site: http://rutgerspress.rutgers.edu

Manufactured in the United States of America

CONTENTS

PART THREE
Genetic Ancestry, Identity, and Group Membership

PART FOUR
Race and Genetics in Public Discourse

FOREWORD

LAWRENCE D. BOBO

Stanford University's Center for the Comparative Study of Race and Ethnicity (CCSRE), which I direct, was pleased to co-sponsor the workshops, public forum, and authors' conference that brought together Stanford faculty and students with the distinguished group of international scholars whose work is presented in this volume. The workshop, "Revisiting Race and Ethnicity in the Context of Emerging Genetic Research," led by Barbara A. Koenig and Sandra Soo-Jin Lee from the School of Medicine and organized by graduate student Sarah S. Richardson, was also supported by a Mellon Foundation grant from the Stanford Humanities Center. CCSRE aims to deepen our understanding of how race and ethnicity shape human thinking, identities, institutions, and general social organization. Our research networks are envisioned not merely as occasions for faculty and students to convene to discuss new research across traditional disciplinary and departmental lines, but as means of actually leading to new research and scholarship and, thereby, setting the intellectual agenda in important domains. The "Revisiting Race" group and its work provide an excellent realization of our hopes for the networks: moving from interdisciplinary dialogue to a public conference held on the Stanford campus in January 2006, and culminating in this scholarly volume.

It is especially fitting that the first faculty network to move forward would be the one focused on how developments in the biomedical sciences, especially the mapping of the human genome and its implications for medical treatment, will affect our thinking about race. There are few topics that raise more fundamental questions about how we do and should understand the very concept of "race" than this one. I am confident that this volume, an informative and lively example of interdisciplinary scholarship, will help set the direction for all those interested in the intersection of race and genetics.

The concept of race itself, of course, is a troubled notion, especially where its relation to, or putative embeddedness in, biology is in question. I am by training a sociologist, in particular, one whose work emphasizes social psychology: how individuals are made into social beings, possessed of both agency and the imprint of the settings, groups, and culture in which they have been reared. Like most social scientists, I have grown comfortable in the thought that we study race without actually believing in race. That is, like most sociologists I adopt a constructionist view of race (Bobo & Tuan, 2006). Although often making an appeal to aspects of physical appearance that may be biologically heritable, such as skin color, eye shape, hair texture, facial features and the like, the practical importance of what we understand as race derives from the human capacity to create and assign meanings. For sociologists, what matters about race is social and concerns the patterns of social organization, institutions, and culture that humans build up around the idea, not what biology or nature has endowed us with (Morning, 2005).

Advances in biological science, for the moment, present something of a double-edged sword. On the one hand, some salutary ideas seem to have strong currency. First, most scholars, and here I would include those working in health-related fields and the physical and biological sciences, recognize that the very concept of race, to borrow a phrase, "reflects a marriage of the social and the biological," rather than something obviously given in the nature of things (Cooper, Kaufman, & Ward, 2003, p. 1166). Second, and even more important, there is general acceptance of the fact that the great bulk of genetic variation occurs within conventionally recognized racial or continental groups, not between them. Third, and perhaps more controversially, to quote from Richard Cooper and colleagues, "variation is continuous and discordant with race, systematic variation according to continent is very limited, and there is no evidence that the unit of interest for medical genetics corresponds to what we call races" (2003, p. 1166). All of these trends are encouraging ones inasmuch as they really do constitute a strong and solid case against falling into thinking of our commonsense or lay racial categories as distinguishing people who are somehow different in kind or in fundamental essence.

On the other hand, and here is the problem, there is much unreflective racialized thinking still going on, including among those working on mapping the human genome. Nowhere is this trend more obvious than in the recent approval by the U.S. Food and Drug Administration of a race-based pharmaceutical targeted to African Americans (Sankar & Kahn, 2005). The recent work of Sandra Soo-Jin Lee (2005) points to three trends that should concern us all. First, conventional racial categories still appear to influence the work and thinking of those who are developing, for example, the DNA repositories used by researchers, which continue to be catalogued by racial and ethnic identifiers. Second, Lee notes, there is plenty of evidence that clinical practitioners rely on lay or commonsense race categories and racial cues in their assumptions about needs and risks for certain conditions and health outcomes. Third, as Jonathan Kahn points out in his contribution to this volume, the pharmaceutical industry sees real market potential in developing lines of race-based drugs. Each of these trends works to legitimate and re-inscribe profoundly biological conceptions of race into popular consciousness and culture.

Ironically, let me say that social scientists in our own fashion are rediscovering and reinserting biology into examinations of racial phenomena as well. For example, former American Sociological Association president and eminent scholar Douglas Massey (2004) has argued that there is a clear *biosocial basis* to a number of negative health outcomes that occur disproportionately among African Americans. That is, he builds a three-part argument. As a result of the conjunction of much greater experience of conditions of poverty and residential segregation by race, African Americans experience far higher rates of exposure to violent crime and other forms of debilitating social disorder (Geronimus & Thompson, 2004). As a normal, routine response to stress, the human body produces higher levels of certain chemicals in the body, such as cortisol. The level of stress and stress response is often referred to as *allostatic load*. Those with persistent or chronic exposure to stress will also have chronic or persistently higher levels of cortisol in the bloodstream, which as a chronic state or condition contributes to a variety of unwanted outcomes like hypertension and elevated risk of heart disease. Now, I mention this not to develop his argument in full, nor to scrutinize it in detail, but rather to say that in some quarters it has been viewed as a dangerous claim, as reinserting a potentially destructive biological logic into analyses of race and health inequality.

What is important about what Massey does, however, is that his eye is always fixed on the *social* construction and *social* causation of what we understand as a race-related social condition. He embraces what philosopher and historian of science Ian Hacking might describe as a view of race as a real social phenomenon that has no basis in true differences of kind (2005).

Hacking has outlined a neat set of distinctions among statistically significant race differences, statistically meaningful race differences, and statistically practical race differences. None of these embrace a view of real biological or genetic difference (indeed, he expressly rejects that basis). And, he goes so far as to reject the terminology of *race-based drugs*, preferring instead *race-targeted drugs* because that vocabulary does a better job of underscoring the social perception side over against the assumption of underlying real biological difference seemingly affirmed by the label *race-based*.

It is not my point here to try to resolve these tough questions. But rather I seek to help set the stage for the chapters in this volume. How do we move forward in an age of pharmacogenomics without reinforcing unwarranted assumptions about race and racial difference? Will new advances lead to the hoped-for disappearance of race as a category of human perception and social organization and thinking? How should we respond to apparent correlations of race with the experience of certain illnesses and responsiveness to certain drugs? These tough, tough questions and the sort of forward thinking and analyses we need are what this volume is all about.

REFERENCES

Bobo, L. D., & Tuan, M. (2006). *Prejudice in politics: Group position, public opinion, and the Wisconsin Treaty Rights Dispute*. Cambridge: Harvard University Press.

Cooper, R. S., Kaufman, J. S., & Ward, R. (2003). Race and genomics. *New England Journal of Medicine, 348*, 1166–1170.

Geronimus, A. T., & Thompson, J. P. (2004). To denigrate, ignore, or disrupt: Racial inequality in health and the impact of a policy-induced breakdown of African American communities. *Du Bois Review, 1* (2), 247–280.

Hacking, I. (2005). Why race still matters. *Daedalus, 134* (1), 102–116.

Lee, S. S.-J. (2005). Racializing drug design: Implications of pharmacogenomics for health disparities. *American Journal of Public Health, 95* (12), 2133–2138.

Massey, D. (2004). Segregation and stratification: A biosocial perspective. *Du Bois Review, 1* (1), 7–25.

Morning, A. (2005). Race. *Contexts, 4* (4), 44–46.

Sankar, P., & Kahn, J. (2005). BiDil: Race medicine or race marketing? *Health Affairs Web Exclusive*, W5.455–W5.463.

ACKNOWLEDGMENTS

This volume was made possible by the generosity and support of many people. Tracing this support allows us to thank those centrally involved and also reveals the intellectual genealogy of the book. Our initial effort to bring together Stanford humanities and social science faculty devoted to scholarship on race with geneticists from "across campus" was supported by an unrestricted gift from the Affymetrix Corporation.[1] In June of 2003 we held a two-day meeting to begin what would become a much longer and deeper conversation. Barbara Koenig was a faculty fellow at the Stanford Humanities Center (SHC) in 2002–03. The SHC and the Greenwall Foundation graciously supported her work—in spite of the fact that work on race and genetics was not part of the original research plan—and hosted the event. Joanna Mountain, an anthropological geneticist at Stanford, played a critical role in organizing this first effort, and we thankfully acknowledge her part in generating the discussions that led to this volume.

John Bender, professor of English and director of the SHC, recognizing immediately the significance of interdisciplinary work on race and new genomic technologies, suggested that we apply for Mellon Foundation Faculty Research Workshop funding to continue the dialogue across the campus and to broaden the discussion by including scholars from around the world. At the conclusion of our two-year workshop the SHC graciously hosted the invitational conference that allowed the authors time to meet face-to-face, read each others' draft papers, make suggestions, and, subsequently, undertake revisions. We are grateful to Professor Bender and his very able staff, including associate directors Elizabeth Wahl and Matthew Tiews for their confidence in our vision and assistance at every step.

The primary support for this interdisciplinary endeavor was provided by Stanford's Research Institute for Comparative Studies in Race and Ethnicity (RICSRE), co-directed by Professor Hazel Markus. The Center for the Comparative Studies of Race and Ethnicity (which sponsors RICSRE) serves as the Stanford campus home for interdisciplinary scholarship on race. Funding from RICSRE supported the two-year workshop, the authors' meeting, and a public conference to examine the implications of race-targeted therapeutics held in January 2006. Professor Markus constantly challenged us with critical questions: Why the "turn" to biological explanations of human difference at this particular moment? How are these efforts different from previous efforts to use technology to "prove" race? Associate Director Dorothy Steele, aided by Chris Queen, was central to the intellectual vision and operational success of the "Revisiting Race" research project.

Other collaborators at Stanford provided support, both financial and moral, for our workshop and authors' conference, including Debra Satz from Stanford's Ethics in Society Program, David Magnus from the Stanford Center for Biomedical Ethics, Henry Greely from the Stanford Center for Law and the Biosciences, and Deborah Rhode from the Stanford Center on Ethics. Stanford undergraduate students Carla Fenves and Catherine Barnard and graduate students Ramah McKay and Sarah S. Richardson enthusiastically helped coordinate the workshops. Individuals who participated in the seminar, thus making key contributions to our understanding of the issues, include Anthony Appiah, Donald Barr, Mildred Cho, Troy Duster, George Fredrickson, Chris Gignoux, Joseph Kaufert, Pilar Ossorio, Paul Rabinow, Neil Risch, Quayshawn Spencer, Michael Thaler, Sally Tobin, and Keith Wailoo. The active involvement of members and speakers in the workshop made this project not only possible but rewarding and enjoyable.

Koenig's move from Stanford to the Mayo Clinic in the midst of completing this complex book project was enabled by expert help from Marguerite Strobel Robinson, Jacqueline Wolf, and Ashley Hicks. Student interns Monica Le and Emma Young helped with final preparation of the manuscript. The Mayo team's able assistance is gratefully acknowledged, as is the skillful editorial work of Alan Harwood, who encouraged us to include this volume in the Rutgers University Press book series he founded, Studies in Medical Anthropology, and Adi Hovav, who shepherded us through the publication process.

Finally, the research of Barbara A. Koenig and Sandra Soo-Jin Lee on the intersection of race and genetics has been supported by several grants from the National Institutes of Health. Initially, Koenig served as a research mentor for Lee under the auspices of a National Human Genome Research Institute (NHGRI) post-doctoral fellowship.[2] That collaboration led to what is arguably the first publication to ask how new genomic conceptualizations of race might affect research on health disparities.[3] Currently, Lee is supported by a career development award from the NHGRI, devoted to studies of race, pharmacogenomics, and distributive justice.[4] Koenig's ongoing work on race and behavioral genetics is supported by the National Institute on Drug Abuse.[5]

NOTES

1. Thane Kreiner and Katie Tillman Buck of Affymetrix have been consistently supportive of serious reflection about race and genetics.

2. National Research Service Award, 5F32, HG00221.

3. S. S.-J. Lee, J. Mountain, and B. A. Koenig (2001), "The meanings of 'race' in the new genomics: Implications for health disparities research," *Yale Journal of Health Policy, Law, and Ethics,* 1 (1), 33–75.

4. NIH/NHGRI K01 HL72465.

5. NIH/NIDA, 1R01DA14577.

Revisiting Race in a Genomic Age

Introduction

Race and Genetics in a Genomic Age

BARBARA A. KOENIG, SANDRA SOO-JIN LEE,
AND SARAH S. RICHARDSON

One specific, unified message accompanied the official announcement of the completion of the Human Genome Project: human beings are essentially the same. Human genetic sequences are 99.9% identical; of the 0.1% of the human genome that varies from person to person, only 3% to 10% of that variation is associated with geographic ancestry (Feldman and Lewontin, this volume). This message was nothing new—for decades, the finding that there is greater genetic variability "within groups" than "between groups" had generally been accepted as evidence that the human species is not divided into discrete races. Geneticists publicly interpreted these results as disproving a biological race concept and voiced their hopes that such scientific findings might help deflate racism. Throughout the second half of the 20th century, most historians, social scientists, and race theorists followed suit, affording biology little status in theories of human difference (Fredrickson, 2002). With the advent of the Human Genome Project, race scholar Paul Gilroy (2000) was even inspired to imagine a future in which race would become obsolete as attention shifted from the body politics of skin color, hair texture, and eye shape to the molecular-level biopolitics of the gene. Contrary to these expectations and hopes, post-genomic science has revived the idea of racial categories as proxies for biological differences. In a recent series of papers, population geneticists argue that the genome holds the key to medically and forensically significant biological differences among human racial and ethnic populations. Increasingly, genetic variation among human populations— races, ethnicities, nationalities—is an object of keen biomedical interest.

Revisiting Race in a Genomic Age documents two years of intensive interdisciplinary conversations on race and genetics at Stanford University. From 2003 to 2005, the workshop "Revisiting Race in the Context of Emerging

Genetic Technologies" hosted a series of discussions that examined how race is revisioned through the lens of emerging genetic findings. Recognizing that the dynamic social meanings of race and the rapidly changing reach of genetic technology outpace the resources of any single discipline or observer, the workshop provided a forum for structured interdisciplinary dialogue. Population geneticists, philosophers, physicians, sociologists, psychologists, historians, legal scholars, and anthropologists shared their research and worked to generate new vantage points from which to interpret and analyze human genomic research on race. The workshop took up a range of questions, including the following:

> Does the global pattern of human genetic variation uncovered by emerging DNA technologies correspond to racial categories as traditionally understood?

> How will understandings of race change as more precise, complete, and predictive genomic information becomes available in the future?

> What ways of describing, categorizing, and communicating about samples collected in the course of human population genetic studies minimize reification of racial categories?

> How might genetic analysis of race, including commercially available genetic ancestry tests, affect notions of personal identity and understandings of social group membership?

> In light of health policy goals to eliminate race-based health disparities, what social harms and benefits arise, or might arise, from research linking racial ancestry with genetic information?

> What are the conceptual tools and interdisciplinary arrangements necessary to interpret emerging genomic research on race?

In 2006, workshop participants returned to Stanford to discuss individual papers in preparation for this volume. In the spirit of interdisciplinary dialogue, humanists, social scientists, and natural scientists were assigned to critically discuss a paper from outside of their field. The chapters in this volume represent the outcome of this three-year experiment in interdisciplinary exchange and register the approaches, questions, and issues that surfaced in the course of this intensive survey of new research on race and genetics.

Why Revisit Race and Genetics Today?

"Race Is Seen as a Real Guide to Track Disease" announced a 2002 *New York Times* article, reporting on a new paper by Risch, Burchard, Ziv, and Tang in

Genome Biology. In that paper, Risch and colleagues argued that genetic differences among populations cluster into five major groups corresponding to a "classical definition of races based on continental ancestry" and boldly made the case for the "validity of racial/ethnic self-categorization" in genetic epidemiology research (2002). In challenging the seemingly unified chorus among scholars and scientists that race is not rooted in human genes, the paper was a pivotal event in redirecting the emerging discourse on race and genetics. It was followed eight months later by a *New England Journal of Medicine* paper (Burchard et al., 2003) that not only advocated the use of race in human genetics research but framed its necessity in terms of the public policy goal of mitigating health disparities among racially identified populations.

The vigorous reassertion of the coupling of race and genes, once seen as antiquated, accompanies the shift of human genetic variation research into the genomic age. The new research revives old debates and polarities over the existence of a biological basis for race. But the new genetic race concept is importantly different than its predecessors; so too is the context of the debate. Race in a genomic age raises new and challenging social, political, and ethical concerns, and, we believe, new opportunities for dialogue. Four distinctive developments distinguish the current debate over race and genetics from its predecessors.

First, the completion of the sequencing of the human genome in 2001 commenced the "genomic" age, instituting a shift from relatively limited gene hunting research to whole-genome analysis. Confronted with a vast pool of largely undifferentiated genomic data, researchers must then find ways to make sense of it. As many commentators have noted, much of the new genomics is non-hypothesis-driven. Researchers query the human genome seeking distinctions and patterns as leads for further research. The data derived from the human genome, like any large, multidimensional database, can be probed, inscribed, and organized in various ways. Race has rapidly become a prominent "search tool." Intensive work of this sort has resulted in the identification of slates of genetic markers common within many racial and ethnic populations. These markers may be useful for diagnostic and etiological research for genetic diseases that show different frequencies in different populations. These carefully constructed, racially inscribed sets of markers in the human genome, however, may then become a point of reference for further research, as the data are analyzed and transformed for use by specialists beyond the fields of their creators. Used uncritically and outside of context, these race-inscribed categories may become naturalized, reified, institutionalized ways of conceptualizing the human genome, with serious implications for all subsequent human genome research.

Second, academic race and genetics research is now entering the marketplace. As several contributors document in this volume, race and genetics research increasingly occurs in a corporate context and is driven toward market applications. There is tremendous financial incentive to package "race" as a genetically underwritten commodity. Pharmacogenomics, genetic genealogy services, and forensics are prominent areas of corporate crossover for academic human population variation researchers. Academic researchers concerned to ameliorate racialist interpretations of their work, for example, by using the term "ancestry" instead of race, nonetheless slip into the language of race in their commercial work. The slippage is transparent and inevitable as human population variation research hits the marketplace. The development of proprietary databases and methods for human population variation research raises further concerns about the soundness of the scientific claims underlying this work and poses a challenge to the self-policing scientific standards of the field.

Third, with the increasing specificity and range of claims about racial ancestry made possible by genetic genealogy services, and inexpensive public access to genetic testing via the Internet, research on race and genetics now enters the politics of identity. Recreational genetics introduces new and challenging frontiers in racial identity formation; as such, it also raises distinctive bioethical questions. Testing services, like any commercial venture, sell both a product and a desire for the product. Marketing literature is laced with the discourse of racial purity and racial mixture, as well as constructs such as blood, kinship, ancestry, and homeland. The implications are as yet unclear. Genetic testing may serve to complicate notions of racial purity or to build them up. As ancestry testing becomes cheaper and more widespread, new configurations of racial and national identity may emerge. At the policy level, genetic race verification services have potentially serious implications for community concepts of kinship and nationhood. In the case of entitlements that are tied to race, such as affirmative action, genetic ancestry testing may inflame long-standing debates about eligibility and the social recognition of race as a class. In all of these areas, the technology of biological race verification will change the terms of debate and analysis.

Finally, today genetic research on race increasingly takes place in a medical context. Throughout much of the 20th century, human population variation genetics was most closely associated with anthropological efforts to reconstruct the history of human migration. This research succeeded in offering impressive corroboration for the "out of Africa" hypothesis of human colonization of the globe and demonstrated the association between time, geographical distance, and genetic variation. In a departure from this anthropological context, today the goals of "personalized medicine" and alleviation of "health disparities" drive social investment in genetic research on human population variation. Research on genetic variation among racial populations is widely pursued as a stepping stone to a future goal of therapies tailored to individual biogenetics, or personalized medicine. Pharmacogenomics, or the search for genomic markers that may help physicians determine safe and effective drug dosage, is the first likely application of personalized medicine. As the pharmaceutical industry seeks marketable technologies to patch over an unexpected post-genomic drought in medical breakthroughs, pharmacogenomics has become a particularly attractive investment. Converging with this trend, increased government interest in alleviating health disparities, which often fall along racial lines, has also directed resources toward research on genetic differences among races. While the discourse of health disparities once focused primarily on differences in health outcomes and access to quality health care, health disparities now fuels investment in research, for instance, on the genetic causes of asthma among African Americans and Hispanics. In the pharmaceutical industry, the promise of remedying health disparities has also been used to lend a politically correct image to efforts to market drugs or genetic tests to racial subgroups. The twin emphases on redressing health disparities and individualizing health care shields race and genetics research from appearing fringe or retrogressive as it once might have. A result is new and unpredictable political alliances around race and genetics, calling attention to the need to appreciate the specificities of the political-discursive context of this research today.

Volume Overview

This volume is designed to be an accessible, comprehensive, interdisciplinary resource on contemporary human population genetic variation research in the United States. The book has four sections. Part 1 offers a general introduction to the history, methods, and key analytical concepts of race and genetics research. Part 2 focuses on race-based therapeutics and the uses of human genetic variation research in clinical practice and drug development. Part 3 treats commercial genetic ancestry tests, examining the methodologies and social implications of this technology. The final section addresses the impact of emerging race and genetics research on public policy, the media, and public discourse at large.

Part I: Concepts of Race

Part 1 provides an overview of key concepts in the race and genetics debates. These chapters introduce the intellectual history of debates around race and biology and characterize the different ways that sociologists, biologists, and philosophers conceptualize race. This opening section makes clear what is at stake in the recent resuscitation of the race and genetics debate. New research challenges old framings of the debate, stretches the boundaries and assumptions of race and ethnic studies, and forces the clarification and reassessment of methods and practices in human population genetics research.

Anthropologist Jonathan Marks opens with a chapter summarizing the history of the biological race concept and profiling the context of contemporary genomic research on race. The 19th-century idea of races as discrete biological types has been replaced today by a clinal (gradual variation across geographical space) understanding of human genetic diversity. This important shift has not, however, done away with biological research on race. As Marks documents, classical race typology persists in contemporary research on genetic variation between geographically distant, historically "non-admixed" populations. For example, sampling and data mining practices frequently focus on the historical populations of East Asia, Northern Europe, and Central Africa and exclude other populations. Both within and beyond specialist discourse, the categories used in this population genetics research are frequently conflated with everyday folk understandings of race as a set of discrete types. Looking forward, Marks predicts that the increasingly corporate context of genetic research on race will make this conflation even more persistent and widespread in popular discourse.

In the next chapter, philosopher of biology John Dupré turns to the concept of the gene. Debates over the genetic basis of race often assume that genes act to produce phenotypic traits in a straightforward way. Dupré urges us to appreciate developments in 20th-century biology that demonstrate the complexity of gene action in human development and phenotypic variability. Race, of course, is not a Mendelian trait. Skin color and other phenotypic racial markers are complex traits of continuous variation. In addition, most markers used for determining racial ancestry are neutral, non-coding genes. What, then, does it mean to say that there are "genes for race"? In a careful review, Dupré distinguishes eight contemporary uses of the term "gene." Dupré shows that a claim of a genetic basis of race fits none of these conventional understandings of genetic causation and explanation, in the same sense that we understand, for example, "the gene for cystic fibrosis." Dupré's contribution underscores the importance of precision and clarity in making any claim of genetic causation—particularly claims of a general nature about race.

Having critically examined the concept of the gene, the next chapter moves to the concept of race. Philosopher Sally Haslanger's chapter outlines a social constructionist framework for conceptualizing race, recommending this framework as the most productive for interdisciplinary debates around race and genetics. While the social constructionist theory of race has been quite influential, its core claims are rarely developed, explicated, and defended. As a result, several misconceptions prevail. A social conception of race is often equated with a denial of the material or somatic reality of racial differences. The claim that race is social rather than biological is frequently interpreted as the claim that race is "not real" or is "merely" a fleeting social fiction. Genetic findings that disprove races as discrete kinds are, on this view, seen as proving that race is not real and, instead, "socially constructed." In light of this, continuing usage of the term "race" is seen as a last vestige of insistent racialist thinking. The paradoxical result of this line of thinking, of course, is that biology remains the arbiter of whether race should be considered real or not.

Haslanger develops and defends a social constructionist analysis of race that assiduously avoids these dead ends. Clearly, there are many meanings and uses of the concept of race in biology, the social sciences, and popular discourse. Rather than seeking to police the meaning of race, Haslanger asks what the most useful working concept of race for an interdisciplinary exploration of new race and genetics research might be. The question is what concept of race best captures its relevant features for a particular purpose and in a particular context. She recommends and develops a social constructionist concept of race as a framework for thinking through the various dimensions of new research on race and genetics. Haslanger argues that any concept of race must take account of the fact that the concept of race continues to identify groups and individuals in everyday life, and that somatic markings are significant in social interaction because they have become proxies for social status and cues for social beliefs and practices. For these reasons, race is both real and social. Race is a somatic phenotype that reflects non-discrete, broad-stroke continental ancestries. The meaning of race, however, is primarily constituted by social and linguistic context of color hierarchy. Of course, this is only true with respect to a particular historical and social context. That context could change, and in that case we may find this concept, or any concept, of race no longer analytically useful. This conception of race figures race as a real social kind and accounts for the salience of biology to race by noting its social importance.

The final chapter in this section, by genetic anthropologist Deborah Bolnick, introduces the models and methods of contemporary human population genetics research and evaluates whether recent research challenges the long-standing view that races are not biological kinds. Bolnick's analysis models how to read new research in population genetics closely, critically, and carefully. Bolnick begins by examining one of the foremost tools of population genetics, the Bayesian population-clustering program called *structure*. Bolnick reviews the uses of *structure* in recent literature and offers a close analysis of the model-theoretic assumptions encoded in the software. She demonstrates how incautious uses of the *structure* program, in combination with problematic sampling designs, may distort data in ways that consistently support a traditional folk race typology and obscure equal or more plausible interpretations of the data. Bolnick then argues more broadly that the claims to precision made by new technologies such as *structure* and tests of racial ancestry are mistakenly taken as a new demonstration that race is "in fact" a biological phenomenon. Yet as Bolnick argues, these findings refine and confirm previous understandings of human population variation, rather than, as widely reported both in the scientific literature and popular press, supporting a revision of the scientific understanding of race and population genetic distribution (see also Feldman and Lewontin, this volume). Bolnick systematically catalogues the ways in which the cutting edge of contemporary human genetic diversity research confirms long-standing findings: humans are overwhelmingly genetically identical; racial ancestry accounts for a very tiny sliver of human genetic diversity; genetic population structure is more variable within than between populations; and human genetic variation is fundamentally clinal rather than discrete.

Part II: Race-Targeted Research and Therapeutics

The chapters in part 2 explore genetic research utilizing racially coded genomic data with the goal of understanding the distribution of disease in the human population, developing population-specific drugs and other therapies, and addressing health disparities across racial and ethnic populations.

Geneticists Marcus Feldman and Richard Lewontin open this section with a detailed review of the relationship between human genetic variation, racial categories, and proposed medical applications. Lewontin is well known for arguing against naïve genetic determinism in biology. Feldman's laboratory at Stanford was among the first to present data showing the geographic variation of human genetic diversity around the globe using over one thousand DNA polymorphisms (Rosenberg et al., 2002). This work demonstrated that an individual's DNA allows assignment of that person to an ancestral geographic region with a high level of accuracy and is often cited as a key paper reinvigorating a genetic race concept. Feldman and Lewontin both argue, however, that classic racial categories are not biologically useful for medical purposes. In a detailed review of recent human population genetics findings, Feldman and Lewontin demonstrate why these findings yield insufficient information about the genetic signature of an individual. They argue, therefore, that the hope that continental ancestry translates back into useful clinical information remains premature.

In their chapter, geneticists Sarah Tate and David Goldstein assess the current state of the movement toward "individualized medicine." Individually targeted therapies have long been promised as the hallmark of a future "molecular medicine." Current pharmacogenomic science takes an interim step toward this goal by probing correlations between individual genetic signatures and drug response. The same drug, at the same dose, works well in some people, has no effect in others, and may lead to serious adverse outcomes in some who take it. In their search for genetic signatures associated with differences in drug response, pharmacogenomic researchers have focused first on differences between racial and ethnic populations. As Tate and Goldstein document, claims that such differences exist are often anecdotal, rarely validated, and lie beyond the scope of Food and Drug Administration (FDA) scrutiny. Clinical practice reflects the trends of pharmacogenomics and has also begun to incorporate assumptions about racial ancestry and genetics. When attempting to individualize drug therapies, a clinician should directly assess the genes controlling expression of that factor rather than relying on either their impression of the patient's continental ancestry or the patient's self-reported race. Since these tests are not yet available, however, Tate and Goldstein predict that doctors will be moved by the new paradigm of research on racial variation in drug response to increasingly rely on hunches about an individual's likely ancestry, assessments that are particularly complex in plural societies like the United States.

In 2005, BiDil became the first drug approved by the FDA for a racialized population (African Americans). In his chapter, historian and legal scholar Jonathan Kahn documents how the U.S. patent system forms part of a complex of institutions and practices that induce drug companies to pursue race-specific applications of their products. Looking closely at the case of BiDil, Kahn examines patents on race-specific drugs within the broader paradigm of "race-based" patent claims, demonstrating that once race is conceptualized as having value for biotechnology products, the patent system provides incentives for using race to maximize patent scope, duration, and robustness. He argues that classification rules in patent law that implicitly recognize race as a genetic category function as bureaucratic practices that have serious implications for the social identities of individuals and groups. The patent system is one of several federal initiatives that have played a central role in producing, classifying, and disseminating human genetic information. Kahn also reviews how certain conceptions of race and ethnicity are built into the biorepositories that support research. Federal guidelines and initiatives, in particular those promulgated by the U.S. Office of Management and Budget, provide specific, targeted incentives to see and use race and ethnicity in relation to biotechnological innovation in a manner that promotes the reification of race as a genetic category.

Duana Fullwiley, an anthropologist, follows the story into the laboratory. Presenting the results of in-depth ethnographic analysis, Fullwiley looks closely at one laboratory engaging in basic genetic epidemiological research on asthma and another researching differences in genetic substrates for cell membrane proteins involved in drug uptake. The community of scientists that Fullwiley follows is itself racially and ethnically diverse and demonstrates a political commitment to accounting for and ameliorating racial and ethnic differences in health outcomes. Nonetheless, the researchers' social and epistemic position commits them to a certain approach to "accounting for race" in their research. Fullwiley demonstrates how the investigators use a preexisting racial logic to drive hypothesis development, organize DNA samples, and explain research findings. The researchers' commitments function in a way that obscures the tautologies at the core of their methods and practices—building race "in" to the analysis leads inevitably to finding race as a salient "outcome" variable.

Sociologist Molly Dingel and medical anthropologist Barbara Koenig present another detailed case study of the use of racial categories in biomedical research, moving into the charged area of behavioral genetics. For decades, research seeking the etiology of addiction has used difference across human populations as a starting point for analysis. Addiction—a biobehavioral trait—provides a useful case study for considering the policy implications of behavioral genetics research targeted by race and linked with differential health outcomes. Use of illicit substances often tracks the boundaries of racialized population groups. Certain Native American communities have high rates of alcoholism and U.S. minority populations are addicted to nicotine at higher rates than middle-class whites. Why do genetic explanations for these phenomena have such appeal and salience? Dingel and Koenig examine recent studies of genetic variants associated with addiction and interrogate the problematic and inherently unstable classification of human populations in this field of research. They document how the molecularization of race and addiction research leads directly to pharmacological intervention, shifting focus from the social world to the gene and discrediting interventions designed to change the underlying social conditions supporting higher rates of addiction in those from racialized minorities.

Part III: Genetic Ancestry, Identity, and Group Membership

The chapters in part 3 treat the rapidly developing technology of genetic ancestry testing. Commercial ancestry testing services sell information of varying levels of specificity about the racial and ethnic background of an individual. The services determine ancestry estimates by scanning an individual's genome for allelic markers found in high frequency in particular national, geographic, ethnic, or tribal populations. Specialized services are available for African American, Native American, Irish, East Asian, and Jewish markets, among others. Some tests are marketed as a means of gaining eligibility for social benefits or membership status. Others are marketed to amateur genealogists seeking to investigate some aspect of family history.

Rick Kittles and Mark Shriver, academic geneticists who are also active in the commercial ancestry testing industry, open this section with an accessible review of the sampling and analytical techniques of this field, the potential applications of the tests, and the risks and limitations of ancestry testing. In particular, Kittles and Shriver explain the development and application of Ancestry Informative Markers (AIMs) technology to personalized genetic ancestry testing. They argue for the value of genetic ancestry testing across many fields and for the social and political importance of the technology. For instance, they cite the present-day slave descendents of Thomas Jefferson who used a Y-chromosomal marker to confirm their paternal ancestry, African Americans enabled to trace African tribal ancestry despite the shattering of families during the slave trade, and the Lemba people of South Africa who used genetic tests to substantiate their claim to Jewish heritage. Kittles and Shriver also review potential methodological pitfalls and social risks of current technology and, significantly, call for the development of a code of ethics and laboratory accreditation for the genetic ancestry testing industry.

In the next chapter, legal scholar and bioethicist Henry Greely profiles the enormous growth of commercial genetic ancestry testing over the past five years and surveys the variety of applications and claims to accuracy of such testing across the industry. Greely describes the different scientific approaches taken by commercial testing companies, including Y-chromosome patrilineal testing, mitochondrial DNA matrilineal testing, and the AIMs methods discussed by Kittles and Shriver. He argues that clear and accurate information about the composition of the databases, the statistical methods used, and proper interpretation of results must be available to the consumer of these tests. There are several areas in which genetic testing companies overstate the scope and precision of their services. For example, consumers are frequently led to believe that a unique Y-chromosomal or mitochondrial DNA finding is fully representative of their ancestry, when in reality it represents only one ancestral line out of thousands. As a legal scholar, Greely's principal concern about these tests is the need for disclosure of test limitations in a consumer-driven market system. He notes that despite Kittles and Shriver's call for a code of ethics for genetic testing, no movement has been made toward its development. In addition, most companies employ proprietary DNA databases and statistical algorithms; the industry operates without regulation or standardized laboratory practices.

Chapters by Kimberly TallBear and Alondra Nelson analyze the unfolding implications of genetic ancestry testing in two communities: Native Americans adjudicating blood quantum requirements for tribal enrollment and African Americans seeking to connect with their African ancestry. Both groups have become prime target markets for genetic ancestry testing companies. American Indian studies and science studies scholar Kimberly TallBear surveys six companies that market tests for Native American ancestry. As TallBear argues, the construct of "Native American DNA" faces serious empirical and methodological problems. The comprehensiveness and historical accuracy of the proprietary databases that are used to produce estimates of Native American ancestry are difficult to assess, and there are no good examples of DNA that is exclusive to all or some Native Americans. Despite these problems, genetic testing is gaining traction in Native America. Tests are marketed to individuals seeking membership in a tribe (or social benefits tied to Native American ancestry) and to tribal nations for use in tribal governance and enrollment. Since genetic testing reveals only lineal descendency, it challenges long-standing tribal paradigms of kinship that emphasize social as well as blood relations within the community. Thus, as TallBear explains, genetic testing could represent a significant reconfiguration of community standards of membership as well as narratives of family, blood ties, and origins.

African diasporic subjects who use genetic testing services to locate pre-slavery African tribal ancestry raise a distinctive set of questions about group, national, and ethnic identity in a genomic age. Sociologist Alondra Nelson analyzes the contemporary African American search for genetic ancestry as a new form of "roots tourism," part of a long-standing effort by many diasporic Africans to trace ancestry beyond the Middle Passage and to reconnect to Africa. Nelson finds that the new scientific authority and specificity of claims about ancestry offered by genetic testing are refiguring and reinvigorating the roots journey, producing new forms of racial and diasporic subjectivity. Drawing on ethnographic work with genealogy hobbyists, Nelson follows the experiences of three individuals who turn to genetic testing in their search for African roots and then integrate new information provided by the genetic test into existing conceptions of identity and ancestry. Significantly, Nelson observes that the test results rarely satisfy the subject's quest for certain and final validation of identity. Rather, they lead the subject in new directions and raise fresh questions about blood, identity, family, and nation. Contrary to expectation, genetic ancestry technologies offer a notion of ancestry that is "admixed," stochastic, and located in networks of ancestors connecting to historical-geographical populations. They challenge homogenous and fixed conceptions of identity as much as they seem to support them. The provocative questions raised by Nelson about identity in an age of genetic ancestry testing frame a rich research field only just taking shape.

Part IV: Race and Genetics in Public Discourse

The final part of the volume opens a discussion of the ethical and policy implications of research linking race and genomics, with special attention to the changing conceptual landscape and context of debates over race and genetics. The chapters in this section consider what constitutes effective scholarly and public engagement with issues related to human genetic variation research, how the media translates information on race and genetics for public consumption, how institutional practices that regulate biomedical research in the United States affect the meaning of race for scientists, and how political language on the relationship between race and genetics is deployed in addressing systemic inequalities in health status.

Bioethicist Pamela Sankar opens the discussion by profiling the divide between social scientists and geneticists in scholarly discourse on race. Examining citation practices and keywords in human population genetics publications, Sankar finds that despite offering sophisticated tools for social analyses of race since the 1970s, the social sciences have failed to keep pace with changes in the field of human population variation research. The result is that they remain too quick to dismiss genetic claims today. Meanwhile, the context of the debate has shifted. Many geneticists, Sankar observes, now appear to accept the thesis that race is a social construction; geneticists are now engaged and interested in the social dimensions of race. Sankar suggests that today there is opportunity for productive cross talk, including active collaboration between genetic researchers and social scientists to transform and clarify the concepts and practices of human population genetic variation research.

Misinterpretation or simplification of scientific research often occurs when journalists translate complex findings for the public. Yet the media plays an inevitably critical role in shaping public understanding of research findings of genetic population differences. In her chapter, journalist and science writer Sally Lehrman characterizes the news environment that makes it difficult for reporters to decouple race and biology. Conventions in story framing and source selection often result in reductionist thinking about race and genetics. The fact that reporters are assigned to specialized "beats," such as sports or crime, increases the possibility of mistranslation. The exclusion of Latino, African American, Asian American, and American Indian voices from the news is yet another factor. Lehrman suggests several strategies for providing a broader understanding of the meaning of race and genetics. Such work is critical, she contends, because journalists serve as the information connectors and mediators through whom findings linking race and genetics are interpreted and conveyed to the public.

Over the past decade, large-scale DNA sampling projects that involve racially and ethnically diverse populations have sought dialogue with communities about their participation in such research. Drawing upon case studies from the Human Genome Diversity Project and the International Haplotype Map Project, science studies scholar Jenny Reardon argues that far from resolving ethical concerns, these initiatives generate a paradox: the populations that scientists and policy makers seek to involve in research in an ethical and democratic manner are not pre-formed; rather, they form in the very act of constituting genomic research projects. Reardon suggests that this "paradox of participation" highlights the need to move beyond the conceptual structures that underlie modern liberal democratic institutions—structures that seek to separate science and society, objects and subjects. Analytic frameworks that call into view the inextricable entanglement of scientific objects and societal arrangements, she argues, are necessary to make sense of contemporary human population variation research.

Turning to the question of research oversight, political scientist Jacqueline Stevens challenges science-funding agencies to confront the possibility that population genetic research may endanger the pursuit of racial equality and undermine public health research goals. Stevens argues that the National Institutes of Health, the largest funder of genomic research related to human health in the United States, must provide greater oversight of the use of race as a variable in scientific research. Stevens outlines specific measures that would establish systematic review of research protocols that use race and racial categories. Stevens addresses several possible counterarguments to such institutional measures and weighs the harms and benefits of allowing the current ad hoc approach of uncritical use of race as a proxy for biological relatedness to continue.

In the final chapter, medical anthropologist Sandra Soo-Jin Lee examines the contemporary political-discursive context of race and genetics research. Lee profiles the rise of racial realism in recent struggles over the use of race in drug design and documents how the endorsement of racially targeted therapeutics in the name of social justice has been enthusiastically incorporated into the platforms of racial realists. Racial realists leverage recent population genetics research to assert that racial difference is a hard biological reality discernible in the human body. Racial realist discourse allies with policy that deemphasizes social solutions to problems such as disparities in health outcomes among racially identified groups. This is, in turn, tethered to the moral-political discourse of "personal responsibility." By these means, she argues, genetic research on race is mobilized as a response to liberal models of race that focus on environmental and historical conditions as the root causes for group differences. Using interviews, transcripts of congressional hearings, and other sources, Lee characterizes an emergent discursive formation drawing on genetic research on race, racial realism, and personal responsibility. Lee's analysis shows that how scientists (and others) frame and explain genetic difference filters into political discourse and influences the range of perceived solutions to questions of inequality and justice.

Challenges and Rewards of Interdisciplinary Dialogue

At the authors' conference for this volume, a geneticist charged a humanist with "not knowing the statistics"; a science studies scholar countered that "geneticists are getting by on a narrow understanding of how meaning is produced." Exchanges like this one demonstrate that interdisciplinary dialogue can be potentially productive and revealing, but also raise questions about the practice of interdisciplinarity. The exchange clarified genuine differences—in epistemic and interpretive practices, in concepts of race—in the humanities, social sciences, and genetic sciences. In the humanities and social sciences, the view that race reflects social hierarchies rather than biological or genetic difference is a starting point of analysis. Visible differences in skin color or hair form, while constitutive of concepts of race and self-evidently "genetic," do not explain the meaning of race in social life—that is, how race functions as an embodied identity and an empirically measurable determinant of social status. Among researchers studying human genetic variation at the molecular level, however, the story is different. Genetic scientists disagree on the utility of racial classification for studying common disease, but they generally accept new data showing the range of human genetic diversity and its global patterning. Many are inclined to see the possibility of biological determinants of a range of differences between races as an open scientific question.

These intellectual differences, however, are made more difficult to bridge by structural dimensions of academic knowledge production that hamper constructive interdisciplinary dialogue in race and genetics studies. Successful interdisciplinary dialogue demands more than speaking in the same vernacular where the meanings of concepts are shared. It requires recognizing fields of expertise and embracing the strength of a combined perspective. Our experience suggests that interdisciplinary exchange must be actively nurtured, structured, and rewarded in order to be successful in a disciplinary environment. We found that even well-meaning interlocutors have few models for listening and communicating in true interdisciplinary dialogue. A "Noah's Ark" model is the default: representatives from several different specialties assemble around a seminar table, each responsible for a certain body of knowledge. Reproducing disciplinary boundaries of legitimate expertise, each interlocutor educates the others about the current perspective of his or her particular field. Each speaks on his or her own specialty and consults with others to learn their perspective, but critical exchange, in which one individual questions fundamental methodological presuppositions of another, is rare, and frequently received poorly.

In race and genetics research, the commitment to interdisciplinarity is especially lopsided. Science analysts need the geneticists, but the geneticists do not need them. Geneticists work in a field of high epistemic and social authority. They can continue to work successfully and productively without taking the time to discuss their work with social scientists and humanists. Interdisciplinary work of this nature has low prestige in their field and few rewards. While many understand its value as a community service, it is also often looked upon as necessary for reasons of political correctness and as falling within a category of activity that forces scientists to attend to politics rather than science. Indeed, as interdisciplinary exposure often involves the questioning of methodological assumptions and requires geneticists to defend how and what they do, it can be an unpleasant experience. In contrast, those who study science and medicine rely on scientists and clinicians as informants. Social scientists are accustomed to approaching engagements with geneticists as opportunities to extract information from their objects of study, rather than as an open, critical exchange of ideas. Additionally, without the compliance of geneticists, social scientists' work would lack legitimacy. The delicate relationship of science analysts to their science informants results in a distinctive power dynamic in interdisciplinary collaboration. Indeed, at our authors' conference, some participants were research subjects of others at the table. Assuming that it is a cost of getting geneticists to the table, critical scholars will frequently tone down and soften their critique, resulting in discussion that yields few illuminating or generative exchanges. Both geneticists and critical scholars need to suspend their dogged focus on their objects of study and consider what ends might be achieved through considering each other's perspectives.

A challenge for the future is to build an ethics of interdisciplinary exchange and clarify the aims of interdisciplinary dialogue in race and genetics studies—this volume is offered in part as an experiment in productive, meaningful interdisciplinary collaborations involving social scientists, biologists, and humanists. Some of the best contemporary models of interdisciplinarity come from engineering and the biosciences.[1] It is common for interdisciplinary (sometimes multisited and multinational) teams of researchers to collaborate on a research problem for extended periods of time, sharing knowledge and resources, debating methodological approaches, and pursuing a problem in many dimensions. These scientific collaborators have more resources, greater similarities in professional training, and a narrower and more explicit set of shared core suppositions than the broad interdisciplinary collaboration we have attempted. Nonetheless we have drawn on their example in our experiment in results-oriented, structured interdisciplinarity. We have sought, with the "Revisiting Race" workshop and now this volume, to develop and model the kind of ongoing, interdisciplinary engagement between scientists and science analysts that will be vital as genetic research expands its reach in the genomic age.

NOTE

1. The Bio-X initiative at Stanford is one instance of an ambitious, large-scale interdisciplinary enterprise bridging biology, medicine, and engineering. (See http://biox.stanford.edu/about.)

REFERENCES

Burchard, E. G., Ziv, E., Coyle, N., Gomez, S. L., Tang, H., Karter, A. J., et al. (2003). The importance of race and ethnic background in biomedical research and clinical practice. *New England Journal of Medicine, 348*, 1170–1175.

Fredrickson, G. M. (2002). *Racism: A short history*. Princeton, NJ: Princeton University Press.

Gilroy, P. (2000). *Against race: Imagining political culture beyond the color line*. Cambridge, MA: Harvard University Press.

Risch, N., Burchard, E., Ziv, E., & Tang, H. (2002). Categorization of humans in biomedical research: Genes, race, and disease. *Genome Biology, 3*, comment 2007.1–2007.12.

Rosenberg, N. A., Pritchard, J. K., Weber, J. L., Cann, H. M., Kidd, K. K., Zhivotovsky, L. A., et al. (2002). Genetic structure of human populations. *Science, 298* (5602), 2381–2385.

Schwartz, R. (2001). Racial profiling in medical research. *New England Journal of Medicine, 344* (18), 1392–1393.

Wade, N. (2002, July 30). Race is seen as real guide to track roots of disease. *New York Times*, p. F1.

Wilson, J. F., Weale, M. E., Smith, A. C., Gratrix, F., Fletcher, B., Thomas, M. G., et al. (2001). Population genetic structure of variable drug response. *Nature Genetics, 29*, 265–269.

PART ONE

Concepts of Race

1

Race

Past, Present, and Future

JONATHAN MARKS

The relationships among anthropology, genetics, and race are far more complex than commonly presented. In this chapter I will explore the development of the race concept as a scientific theory in the 18th and 19th centuries and the insights provided over the course of the 20th century by genetics and anthropology. This chapter will show how race as an idea has evolved over the last century, so that even people who swear to its existence would not agree on its nature; how race involves a constant negotiation between difference (a biological state) and otherness (a cultural state), creating groups of people whose biological diversity is commonly at odds with their sense of unity; and, finally, how the complexities of the usage of race belie the possibility of it beneficially augmenting medical practice.

Origins of the Race Concept

The concept of race arose at the conjunction of two historical moments: the Scientific Revolution (privileging the study of nature and particularly its classification) and the Age of Colonialism (establishing hegemonic economic relations with unfamiliar, and commonly fluid, political and social entities).

Everyone everywhere (including pre-modern Europeans) has always identified themselves as distinct from and commonly in opposition to their neighbors. But the distinctions are rarely considered natural, immutable, or global—the core features of the race concept. Thus, although Josiah Nott and George Gliddon (before the Civil War) invoked a tomb painting from 18th Dynasty Egypt as putative evidence that the ancient Egyptians depicted and conceptualized race roughly as we do today, this is a misrepresentation. The painting simply shows that they noticed physical differences in relation to the people north of them and south of them—in other words, they weren't blind (see Isaac, 2004).

Race as we understand it today is a result of the encounter of early modern Europeans with exotic peoples and the new esteem accorded to naturalistic, scientific explanations of the meaning of the diversity represented by those peoples in relation to their "discoverers." Consequently, it is difficult to identify real intellectual antecedents for the works of François Bernier in 1684 and Carl Linnaeus in 1735 (see Stuurman, 2000). Comte de Buffon introduced the word "race" into the study of human diversity in 1749, but he used it in a casual and non-technical sense, meaning a "local strain." By the turn of the 19th century Buffon's term ("race") had become wedded to Linnaeus's concept ("subspecies"), and the modern scientific idea of race emerged—a fundamental taxonomic division of the human species. Racial classifications of the 19th century are noteworthy principally for their inconsistency and instability—one scholar emphasizing complexion; another, hair form; another, skull shape; and still another, geography. While there might be crude agreement in, for example, separating the peoples of Northern Europe, West Africa, and Southeast Asia from one another, most of the rest of the peoples of the world seemed to be in some degree of perpetual disarray.

Race in the 20th Century

The fruits of racial classification were descriptions of types, like Platonic essences. Linnaeus described *Homo sapiens Europeaus albus* in 1758 as having blue eyes and long, flowing blond hair ("Pilis flavescentibus prolixis[;] Oculis cæruleis"), while knowing full well that the vast majority of Europeans possessed neither of those features. His purpose was not empirical, but ideal; he was describing what *Homo sapiens Europeaus albus* was *supposed* to look like.

Such a concept was difficult to assimilate to Darwinian biology, and Mendelian genetics made race-as-essence even more untenable. Mendelian inheritance is quantitative and probabilistic; an individual has a one-eighth chance of inheriting a particular gene from any great-grandparent, and that holds across all genes and across all great-grandparents. At the same time that "one drop of blood" sufficed to make a person non-white, it was becoming unclear just what could be transmitted genetically in such a fashion. "Passing," the scourge of racial classifiers, was a simple lie under essentialist logic, but an ambiguous fact of complex ancestry under Mendelian logic.

By the early 20th century, racial essences were being supplanted by races that were conceptualized quite differently, as geographical subdivisions. William Z. Ripley (1899) formalized the peoples of Europe not so much into types as into regions: Nordic, Alpine, and Mediterranean. That each of these divisions might also embody an average physical appearance created much confusion between race-as-essence and race-as-geographical-type.

In the first, however, a race was embodied in a person; while in the second, a person was a member of a race. The very locus of race had been altered, and its relation to the individual organism had been reversed. A series of book-length works proceeded to explain what race "really" was, stretching from Huxley and Haddon's *We Europeans* (1936) through Ashley Montagu's *Man's Most Dangerous Myth* (1942) and William C. Boyd's *Genetics and the Races of Man* (1950). Of considerable influence in this regard was Theodosius Dobzhansky's magisterial synthesis, *Genetics and the Origin of Species* (1937), which used "race" zoologically, in a strictly geographical sense.

No longer abstract forms, human races were now groups of real people, that is to say, populations. They were the explicit focus of the 1950 Cold Spring Harbor Symposium, organized by Dobzhansky and physical anthropologist Sherwood Washburn. At the end of the symposium, Harvard's race expert, Earnest Hooton, sighed to the co-organizer, his former student, "Sherry, I hope I never hear the word 'population' again!"

The political environment of the 1930s had engendered a crisis in racial studies, which culminated in a paradigm shift (in Kuhn's original [1962] sense, with key terms being used incommensurably). The shift from a Platonic to a geographical concept of race, or from an essentialist to a populational concept, reflected a change in ideas about what race is, but not a negation of it (e.g., Krogman, 1943). Cultural anthropologists, while downplaying the possibility of racial endowment as a significant cause of behavior or history, accepted the existence of such taxonomic divisions as a scientific fact. That our species could be divided taxonomically was separated from questions of behavior and civilization by the progressive students of Franz Boas, the founder of modern American academic anthropology. Alfred Kroeber's 1923 textbook, *Anthropology*, for example, divided the peoples of the world into three branches and sixteen sub-branches. Ruth Benedict and Gene Weltfish's *The Races of Man*, a pamphlet commissioned by the USO in 1943 for distribution to soldiers, and later declared subversive for its egalitarian message, nevertheless proceeded from the assumption that there indeed exists a small number of different kinds of people. Likewise, Ashley Montagu's 1950 "UNESCO Statement on Race" specified three "divisions" of *Homo sapiens*, corresponding to European, Asian, and African, with an unspecified number of subdivisions.

Perhaps the last major exposition of this kind of taxonomic activity was William Boyd's genetic synthesis, published in *Science* in 1963, identifying 13 races, clustered into 6 larger groups (5 of the 13 hailing from Europe, but only 1 each from Africa and Asia—testifying eloquently to the cultural aspects of the endeavor). Meanwhile, the political environment of the early 1960s—the civil rights era—created a second crisis in racial studies and engendered a second paradigm shift. With the sitting president of the American Association of Physical Anthropologists (Carleton S. Coon) giving clandestine assistance to the segregationists, Sherry Washburn (1963) proclaimed the study of race to be an obsolete research program. Frank Livingstone wrote, "There are no races, there are only clines" (1962, p. 279)—invoking a term coined by Julian Huxley in 1938 to describe geographical gradients of features in natural populations, which seemed to be a better description of human diversity (see Serre and Pääbo, 2004). Local populations could be taken as the closest thing to natural units of the human species, but higher-order clusters were now recognized as arbitrary and ephemeral (Thieme, 1952; Hulse, 1962; Johnston, 1966). The British physical anthropologist J. S. Weiner described the human species "as constituting a widespread network of more-or-less interrelated, ecologically adapted and functional entities" (1957, p. 75).

In 1972, geneticist Richard Lewontin gave a quantitative basis for dismantling the race-as-population, calculating that there was about sixfold more within-group variation than between-group variation detectable in the human gene pool. This implied that the idea of races as discrete, geographically localized groups was unsustainable. Eric Wolf's *Europe and the People without History* (1982) refocused attention on the exchange and contact (economic and genetic) that has always existed among human populations.

By the early 1980s, physical anthropology had withdrawn from the study of race, which had defined the field's scope for much of the century. The principal aspect by which humans differed from one another was cultural; what was not cultural was principally polymorphic; what was not polymorphic was principally clinal; what was not clinal was principally local; and what was left of human diversity—that which was not cultural, polymorphic, clinal, or local—was not much. Race was therefore abandoned to the cultural anthropologists on one side (as a major set of social facts), and to the geneticists on the other (as a minor set of natural facts).

This left a small window remaining open, however, for yet another shift in the conceptualization of race. If biological variation could be seen as a Venn diagram, with 85% overlap and a small amount of unique diversity in each circle, then one could direct one's focus very closely to the small amount of localized variation, essentially contrasting the most extreme members of any circle.[1]

This approach focuses on the extremes, however, at the expense of the majorities. It fails to tell you how many natural groups there are (which is decided a priori, as Bolnick details in this volume), fails to classify properly most of the world's living peoples (who do not hail from the geographic extremes), and fails as well to classify diachronic populations. An archaeological population from East Africa can be allocated heuristically to the corners of the globe in this fashion (Williams, Belcher, and Armelagos, 2005), and a 9,500-year-old skeleton from Kennewick (in America) is infamously allocated first to France, and then re-assigned to Polynesia (Chatters, 2001; Powell, 2005).

The word remains the same (yes, "race"), but the concept underlying it has again been radically reformulated. It is now a very careful look at the most divergent qualities of the most geographically separated peoples, so as to maximize detectable differences between groups. This is exemplified by the study of Ancestry Informative Markers, or AIMs—an analysis of the residuals of human genetic variation once the major features of that variation are dismissed. Here, an individual's African ancestry is inferred from genetic similarity along a few dozen genetic markers (sampled across 25,000 genes and 3 billion DNA bases in a human genome), derived from a few dozen cell lines from Central-West Africa, carefully chosen to be maximally different from a comparable sample of East Asians and Northern Europeans. While this is not "race" in any previously familiar sense of the term, it is readily conflated with such notions, whether ingenuously or not.

To enter into a discussion of whether race is "real" today, then, is to miss the fact that race has itself evolved over the last century, from a qualitative judgment of "what you really are" to "where you really come from," and finally to a crude quantitative assessment of your similarity to the gene pools of the most divergent peoples.

Race as a Theory of Kinship

Anthropologists know that we "*make* sense" of human diversity; its sense or meaning does not come ready-made for us. In Claude Lévi-Strauss's famous phrase, race can be best understood as a scheme of classification that is "good to think with," in that it imposes an order upon nature, which helps to confront, understand, and manipulate it. That is true of many things that are not tangible, but which help us make sense of the world and stimulate social action (Lévi-Strauss, 1962).

Race—in any guise—is a theory of kinship. It tells you who you are and what you are. Kinship is a system of classification, in which complex relationships of biology (procreation) and law (marriage) are organized into a coherent framework (Bohannon 1963; Fox 1969; Carsten 2003; Franklin and McKinnon, 2002). This framework provides the basic social orientation for an individual: it defines who is a member of your family and who is not. As new forms of marriage and new forms of procreation appear, they are perceived as threatening and potentially disorienting because they challenge this social order.

While procreation is biological, descent is not. Descent depends on the assignment of a status based on the facts of birth. This is why "bastardy," for example, is a cross-cultural insult. "No primate other than man can remember his ancestors up to the thirteenth and fourteenth generations, nor can he conceptualize 'second cousin once removed,' even though he *has*, biologically, such a relative. And certainly he could not attach legal, political or economic significance to such a relationship" (Fox, 1969, p. 30). The facts of nature are quickly overtaken by the facts of culture. Kinship systems are codes of meaningful similarity: Your mother's sister and your father's brother's wife may be equivalent relations to you ("aunts"), or they may not be, given that the former is a blood relative and the latter is not. Since marriage is a political status and coexists as an economic, social, sexual, reproductive, and residential unit, it can take many diverse forms—each of which has a direct impact upon one's ideas of what the family is. The boundaries of the family are consequently symbolic and powerful: a first cousin may be a preferred spouse in one setting (like Charles Darwin's) or an incestuous mating in another (like Jerry Springer's).

Kinship, then, is a symbolic system of classification that adopts some aspects of natural relationships and suffuses them with arbitrary cultural distinctions. In this way, one's identity is established and the appropriate expectations for particular social interactions can be formalized and learned. One knows on whom one can count, who can be approached by virtue of sharing a common bond (that of ancestry), and who is alien. The distinction between "ingroup" and "outgroup" is understood as hierarchically inclusive and oppositional: two rival groups may readily unite against a common outsider (Leach, 1968).

Most significantly, however, kinship uses metaphors of nature to express social norms—blood is thicker than water, incest is a crime against nature, blood will tell, and the like—and thus commonly misrepresents itself as a natural system by obscuring or concealing its historical or constructed aspects (Schneider, 1968).

This is the intellectual context in which to understand race. Race is simply one aspect of the complex system of identity formation. Identity is itself a negotiation among statuses rooted in genetics (dwarf? diabetic?), other aspects of biology and behavior (teenager? transgender?), sociopolitical categories (Catholic? Cuban? doctor?), and what slots are even available to be occupied (witch? Hispanic? mulatto? Brooklyn Dodgers fan?). These run a gamut from mostly naturally constituted to mostly historically constituted, but what they share is that they orient you in a social universe, give meaning to life, and regulate behavior and social interactions by indicating just who and what you are and to whom you may be fundamentally akin.

Corporate Genetics

It is thus not surprising that kinship would emerge as a potentially lucrative market for genetics companies. Certainly, the most noteworthy aspect of the life sciences over the last generation has been the transformation of genetics. The development of technologies from 1960 to 1985—gel electrophoresis, the isolation and production of restriction enzymes, DNA sequencing, and the Polymerase Chain Reaction—made it possible to conceive of a Human Genome Project. But the project could not be justified solely as a chance to apply these technologies on a large scale. To secure the billions in federal funding it would take, the Human Genome Project had to convince the public of its value. In large part, the Human Genome Project was justified by recourse to familiar hereditarian social philosophies. Said James Watson famously in *Time* magazine, "We used to think our fate was in the stars. Now we know, in large measure, our fate is in our genes" (Nelkin and Lindee, 1995). Finally, through the investment of capital, biotech start-up companies began to transform the Human Genome Project into profit ventures. This is not necessarily bad, but it is not the science that we grew up with, nor that our professors taught us. This is not the science of Theodosius Dobzhansky or of James Neel—which was by no means "pure," but whatever it was, it was science. And the nonscientific issues it grappled with were relatively simple: notably, ideological prejudice. That, of course, is still with us, but now there is a profit motive; there are products to sell and a market to create and maintain.

Capital has transformed genetic science, but into what? There is not even a word for a science in which the production of capital is so intimately associated with the production of knowledge, in close synergy with the creation of a market for that knowledge. You can't call it pseudo-science because the people doing it have impeccable scientific credentials.[2] Whatever modern human genetics has become, it is not the science of a generation ago, and it now uses technology, professional expertise, and the authority of the scientific voice in support of goals that are not classically or normatively scientific.

An unprecedented amount of it is taking place outside of academic contexts, producing unprecedented questions of conflicting loyalties, interests, and motivations. And it need hardly be pointed out that it is in the interests of this transformed science to overvalue the significance of genetics in life. The more of life you think genetics controls, the larger the potential market for its products. But how can you possibly then know what to believe amid the conflicted interests, intellectual prejudices, and cultural naïveté of the practitioners?

In 1937, Earnest Hooton struggled (largely unsuccessfully) to differentiate his own racial science from that of his German colleagues. "There is,"
Hooton wrote, "a rapidly growing aspect of physical anthropology which is
nothing less than a malignancy. Unless it is excised, it will destroy the science. I refer to the perversion of racial studies and of the investigation of
human heredity to political uses and to class advantage. . . . [T]he output of
physical anthropology may become so suspect that it is impossible to accept
the results of research without looking behind them for a political motive"
(1937, pp. 217–218). To the extent that genetics now leads the way in the study
of human variation, the same class values, naïveté, and enlightened self-
interest remain—witness the reluctance of the American Society of Human
Genetics to repudiate the folk ideas about heredity popularized by R. Herrnstein and C. Murray's *The Bell Curve* in 1994 (Andrews & Nelkin, 1996) and
the confusion over the analysis of the human gene pool with the computer
program *structure* and its relation to the ontology of race (Rosenberg et al.,
2002; Wade, 2002; Feldman & Lewontin, this volume; Bolnick, this volume).
But the financial conflicts introduced in the last generation by privatized
interests make it even more difficult to gauge the truth value of any claims in
human genetics, as evidenced by the public discussion of race precipitated
by BiDil (Kahn, this volume) and the boom in scientific genealogy that tends
to gloss over the fine print (Greely, this volume).

This confusion has proved to be easily exploited. There are several areas
of contemporary racial/genetic science where Hooton's worst fears, it would
seem, have come true.

Racial Taxonomism

The first claim centers on the racial taxonomists, those who reify race and
claim that human beings do, indeed, fall into a small number of largely
discrete natural categories. Spurning the last half century of research into
patterns of human variation, the argument here is that left-wing cultural
prejudice has suppressed an obvious fact of biology (Entine, 2000; Sarich
& Miele, 2004; Leroi, 2005).[3] The precise composition of these ostensibly
natural categories, or what they represent, is rarely made explicit; and where
it is, the assertion is easily refutable.

Philosopher Robin Andreasen (2000, 2004) has recently attempted to
resuscitate the race concept formally, by reference to what she calls a "cladistic" race. Cladistics is a philosophy of biological classification that privileges
recent common ancestries in the establishment of taxonomic categories
(Eldredge & Cracraft, 1980). Race here is a node of the tree of population
bifurcations, popularized in the work of Cavalli-Sforza, Menozzi, and Piazza
(1995). Actually, however, the tree itself is a statistical reification and not

at all a series of historical events, upon which cladistics relies (Templeton, 1998). These trees are phenetic (based on similarity) rather than cladistic (based on monophyletic descent, that is, from a series of unique ancestors) and incorporate many kinds of groups (Moore, 1994), often subsuming recently re-merged lines of ancestry of varying degrees (see fig. 1-1).

Real human populations are connected genetically, and the labels that Andreasen takes as racial generally do not satisfy the cladistic stricture of monophyly (that is, unique single ancestry). What is most confusing here is that race and cladistics are both about classification. If "cladistic races" are real, then they imply a system of classifying people into natural monophyletic groups; but such groups do not exist and are simply being inferred from a specious interpretation of the genetic literature. If a cladistic race concept contains such internal contradictions, and must mythologize human history in order to be applicable, it is difficult to see much anthropological value in it.

The study of human diversity would be better served by the recognition that human populations are in reality variably sized bio-cultural units. Each has its own genetic idiosyncrasies, the result of adaptation to local circumstances and to the historical accidents of survival and proliferation. Each has the properties of being (1) historically ephemeral (as Hittite or Anasazi); (2) hierarchically organized (as in the nested categories Caucasoid, Nordic, Slavic, Baltic, and Latvian); (3) genetically porous (in manners that run a gamut from exogamy and adoption to kidnap and mass rape); and (4) culturally bounded (as a species with little genetic variation relative to apes and

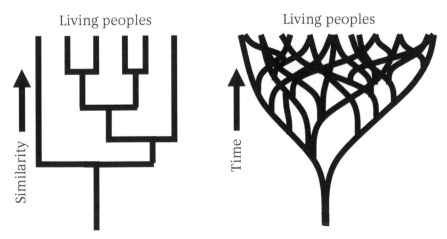

FIGURE 1-1. *Left*, a bifurcating tree, the product of applying clustering methods to genetic data derived from discretely defined populations, to estimate the patterns of similarity among gene pools. *Right*, a reticulating tree, depicting the complex biological histories of human populations.

with a concomitant tendency to identify our group by adornment, speech, movement, and custom, whether or not biological distinctions exist).

This is what human variation is about. *Race*—that is to say, the existence of a fairly small number of naturally distinct or different kinds or types of people—is not.

Population Genetics

The second area in which conflicts of interest of one sort or another have led to confusion, and to an erosion of confidence in its claims, is population genetics. At the same time that Lewontin was undermining the existence of geographical race, other population geneticists were using it unproblematically as an analytic category. As late at the 1990s, population geneticists were coming forward with claims such as "ancestral Europeans are estimated to be an admixture of 65% ancestral Chinese and 35% ancestral Africans" (Bowcock et al., 1991, p. 840), duly publicized as "All Europeans are thought to be a hybrid population, with 65% Asian and 35% African genes" (Subramanian, 1995).

This ambiguity recapitulates earlier states of affairs in the genetic analysis of human diversity. Utilizing the earliest genetic data on ABO blood groups, Laurence Snyder proceeded to identify basic races; but since the ABO allele frequencies only vary within a fairly narrow range, it is not uncommon for distantly related peoples to have similar ABO frequencies. Thus Snyder (1926) was obliged to associate (spuriously) the peoples of Poland and China, for example. In other words, the available genetic data were largely useless for studying race, as Hooton (1931, p. 490) dismissively noted. The argument was not about what the basic patterns of human variation are, but rather, how genetic data reveal the basic patterns more clearly than other classes of data. The self-serving assumption that genetic data superseded all others in this area prompted a heated exchange at the Royal Anthropological Society in 1932 (discussed in its journal, *Man*) and similar arguments in the journal *Science* in the 1940s and the *American Journal of Physical Anthropology* in the early 1950s (Marks, 1996).

In retrospect we can see that, ironically, the genetic data did have it right—human variation, indeed, just does not sort out racially—but since race was understood to be the organizing principle of human variation at the time, it was crudely assumed that the genetic data *had* to show it and would necessarily reveal it more compellingly than other classes of data. In other words, the geneticists were finding what they were looking for and were finding it with an astounding lack of introspection or intellectual breadth.

The sad case of the recent Human Genome Diversity Project reveals the ambiguities of contemporary human population genetics. The project failed to grapple adequately with the issues that large-scale collection of indigenous people's cells might incur: the patenting of cell lines, the manipulation of rights and identities, the meaning of voluntary informed consent in a cross-cultural context, the social and political legacy of colonialism, ownership of the body, and personhood, much less the widespread fear of witchcraft by blood theft and the possible development of weapons of mass genetic destruction. The best the Diversity Project could do was to avow that at least they weren't racists and that they didn't care about making money from the venture. Rather than resolve the problems, however, human population geneticists instead turned to private sources of funding; and now the Genographic Project makes much the same claims about their good intentions, with even less attention given to the actual issues that undid the Diversity Project (Small, 2006; Harmon, 2006).

Today, with massive federal funding failing to materialize for the Human Genome Diversity Project (Reardon, 2004), pharmaceutical money may be poised to take its place. The niche marketing of BiDil as a race-specific drug (Kahn, 2004) violates the patterns of human genetic diversity as surely as a perpetual-motion machine violates the laws of physics. Nevertheless, it represents the foot in the door and can readily be complemented and rationalized through the reification of race. Not surprisingly, there is considerable overlap between the scholars advocating racial taxonomism and those touting BiDil and racialized medicine (Satel, 2002; Leroi, 2005).

Behavioral Genetics

In behavioral genetics we find perhaps the greatest amount of confusion. The broad convergence of interest between innatist theories of human behavior and conservative social politics has been acknowledged for close to a century (Davenport, 1911; Darrow, 1926; Lewontin, Rose, & Kamin, 1984). This was the crux of the controversy over *The Bell Curve*, co-authored by a political scientist (Herrnstein & Murray, 1995).

In principle, behavioral genetics should not be politically threatening since the basic patterns it seeks to study are well known. Human genetic variation is, in the main, clinal and polymorphic; human behavioral variation is, in the main, discrete and polytypic and is the product of historically and socially produced differences. That is to say, in the grand scope of what human beings think and do, the major differences are cultural. Some people may be genetically happier or more impulsive or better spellers, but such people dress the same, speak the same language, eat the same kinds of foods, and pray to the same gods as the people with different alleles.

In other words, to the extent that people differ in temperament or brainpower, and that such differences may be genetic in origin, those differences can only explain at best why two people in the same population do slightly different things. For example, there might well be an imaginary brain hormone receptor with an allele that makes you a little happier. We should expect its distribution, however, to be similar to other genetic variation: that is, clinal and polymorphic.

This has an important correlate, however. The same brain hormone receptor allele in a Harvard professor and a Yanomamo would not make their lives any more similar. They might get through the day a little more happily than other Harvard faculty members and other Yanomamo, but no significant aspect of their lives would thereby converge—for what makes their lives so different lies not in their brain genes, but in their social histories and circumstances. Thus, to the extent that there exists the possibility of studying the effects of genetic variation upon human behavior, that opportunity exists for studying only those features of human behavior that vary as genetic differences tend to—within populations. That is to say, only the most minor features of the diversity of human thoughts and behaviors.

The interests of the behavioral geneticist and the racist differ, but they converge if the discordant patterns of human behavioral and genetic variation are ignored. Then, genetic variation in behavior, which may be useful in explaining within-group differences, can casually be extrapolated to explain between-group differences in mind and deed. Such an extrapolation is entirely illegitimate; yet every generation seems to be confronted with it (Lewontin, 1970; Marks, 2005). Thus, the arguments of Herrnstein and Murray (1995) are identical in structure to those of Arthur Jensen (1969)—including especially the misapplication of "heritability" (a within-group calculation misapplied to explain between-group differences).

Interestingly, some of the strands are connected through the philanthropic activities of a wealthy textile magnate named Wycliffe Draper. Draper sponsored projects ranging from the deleterious effects of race-crossing in the 1920s, through the demonstration of IQ differences between whites and blacks, and up to activism against school integration in the early 1960s. The focus of his philanthropy gradually expanded to include research committed to demonstrating innate variation in behavior and personality, accompanied by the seductive (if invalid) extrapolation from the causes of within-group variation to the causes of between-group variation. Draper's money also subsidized the publication of the *Mankind Quarterly*, an ostensibly scholarly journal that scandalized the scientific community with its racist orientation in the early 1960s, and went on to fund Arthur Jensen and William Shockley generously in the 1960s and 1970s.

In the 1980s, Draper's Pioneer Fund sponsored the research of the psychologist Thomas Bouchard on identical twins separated at birth. These well-publicized studies, invoking the genetic basis of intelligence and personality, are usually intertwined with stories about the psychic abilities of these twins, apparently on equally firm scientific footing. And in the 1990s, it sponsored the racial research of J. Philippe Rushton, currently the Pioneer Fund's president.

The Pioneer Fund did not subsidize *The Bell Curve*, but *The Bell Curve* cited many of the researchers it had sponsored as well as several articles published in the *Mankind Quarterly*. Philippe Rushton's bizarre ideas—that Africans evolved to be dumb and promiscuous; Asians, to be smart and under-sexed; and Europeans, to be the happy medium—are also cited in *The Bell Curve*, along with a pre-emptive appendix defending them.

Even work in human behavioral genetics that is not so overtly political usually turns out to be unreliable, a path to headlines and best-sellers with little scientific merit—such as the genes for male homosexuality, novelty seeking, and religiosity (Hamer, Hu, Magnuson, Hu, & Pattatucci, 1993; Hamer & Copeland, 1998). What research can actually be believed in this area is frankly anybody's guess; and its connections to race are even more tenuous, but those inferences are readily drawn. As Hamer and Copeland nonchalantly explain, "There is still such great variation in the [novelty-seeking] gene in modern day humans. . . . [D]ifferent ethnic and racial groups, who evolved under different environmental circumstances, have noticeably different frequencies of the different variations" (1998, p. 49).

Health Care, Race, and Genetics

Anthropology has productively studied human diversity for decades *without* the underlying belief that a small number of natural divisions of the species exists, and it certainly will continue to do so. It stands to reason that health care would benefit more from the scientific knowledge of how human variation is patterned than from racial mythology.

Let us consider some nonracial research strategies. First, we can contrast groups as social units. The health differences between blacks and whites can be contrasted as readily, without reifying them, as one might contrast the health of a sample of tailors and coal miners. Second, we can contrast groups as phenotypic units: darkly versus lightly complected, tall versus short, blond versus brunet; fat versus thin. Third, we can study geographical or ecological units: high-altitude dwellers in the Andes, circumpolar peoples, rainforest peoples, industrialized peoples in temperate climates. And fourth, we can contrast human populations in terms of their local histories: adapted to their conditions or disproportionately expressing the genetic legacy of an ancient founder by "accident."

It is critical to distinguish also between risk assessment and diagnosis (on the one hand), and intervention (on the other). Knowing the identity or identities of a patient can be of obvious value, as the health risks associated with being a prostitute, Ashkenazi Jew, Pima Indian, or computer programmer may be quite specific (for example, gonorrhea, Gaucher's disease, diabetes, or carpal tunnel syndrome, respectively). Nevertheless, it seems hard to imagine that the same health problem would necessitate distinct kinds of treatment in different groups (unless the difference involved adapting the intervention to particular sociocultural norms).

Even the principal and most familiar genetic risks do not sort themselves out "racially" in any ordinary meaning of that term. Human populations have their own particular genetic risks, the result of their biosocial histories. Sickle-cell anemia, so familiar for its prevalence in Equatorial Africans, is rare in Southern Africans and more common in Saudi Arabians. Tay-Sachs disease is more common in Ashkenazi Jews, but what have they to do with race? Porphyria variegata is common among South African whites of Dutch ancestry (by virtue of founder effect) and also among Coloureds (by virtue of gene flow). Ellis-van Creveld Syndrome is famous for its incidence among the Pennsylvania Amish, but again, what have they to do with race? Alpha-thalassemia is found commonly among East Asians and Melanesians, apparently adaptive à la sickle-cell. Cystic fibrosis is most prevalent in Northern Europeans, and lactose intolerance is common in everybody else. The appropriate model here is not a racial one, but a biosocial one.

Health care can benefit by knowing something of the self-identification of the subject, given that different groups have different risks, due to their histories or life circumstances. But race is not the cause of it; in fact, race will positively obscure it. So of what benefit would racialized medicine possibly be? A therapeutic intervention would have to be based on the genotype, not on any racialized identity. And if there is anything that is secure as positive knowledge in the study of human variation over the last few decades, it is that race is a very poor surrogate for genotype.

Conclusion

Perhaps the most extraordinary aspect of the revival of race in biomedicine is its explicit rejection of decades of professional scholarship on the subject of human variation and the acceptance instead of common, or folk, knowledge. It has an interesting parallel elsewhere on the ideological spectrum. Anthropologists have grown accustomed to hearing that they are pawns in a conspiracy to deceive the public and to indoctrinate them into the counterintuitive dogma of evolution. That the creationists and the race reifiers would find a common enemy in the anthropologists, and adopt parallel rhetoric against them, is a great irony. The race reifiers, after all, generally claim to be speaking in Darwin's name, just as the creationists claim to be speaking in Jesus' name, against their enemies, the anthropologists.

There seems to be little else binding the creationists and the scientific racists together, aside from their common hostility toward the accumulated scientific knowledge of anthropology. This bond of anti-intellectualism would be unremarkable in American society (Hofstadter, 1963) were it not for one crucial difference between the anthropology-bashing scientific racists and the anthropology-bashing creationists: namely, that racialized biomedicine puts actual lives at risk by ignoring natural patterns of variation and mistaking the confabulations of social history for authoritative knowledge about the world, while creationism only threatens minds.

NOTES

1. This is in fact what forensic anthropologists had been practicing, although with very little reflection. If you know that the extremes of human groups can be crudely distinguished on the basis of metrics describing nasal breadth, facial conformation, cranial shape, and femoral curvature, it is not difficult to apply those measurements to make a better-than random guess about the origin of an unknown skeleton.

2. You could call it "schmience," I suppose.

3. This charge, denying the actual empirical patterns of human variation and claiming that scholarly determinations about race had been ideologically hijacked by left-wing anthropologists, was leveled initially by segregationists a half-century ago (Putnam, 1961).

REFERENCES

Andreasen, R. O. (2000). Race: Biological reality or social construct? *Philosophy of Science, 76*, 653–666.

Andreasen, R. O. (2004). The cladistic race concept: A defense. *Biology and Philosophy, 19*, 425–442.

Andrews, L. B., & Nelkin, D. (1996). The Bell Curve: A statement. *Science, 271*, 13–14.

Bohannon, P. (1963). *Social anthropology.* New York: Holt, Rinehart.

Bolnick, D. A. (2008 [this volume]). Individual ancestry inference and the reification of race as a biological phenomenon. In B. A. Koenig, S. S.-J. Lee, & S. S. Richardson (Eds.), *Revisiting race in a genomic age* (pp. 70–85). New Brunswick, NJ: Rutgers University Press.

Bowcock, A. M., Kidd, J. R., Mountain, J. L., Hebert, J. M., Carotenuto, L., Kidd, K. K., & Cavalli-Sforza, L. L. (1991). Drift, admixture, and selection in human evolution: A study with DNA polymorphisms. *Proceedings of the National Academy of Sciences, USA, 88*, 839–843.

Boyd, W. C. (1950). *Genetics and the races of man.* Boston: Little, Brown.

Boyd, W. C. (1963). Genetics and the human race. *Science, 140*, 1057–1065.

Carsten, J. (2003). *After kinship.* London: Cambridge.

Cavalli-Sforza, L. L., Menozzi, P., & Piazza, A. (1995). *The history and geography of human genes.* Princeton, NJ: Princeton University Press.

Chatters, J. C. (2001). *Ancient encounters: Kennewick man and the first Americans.* New York: Simon and Schuster.

Darrow, C. (1926). The eugenics cult. *American Mercury, 8,* 129–137.

Davenport, C. (1911). *Heredity in relation to eugenics.* New York: Henry Holt.

Dobzhansky, T. (1937). *Genetics and the origin of species.* New York: Columbia University Press.

Eldredge, N., & Cracraft, J. (1980). *Phylogenetic patterns and the evolutionary process: Method and theory in comparative biology.* New York: Columbia University Press.

Entine, J. (2000). *Taboo: Why black athletes dominate sports and why we're afraid to talk about it.* Washington, DC: Public Affairs Press.

Feldman, M. W., & Lewontin, R. C. (2008 [this volume]). Race, ancestry, and medicine. In B. A. Koenig, S. S.-J. Lee, & S. S. Richardson (Eds.), *Revisiting race in a genomic age* (pp. 89–101). New Brunswick, NJ: Rutgers University Press.

Fox, R. (1969). *Kinship and marriage.* London: Pelican.

Franklin, S., & McKinnon, S., eds. (2002). *Relative values: Reconfiguring kinship studies.* Chapel Hill, NC: Duke University Press.

Greely, H. T. (2008 [this volume]). Genetic genealogy: Genetics meets the marketplace. In B. A. Koenig, S. S.-J. Lee, & S. S. Richardson (Eds.), *Revisiting race in a genomic age* (pp. 215–234). New Brunswick, NJ: Rutgers University Press.

Hamer, D., & Copeland, P. (1998). *Living with our genes: Why they matter more than you think.* New York: Doubleday.

Hamer, D., Hu, S., Magnuson, V. L., Hu, N., & Pattatucci, A.M.L. (1993). A linkage between DNA markers on the X chromosome and male sexual orientation. *Science, 261,* 321–327.

Harmon, A. (2006, December 10). DNA gatherers hit a snag: The tribes don't trust them. *New York Times,* p. 1.

Herrnstein, R., & Murray, C. (1995). *The Bell Curve.* New York: Free Press.

Hofstadter, R. (1963). *Anti-intellectualism in American life.* New York: Alfred M. Knopf.

Hooton, E. A. (1931). *Up from the ape.* New York: Macmillan.

Hooton, E. A. (1937). *Apes, men, and morons.* New York: Macmillan.

Hulse, F. S. (1962). Race as an evolutionary episode. *American Anthropologist, 64,* 929–945.

Huxley, J. S. (1938). Clines: An auxiliary taxonomic principle. *Nature, 142,* 219–220.

Huxley, J., & Haddon, A. C. (1936). *We Europeans.* New York: Harper and Brothers.

Isaac, B. (2004). *The invention of racism in classical antiquity.* Princeton, NJ: Princeton University Press.

Jackson, J. (2005). *Science for segregation.* New York: New York University Press.

Jensen, A. (1969). How much can we boost IQ and scholastic achievement? *Harvard Educational Review, 39,* 1–123.

Johnston, F. E. (1966). The population approach to human variation. *Annals of the New York Academy of Sciences, 134,* 507–515.

Kahn, J. (2004). How a drug becomes "ethnic": Law, commerce, and the production of racial categories in medicine. *Yale Journal of Health Policy, Law, and Politics, 4,* 1–46.

Kahn, J. (2008 [this volume]). Patenting race in a genomic age. In B. A. Koenig, S. S.-J. Lee, & S. S. Richardson (Eds.), *Revisiting race in a genomic age* (pp. 129–148). New Brunswick, NJ: Rutgers University Press.

Kroeber A. L. (1923). *Anthropology.* New York: Harcourt, Brace.

Krogman, W. M. (1943). What we do not know about race. *Scientific Monthly, 57*, 97–104.

Kuhn, T. (1962). *The structure of scientific revolutions.* Chicago: University of Chicago Press.

Leach, E. R. (1968, October 10). Ignoble savages. *New York Review of Books, 11* (6) (http://www.nybooks.com/articles/11540).

Leroi, A. F. (2005, 14 March). A family tree in every gene. *New York Times.*

Lévi-Strauss, C. (1962). *La pensée sauvage.* Paris: Librairie Plon.

Lewontin, R. C. (1970). Race and intelligence. *Bulletin of the Atomic Scientists, 26*, 2–8.

Lewontin, R. C. (1972). The apportionment of human diversity. *Evolutionary Biology, 6*, 381–398.

Lewontin, R. C., Rose, S., & Kamin, L. J. (1984). *Not in our genes.* New York: Pantheon.

Livingstone, F. B. (1962). On the non-existence of human races. *Current Anthropology, 3*, 279–281.

Marks, J. (1996). The legacy of serological studies in American physical anthropology. *History and Philosophy of the Life Sciences, 18*, 345–362.

Marks, J. (2005). Anthropology and the Bell Curve. In C. Besteman and H. Gusterson (Eds.), *Why America's top pundits are wrong: Anthropologists talk back* (pp. 206–227). Berkeley: University of California Press.

Montagu, A. (1942). *Man's most dangerous myth: The fallacy of race.* New York: Columbia University Press.

Moore, J. H. (1994). Putting anthropology back together again: The ethnogenetic critique of cladistic theory. *American Anthropologist, 96*, 925–948.

Nei, M., & Roychoudhury, A. (1974). Genic variation within and between the three major races of man, Caucasoids, Negroids, and Mongoloids. *American Journal of Human Genetics, 26*, 421–443.

Nelkin, D., & Lindee, M. S. (1995). *The DNA mystique: The gene as cultural icon.* New York: Freeman.

Powell, J. F. (2005). *The first Americans: Race, evolution, and the origin of Native Americans.* New York: Cambridge University Press.

Putnam, C. (1961). *Race and reason.* Washington, DC: Public Affairs Press.

Reardon, J. (2005). *Race to the finish: Identity and governance in an age of genomics.* Princeton, NJ: Princeton University Press.

Ripley, W. Z. (1899). *The races of Europe.* New York: D. Appleton.

Rosenberg, N. A., Pritchard, J. K., Weber, J. L., Cann, H. M., Kidd, K. K., Zhivotovsky, L. A., et al. (2002). Genetic structure of human populations. *Science, 298*, 2181–2185.

Sarich, V., & Miele, F. (2004). *Race: The reality of human differences.* New York: Westview.

Satel, S. (2002, May 5). I am a racially profiling doctor. *New York Times.*

Schneider, D. M. (1968). *American kinship.* Chicago: University of Chicago Press.

Serre, D., & Pääbo, S. (2004). Evidence for gradients of human genetic diversity within and among continents. *Genome Research, 14*, 1679–1685.

Small, M. (2006). First soldier of the gene wars. *Archaeology, 59*, 46–51.

Snyder, L. (1926). Human blood groups: Their inheritance and racial significance. *American Journal of Physical Anthropology, 9*, 233–263.

Stuurman, S. (2000). François Bernier and the invention of racial classification. *History Workshop Journal, 50*, 1–21.

Subramanian, S. (1995, January 16). The story in our genes. *Time Magazine*, 54–55.

Templeton, A. R. (1998). Human races: A genetic and evolutionary perspective. *American Anthropologist, 100*, 632–650.

Thieme, F. P. (1952). The population as a unit of study. *American Anthropologist, 54*, 504–509.

Tucker, W. H. (2002). *The funding of scientific racism: Wickliffe Draper and the Pioneer Fund.* Urbana: University of Illinois Press.

Wade, N. (2002, December 20). Gene study identifies 5 main human populations. *New York Times.*

Washburn, S. L. (1963). The study of race. *American Anthropologist, 65*, 521–531.

Weiner, J. S. (1957). Physical anthropology: An appraisal. *American Scientist, 45*, 79–87.

Williams, F. L., Belcher, R., & Armelagos, G. (2005). Forensic misclassification of ancient Nubian crania: Implications for assumptions about human variation. *Current Anthropology, 46*, 340–346.

Wolf, E. R. (1982). *Europe and the people without history.* Berkeley: University of California Press.

2

What Genes Are and Why There Are No Genes for Race

JOHN DUPRÉ

Talk of the genetic basis of race has resurfaced in the aftermath of spectacular progress in the development of genetic technologies, most especially technologies that provide genetic tests, allegedly "for race," that coincide quite closely with self-reported racial identities. In a much discussed Op-Ed piece for the *New York Times* (March 14, 2005), Armand Leroi, an evolutionary biologist at Imperial College, argued that the classic claim by Richard Lewontin (1972) that human variation was overwhelmingly within rather than between races, so that traditional racial categories could be seen as socioeconomic constructs, was based on an elementary statistical error.[1] Whereas, taking all genes separately, Lewontin's claim was true, the clustering together of genes characteristic of particular groups was able to show a distinctive genetic inheritance to traditional racial groups. These developments carry a significant danger of lending new respectability to controversial speculations about racial differences in such politically charged characteristics as IQ.

Unfortunately, the further these discussions move away from the technical contexts in which these genetic tests originate, the more misunderstandings appear. Other chapters in this volume (see Bolnick, Fullwiley) explain in detail the statistical procedures to which Leroi is referring and the limitations of conclusions drawn from them. As I shall try to show in this chapter, insights from recent genomic science have helped clarify and highlight the ambiguities and misunderstandings that threaten incautious interpretations of genetic data and even of the very concept of a gene. I shall also show how misunderstandings of these concepts can and do lead to spurious conclusions of the kind just outlined with regard to race and that, in fact, the kinds of tests just referred to do nothing to underwrite traditional racial categories.

The topic can be approached by considering the apparently quite straightforward claim that there are genes for race. This may seem banal and obvious; physical characteristics such as skin color presumably have a genetic basis. Far from being banal, the claim just mentioned is so difficult to interpret as to be close to unintelligible. This is because of the great difficulty of making clear sense of its main terms. The difficulty in defining race is familiar: although it is quite widely accepted by relevant experts that race is primarily a socially constructed concept, as the present volume documents there are still many who think of the concept as fundamentally biological. And whether it is a social or biological concept, there is no agreement as to how many races there are. Other contributors to this volume will explain many of the issues here. Difficulties with the concept of gene may be less familiar. The first task of this chapter will be to explain why this concept is so problematic. This will make possible the differentiation of various interpretations of the claim under consideration about genes and race and examination of the plausibility and implications of each of them.

Are There Genes?

To explain the complexities of the various uses of the term "gene," it will be helpful to provide a very condensed and somewhat Whiggish history. Genetics is generally thought of as starting with the work of the Austrian monk Gregor Mendel. Mendel's famous experiments in the 1860s involved crossing varieties of peas which, after generations of work by plant breeders, consistently produced plants with known phenotypes. For example, he interbred lines of peas with yellow- and green-colored seeds. The first generation produced by this crossing was found to have uniformly yellow seeds. When these hybrid peas were crossed with one another, the second generation was found to have 75% of plants with yellow seeds and 25% with green. Similar results were claimed with other observable features.

The (mildly Whiggish) interpretation of these results goes as follows. Plants are assumed to contain factors that produce the observable traits. Each plant contains two such factors, one of which is passed on to the offspring. If we call the factor that produces yellow seeds Y and the other G, we assume that the true-breeding lines have two copies of the same factor, and we refer to the lines as YY and GG. The first-generation hybrids will therefore have one factor of each kind, which we refer to as YG. The observation that the first-generation hybrids are all yellow, hence that YG plants have yellow seed, is interpreted as showing that the Y factor dominates. In the second generation, assuming that parents are equally likely to pass on either of their factors, the plants will be divided between YYs, GGs, and YGs in the ration 1:1:2. Since only the quarter that are GG will produce green seeds, this explains the quantities found in the classic experiments.

Also famously, Mendel's results were largely ignored until they were taken up by several scientists independently at the beginning of the 20th century. In 1909 the Danish biologist Wilhelm Johannsen named these factors *genes*. And in the first few decades of the 20th century a highly successful research program, now referred to as Mendelian genetics, greatly expanded empirical knowledge of the transmission of traits from organisms to their descendants. The most influential embodiment of this program was the work of Thomas Hunt Morgan and his students and collaborators on the fruit fly *Drosophila melanogaster*.

The most crucial thing to note about Mendelian genes, the objects of study in this episode of scientific history, is that they were causes of differences. No difference, no genes. In this strict Mendelian sense, there are no genes for traits that are universal in a population. This is a concept suited to evolutionary theory, where selection can only work on differences, and one that remains prominent in medical genetics, since medicine is centrally concerned with deviations from the norm—and hence with genetic peculiarities that cause differences. In light of this general point we can easily see that the idea of genes for race is highly problematic.

Even supposing, for the sake of argument, that belonging to a particular race is a biological trait at all, it is certainly not the kind of trait that could be the subject of a Mendelian experiment. We could not, for example, examine the offspring of two black people, or a black person and a white person, and decide how many of them were black. Whatever the phenotypic criteria are for these categories, they are not immediately accessible to inspection. This simple observation, incidentally, is already enough to throw serious doubt on the idea that race might be a biological trait, but we will continue for the moment with the counterfactual assumption that it is. Minimally, the problem is that race, even if it were a biological trait, would be far too complex a trait. Perhaps there are Mendelian genes for dark skin, hair texture, the shape of facial features, and so on, but race is at least a matter of there being many such traits. Some people have some but not all of the relevant set of traits. There is quite certainly no gene that makes the difference between being black or white, even ignoring, for the moment, the fact that there is an important social aspect to many racial categorizations.

The simple preceding point is important because a lot of talk of genes is still firmly embedded in the tradition of Mendelian genetics, a tradition that, first, explicitly licenses the idea that genes are *for* a phenotypic trait, the trait to which they make a difference, and, second, almost inescapably suggests the erroneous inference that the trait is caused by the gene. Medical genetics, as just noted, is still very largely Mendelian, in that its traditional and continuing central concern is with genes that make a specific difference, resulting in sometimes devastating pathologies. Medical genetics, it is true, is now moving rapidly toward a concern with predisposing genes, specific alleles that make it more probable that a particular pathology will develop. Good examples are the BRCA-1 and BRCA-2 genes, which strongly predispose women to developing breast cancer. But this hardly brings us nearer to a promising line for understanding genes for race. The idea of an allele that increases the probability of belonging to a particular racial group is a nonsensical idea. Many of the hypothesized genes for complex properties—intelligence, sexual orientation, violence, and so on—that appear regularly in the popular press raise a similar problem of spurious assimilation to traditions of Mendelian research on heredity, a further reason to emphasize the conceptual pitfalls. But to get to some slightly more plausible lines of thought, we should briefly move to some more recent history.

From the early stages of this program it was widely, but by no means universally, assumed that Mendelian factors, or genes, would eventually turn out to be specific material entities. Quite quickly, a consensus emerged that these were located on chromosomes—threadlike structures that were observable with the microscopes of the time. This consensus was reinforced as techniques were developed that enabled the order of the genes along the chromosomes to be ascertained, techniques which also made the hypothesis that genes were physical entities increasingly hard to resist.

A turning point in the attempt to convert hypothetical Mendelian genes into something solidly material was the determination of the structure of DNA by James Watson, Francis Crick, and others in 1953. The molecule had a number of features that seemed essential for anything that could be the bearer of Mendelian genes. The very long sequence of varying components making up a DNA molecule could be seen to have the information-carrying capacity, by varying the order of its nucleotide components, to specify the many traits for which there were genes. The double helical structure, allowing the possibility of separating into two strands, each of which could provide the template for a new double helix, provided a mechanism for the indefinite transmission of this information. And DNA was also a sufficiently stable molecule to maintain with some reliability the information it carried. It was naturally hoped that Mendelian genes would turn out to be specific sequences of nucleotides in the DNA that made up the chromosomes. When, a few years later, the "code" through which triplets of nucleotides "represented" specific amino acids was discovered, revealing the way in which sequences of DNA could provide information for the production of functional proteins, such a hope seemed to some even closer to realization.

However, the last fifty years of molecular genetics can also be seen as a gradual unraveling of this attractive vision. First of all, functional proteins correlate very poorly indeed with the phenotypic traits that are of interest to the student of whole organism inheritance. Even quite simple traits turn out to be the result of developmental processes involving many different protein products and much else besides, and for more complex traits the number of proteins involved might be hundreds or thousands. Hence, a particular gene, conceived as the sequence of DNA coding for a particular protein, would typically have no very specific phenotypic upshot. Not only did traits turn out to have many genetic causes, but a particular gene would generally contribute to the development of many traits.[2]

More recently, the situation has proved to be far more complex still. First, the assumption that identifiable bits of DNA sequence are even "genes for" particular proteins has turned out not to be generally true. Alternative splicing of fragments of particular sequences, alternative reading frames, and post-transcriptional editing—some of the things that happen between the transcription of DNA and the formation of a final protein product—are among the processes the discovery of which has led to a radically different view of the genome. The relationship between stretches of DNA and protein products is already many/many. Coding sequences in the genome are therefore better seen as resources that are used in diverse ways in a variety of molecular processes and that can be involved in the production of many different cellular molecules than as some kind of representation of even a molecular outcome, let alone a phenotypic one.

Moreover, most of the genome doesn't code for anything. When it was still assumed that the function of the genome was to code for proteins, this non-coding sequence came to be known as "junk DNA." As a more complex view of the genome is emerging, it has become an increasingly more plausible project to look for different functions of this material. It is understood that much of the non-coding DNA is nevertheless transcribed into RNA, and the list of identified functions of these RNA molecules is growing rapidly. DNA sequences at other sites attach to various chemicals in the cell which in turn affect the rate of transcription at related loci. And it is plausible that even parts of the chromosomes that do not have specific chemical functions may have structural importance. The structural configuration of the chromosomes will affect, for instance, which parts are accessible to the transcription machinery. The assumption that the genome merely stores information is becoming untenable, and it now appears rather as an object in constant dynamic interaction with other constituents of the cell.

A problem that emerges from all this and that is exercising a growing number of philosophers of biology is whether any coherent interpretation of the concept "gene" can be recovered from these complexities. The first part of an answer is well captured by the proposal of Lenny Moss (2003) to distinguish two kinds of usage, which he calls genes-P and genes-D. Genes-P are related to phenotypes and the biological tradition of preformation. They are most obviously the genes for this or that phenotypic trait found in the Mendelian tradition. Genes for cystic fibrosis or Huntington's disease are genes-P, as are the red-eye or wingless genes in *Drosophila*. Mendelian genes retain important, but highly circumscribed, uses, but they are quite unsuited to general characterization of the genome. Genes-D are understood in relation to development. Genes-D are defined not by their phenotypic outcome but by their molecular sequence. Though they are often referred to by means of a protein for which they "code," it is important to be aware that the sequence is fundamental. So, for instance, the N-CAM gene, named for the neural cell adhesion molecule for which it codes, can actually produce perhaps 100 different isoforms of this molecule in different tissues and at different developmental stages. Genes-D are the functional constituents of the genome as these constituents need to be distinguished in order to understand molecular function and, more specifically, the way genomes contribute to the differentiation of cell function in development. The philosophical issue here is whether there is any canonical division of the genome into genes-D or whether, as I and a number of commentators suspect, this is just a name for any sequence of nucleotides that may, for a particular investigation, be of interest to a particular group of researchers.[3] But what is clear is that genes-D cannot, in general, be identified with relation to their outputs even at the level of functional proteins, let alone at the level of phenotypic traits.

Kinds of Genes

To get a better sense of the diversity of contexts and uses in which the term "gene" appears, I will now summarize some of the more prominent uses of the term with a view to exploring what relevance, if any, they might have to our understanding of human race. I don't claim that this is a complete list of such uses or the only way in which this concept could be conceptually divided. I hope this list will, however, illustrate the diversity of uses across the Mendelian/molecular and gene-D/gene-P divides and, consequently, make it clear how hazardous it is to talk about genes, or the genetic, without a good deal of clarity about what is intended.

1. *The hypothetical cause of a phenotypic difference.* This is the original meaning of the term "gene" in classical Mendelian genetics and the standard gene-P. I mentioned, as an example, the gene for cystic fibrosis.
2. *The physical cause of a difference.* There are, of course, physical features

of the genome corresponding to traits with Mendelian inheritance patterns. These may be point mutations, deletions, insertions, inversions, duplications, and so on. They are of continuing interest mainly in the study of genomically based pathologies and also in the application of genomic knowledge to the improvement of techniques for plant and animal breeding. But they are not the objects of primary interest in the study of normal development. "The gene for cystic fibrosis" actually refers to a large number of possible mutations in a particular part of the genome, so the relation of this concept to the previous one is not straightforward correspondence.

3. *The physical cause of a trait (the gene for X).* Such a thing can only be assumed to exist at all in so far as there is a Mendelian trait; in which case it will be the kind of thing described in 2. For most X, there is no genomic feature or set of features that can be distinguished as the gene or genes for X. This is very likely the case for most of the Xs for which genes are regularly announced in the popular press, contributing to massive popular confusion on the general nature of genetics. The expression "gene for X" is, according to Moss (2003), the canonical expression of the failure to distinguish genes-P from genes-D and very often signals a conflation of these two concepts.

4. *Quantitative trait loci.* This category constitutes a technical qualification of the negative remark at the end of the last category. Breeders interested in, for example, leaner cattle or bigger cabbages, can locate particular genomic loci that have particular relevance to such features and use this information to improve breeding programs. These are "genes for big cabbages" only in the sense that changes in these loci tend to affect the size of cabbages more than other loci. These loci may have countless effects, and their effects on cabbages may be dependent on interactions with numerous further genetic and environmental factors. Quantitative trait loci can be identified for breeding purposes through genetic markers (see below).

5. *An open reading frame (ORF).* This is the sense of "gene" intended, more or less, when we are told that there are only 13,000 genes in the human genome. It is a bit of sequence that is sometimes transcribed as a block into RNA. These transcripts are then subject to the processes of alternative splicing, editing, and so on that result in the much larger number of proteins, probably at least 10 times the number of ORFs, though still a very speculative quantity. This is a concept located firmly in genomic studies, and it is unlikely to be confused with a "gene for" a specific trait.

6. *A functional part of the genome.* This very loosely defined concept may well be the way the concept is heading in technical molecular biology.

This, I take it, is the paradigmatic gene-D. I gave the example above of the N-CAM gene, which plays a central role in the production of a set of closely related proteins. These can coincide with the ORF or be much smaller genomic elements.

7. *An error in the genome.* This is the molecularized Mendelian concept in much of medical genetics. It refers to any peculiarities of the genome with pathological consequences. For example, the cause of a particular case of cystic fibrosis is a particular error, one of the set of mutations corresponding to the Mendelian gene.

8. *A genetic marker.* These will be discussed below. Whether they are properly referred to as genes is questionable, but this is the concept that underlies many recent claims about genetics and race. A genetic marker is a specific bit of sequence that need not have any function or correspond to any natural unit, but which is used to locate a part of the genome, generally because it is close to some functionally interesting part, for which it can serve as the marker.

Race and Genes

I have already noted that the first category on the list has no possible relevance to race and, by implication, neither has the second. Whatever races are, they are not Mendelian traits, traits that are present or absent in any individual and are transmitted in specific ratios across generations. Number 3 deserves a little more discussion. If there are biological features that constitute belonging to a particular race (and I continue to assume this for purposes of the argument), then surely there is some set of factors that causes those features. In considering this suggestion, it is worth recalling that all or most biological kinds encompass a substantial amount of variation.[4] Hence, if there are causes that make something a member of a kind, these causes are themselves likely to be diverse. Anyone who thinks that particular races are objective biological kinds must admit that they are variable kinds with diverse memberships.

It must be admitted that the variability of a trait does not in general prevent genetic analysis of the processes involved in its ontogeny. Examples are the complex diseases—diabetes, Alzheimer's, cardiovascular disease—that are currently undergoing this kind of investigation. Two further points distinguish such cases from the case of race, however. First, it is important to distinguish the basis of correct function from the basis of dysfunction. Various blows on the head can disrupt proper brain function, but there is no correct blow on the head that explains normal brain development. In the case of diabetes, for example, part of the genetic project is distinguishing different classes of genetic failure which call for quite different therapeutic responses. None of these failures is in "the gene for correct blood sugar regulation." But second, and most importantly, the fundamental reply is that race is not a biological trait at all, it is a social classification. So there is no candidate subject for genetic explanation. In some cases this is self-evident, as in the category of "black" defined by the one-drop rule in the United States. More generally, this conclusion is entailed by the failure of races to constitute credible biological kinds. What I mean by this, and the reasons for it, should become clearer in the later stages of this chapter.

Of course, even though races are not kinds, and even if race is not a biological trait, several phenotypic features strongly associated with conceptions of race are. Most obvious is skin color. Skin color is not, however, a Mendelian trait but, like height, the kind of continuously variable character associated with many genetic loci. If one were interested in breeding people with darker or lighter skins, one could probably discover QTLs (qualitative trait loci) that would facilitate a sophisticated breeding program of this kind. But particular alleles contributing to skin color are quite certainly not located exclusively in members of one (socially defined) race. Apart from the fact that people of different races do, of course, interbreed, this is evident in the continuous variability of skin color.[5] And again, of course, there are people who count socially as white who have darker skins than some people who count as black (a well-tanned person from Southern Europe, say, versus an American with one grandparent of African descent).

These elementary observations about variability of race-indicative phenotypes and the regular interbreeding between people socially defined as belonging to different races is enough to show that there is no interesting work to be done in this area by genes-D. One could investigate the developmental processes by which skin color, say, is determined, and this would include investigation of various genes in the sense of number 6, above (genes-D). It is almost certain that a variety of developmental processes might equally be found to lead to the same skin color, and certain that no such precise developmental sequence would be exactly correlated with any particular socially defined race.

Development, the more or less species typical physiological trajectory of an organism, depends on a great variety of factors. Though the genome is, of course, an essential factor, what genomic resources are deployed at any point in the developmental cycle depends on many other factors of many kinds. Familiar metaphors for the genome—blueprint, recipe, program, and so on—suggesting that the genome alone determines the development of an organism, are entirely misleading. The resources required for development are sometimes divided between the genetic and environmental, but within such a division environmental resources will range from parental care and social context to the set of extragenomic factors passed from mother to offspring in the egg cytoplasm. Any of these "environmental factors," up to and including the social, can affect the chemical and physical structure of the genome in ways that will contribute to the determination of which genomic resources are exploited by the developing organism. These complexities of development and of genomic function explain more deeply the point stressed earlier in this chapter: that it is in general quite mistaken to think of bits of the genome having specific functions defined in terms of phenotypic outcomes. So, finally, there is no reason to expect a particular set of genomic features to provide a complete causal explanation of a feature such as skin color. Like other features, we should expect skin color to be the final outcome of various possible developmental pathways, exploiting a range of genomic and other developmental resources.

Genetic Testing for Race

With this background we can address the concerns that have been raised by recent claims that race can reliably be discerned with genetic tests. First, we need to relate the relevant categories of genetic test to the various gene concepts that have been distinguished earlier in this chapter. These genetic tests depend on large numbers of genetic markers; and though these are genes-D, in the sense that they are specific bits of DNA sequence, there is no necessary, or even expected, connection to functional units of DNA. The tests in question are essentially similar to the technologies used in criminal forensic genetics and paternity testing.[6] Human genomes differ considerably in detail, and one current successor project to the Human Genome Project is the cataloguing of single nucleotide polymorphisms, SNPs, particular points in the genome at which different individuals are found to have differing nucleotides. A particular SNP can provide the basis for a specific genetic marker. If one tests an individual for a sufficient number of common SNPs, one can find a profile that becomes more unique the larger the number of SNPs for which one looks.[7] An important thing to note about all these technologies for genetic profiling is that the variation studied is preferentially drawn from parts of the genome that do not have coding functions. This is for the simple reason that the less functionally critical the sequence is, the more variation will accumulate in it. Variation in sequence with important coding function is likely to have deleterious effects on the organism, so that integrity of sequence is maintained both by internal editing processes and, failing that, natural selection.

As well as being clear about the relevant concept of gene here, we need to consider how race is being conceived. The simple answer is that race is being identified with geographical ancestry. Certainly, the concepts are related. Traditional broad racial categories were assumed to coincide with origins in the world's major continents, and more local racial concepts are often identical to concepts based on ancestry: African American means, more or less, having ancestors from (West) Africa. In the United States "black" is often understood as synonymous with "African American," though in the United Kingdom the former term is used much more widely to include more or less anyone with naturally (i.e., not environmentally induced) dark skin and can include South Asians or aboriginals from Australia or New Zealand.

The reason why SNPs or other indicators of genetic variation can track ancestry is clear enough. Since SNPs appear by random mutations and are passed on to offspring, they will, if they are fortunate, diffuse slowly around and away from the populations in which they originally occurred. SNPs will for a considerable time be most common in those areas where they have originated. Thus, and especially where populations are less mobile, particular populations of SNPs will tend to characterize particular geographical regions. We may finally consider the significance of recent reports (Rosenberg et al. 2002; Bamshad et al. 2003), apparently disturbing to some, that a genetic test can give results that very reliably predict whether Americans categorize themselves as African American, white (or European American), or Asian. Such tests indicate whether a person has ancestors who came from a particular geographic area in which particular SNPs originated. Thus, the ability to distinguish, through these tests, those Americans who identify as African American shows that those who so identify tend to have more ancestors from a particular region, presumably in this case West Africa. This should hardly surprise anyone. It should be stressed, though, that all of these SNPs will be found in many people who don't identify as black since racial interbreeding will ensure that they are gradually spreading through the wider population.

If one is trying to specify an individual from among a generally interbreeding population, the desirability of tracking nonfunctional, or minimally functional, parts of the genome is clear. The more functional the locus, the more selection will work to reduce variability. This argument is less straightforward in the case of testing for geographical ancestry. If populations have adapted to local conditions, then genes involved in such adaptations will be the most reliable indicators of that ancestry. Markers linked to such selected genes will be the ones that will tend to spread through the population. As a matter of fact it is unclear whether such functional genes are available.[8] The majority of loci actually applied are variations that occur in all populations and differ only in their frequencies within populations. The best correlation with West African ancestry in U.S. populations is exhibited by the so-called Duffy null gene, a variation that confers almost complete protection against the malaria *Plasmodium vivax*. This gene occurs in a very large majority of sub-Saharan Africans but is rare in Europeans. This degree of bifurcation is not currently known for any other locus. The extent to which the genetic variation characteristic of geographical locations is due to such adaptive histories is unknown, though for reasons that will be briefly discussed in a moment, much of this variation is likely to be a great deal more local than even the geographically restricted racial categories currently being considered. In what follows I shall mainly consider nonfunctional SNPs.

It should be noted that the same process of geographic origin of genetic mutations does explain the concentration of genetic disease in people of particular geographic origins. Many genetic diseases are consequences of simple point mutations in coding sequences that lead to a pathological defect in a functional protein. Echoing the preceding discussion, selective processes such as linkage to a selected locus or heterozygote superiority (as in sickle cell disease) will most effectively spread the deleterious genes, though drift may also be a sufficient explanation. Such localized genetic diseases lend some support to the idea that racial identification may have a use in targeting of genetic medicine, though the weaknesses of such a strategy should also by now be clear. First, ancestry and race are not identical concepts, and only ancestry has any relevance to the incidence of genetic disease. Racial self-identification is at best a rough proxy for a specific ancestry. Second, as with the functionally neutral mutations that are typically sought in ancestry testing, deleterious mutations will not be confined to people who identify as having ancestors in the relevant geographic region of their origin. There is distinct danger that exaggerated correlations will be assumed, and disease will be overlooked in people who do not identify with the right racial proxy groups. We may hope that as genetic tests become rapidly cheaper, diagnosis by racial classification will be a very temporary transition toward more general disease screening or individual genetic testing.

Are There Biologically Distinguishable Human Kinds?

There is a long philosophical tradition of asking whether the categories into which we sort things are in some way given to us by nature or rather imposed on the world. Categories such as the chemical elements or, though more controversially, biological species are often assumed to be given by nature and discovered by us, and these are often referred to as "natural kinds."[9] Pencils, penitentiaries, and philosophy professors, on the other hand, are clearly humanly created categories. A formulation of the question about race that has been partially resurrected by recent genetics is whether human races are natural kinds (see Haslanger, this volume). I take the answer to be an unequivocal no. The procedures just discussed for testing for geographic ancestry in America are effective because the large majority of African Americans have ancestors in a relatively specific geographic region. This does nothing whatever to support broader racial categories. As noted above, in the United Kingdom people are classified as black if they are non-white and, hence, experience discrimination. It includes people with Asian origins as well as those of African and Afro-Caribbean descent. It would also include (no doubt a very small number of) native Australians or New Zealanders. This is about as heterogeneous a group of humans in terms of origin, and therefore genetic profiles, as it would be possible to construct. In the United States the category is more narrowly circumscribed in terms of descent from black populations in West Africa, though to the extent that the one-drop rule is taken seriously this would make the category even more heterogeneous in terms of origin and genetics than the British version. At any rate, no serious scientist thinks these categories, even if the American category is interpreted in terms of some greater predominance of African ancestors, have any biological grounding that could justify any claim to the status of a natural kind.

However, as previously remarked, the human population does have some geographic structure. Relatively isolated populations of humans, as with most species, make minor but specific evolutionary adaptations to their environments quite quickly. Skin color has been mentioned as one superficial characteristic which is notably fluid in human micro-evolutionary history. Following Kaplan and Pigliucci (2003), I have discussed elsewhere (2003, ch. 7) the relevance of local human "ecotypes" to discussions of race. The point is just that while there have been and continue to be numerous very local human types adapted to specific local conditions, this is a vastly finer-grained classification than any standard racial category. This phenomenon very possibly explains such observations as the dominance of Kenyans among international marathon runners as the consequence of local adaptation to a culture involving extensive running at high altitude. (It has also been pointed out, though, that this tradition of success promotes a culture of aspiration in this particular direction.)

Broad racial categories, at any rate, comprise large numbers of ecotypes that are likely to differ in most respects of local adaptation. All equatorial peoples share dark skins and related adaptations to high temperature and strong solar irradiation, but there is little reason to suppose that they share any other adaptations not specifically responsive to climate. This is, of course, why claims such as a correlation between race and IQ are so biologically implausible. Though it is possible that local adaptation may promote subtle differences in cognitive skill sets, no good reason has been

offered why these differences should be common to all ecotypes in low lati-
tudes. Given that race broadly defined is a social kind with no interesting
biological grounding, it is overwhelmingly plausible that familiar social
explanations—less educationally enriched environments, subtly culturally
biased questions, and so on—will be more relevant to explaining prima facie
data of this kind.

Why Does All This Matter?

It is sometimes remarked that it is misguided and even dangerous to engage
in debates about the biological reality of race. By doing so, it is said, one
is offering quite unnecessary hostages to fortune. What if races did turn
out to be biologically significant categories? We would still have no reason
to discriminate against people because they fell into a different biological
kind from our own. After all, male and female are indisputably significant
biological categories, but this provides no justification for treating women
(or men) as inferior. This is all true enough, and it is certainly important,
if only because of the last point about sexual kinds, to insist that biological
difference is no simple justification for social discrimination.

However, I think it is important to engage with the biological issue. First
of all, there is no very serious hostage to fortune involved. We know enough
about race to be quite confident that races will not turn out to be significant
biological kinds, and it is at least worth explaining recent developments in
genetics which are liable to be interpreted as underwriting biological interpretations of race.

Second, and more importantly, although we should not (of course)
unjustly discriminate against people on the grounds of difference, real differences can and do provide reasons for different treatment. The political
consequences of sexual difference are a much more complex issue than
those of racial difference. Minimally, the fact that most women bear children
is a reality that cannot simply be ignored in a just society. The problem, or
one problem, is to make sure that this fact does not lead to systematic and
unjust disadvantaging of women. In the case of race, by contrast, there is no
such difference and therefore no such problem. If, as some racists may once
have thought, black people were an evolutionary experiment somewhere on
the step from apes to white people, there would be a real question as to what
differences in treatment, if any, this justified (as, indeed, there is beginning
to be a debate as to whether we are morally justified in treating apes in
quite different ways from humans). But such a sharp distinction between
human races is, needless to say, biological nonsense. Racial categories group
together highly diverse groups of people on the basis of multiply evolved and
trivial surface characteristics, and it would be miraculous if there turned

out to be systematic biological differences dividing members of socially distinguished racial groups. So there is no question of what differences there should be in the treatment of people of different races: there should be none. The only question is the political one of how we move from racially divided societies practicing racial discrimination to a situation in which race ceases to be a concept of any interest to anyone. Addressing biological misunderstandings doesn't do much to get us there, but it provides a small part of the necessary groundwork.

Conclusion

Contrary to some popular misunderstandings, there are no "genes for" race in any of the various senses of the word "gene." There is a lot of local variation within the human species, as there is for almost any widely distributed species; but as migration, easier travel, and so on make the species increasingly panmictic, this variation is likely to become ever more dispersed. This variation, moreover, provides no grounding whatever for the much coarser classifications that make up traditional racial categories or, indeed, any other comparable higher-level categories. The human species is an unusually genetically homogeneous one, and there are no important natural kinds distinguishable within it. As I have also discussed, genetic techniques make it possible to identify the geographic origins of some of the ancestors of individuals. But this reflects random and insignificant changes that occur in local human populations, or perhaps superficial adaptations to very local conditions, not the discrimination of significantly different kinds. Recent biology has confirmed the conviction of those who have long insisted that racial kinds were social kinds, and it has undermined any possible argument for placing these kinds in the realm of the biological. In its broadest and most common understanding, the concept of race remains little more than the reified residue of racism.

NOTES

The support of the Economic and Social Research Council (ESRC) is gratefully acknowledged. This work was part of the program of the ESRC Centre for Genomics in Society (Egenis). The chapter has benefited greatly from comments on an earlier draft by various colleagues in Egenis, especially Christine Hauskeller, Staffan Müller-Wille, and Maureen O'Malley. I also received very helpful comments from Sarah Richardson. Finally, my understanding of the topic was much improved by attendance at the Authors' Conference in Stanford in January 2006.

1. Leroi attributes the statistical insight to Edwards (2003). Robust replies can be found in Graves (2005).
2. I should note that the many/many relations between genes and traits were not unfamiliar to geneticists in the first half of the 20th century.

3. See various discussions in Beurton, Falk, and Rheinberger (2000). For empirical evidence that biologists do not have a clear consensus on the meaning of "gene," see Stotz and Griffiths (2004).

4. Issues about biological kinds are discussed in several essays in Dupré (2002).

5. I understand that in parts of Latin America social discrimination is based on continuous variations of skin color. Though this may no doubt ground a form of racism, it does not assume the dichotomous view of distinct races that is the present subject of discussion.

6. Individual genetic fingerprinting is most often based on measuring repeated sequences of variable length in non-coding DNA. This is a somewhat simpler technology, but the basic point is the same, namely, an inventory of variable aspects of the genome. It is also worth noting that most tests of this kind use loci on either the Y chromosome or the mitochondria, thereby restricting their relevance to only male or female ancestors. This has the rather striking consequence that claims to ancestry at, say, 10 generations in the past will actually be based on the genome of just one of the 1,024 ancestors in that generation.

7. Actual testing, both for ancestry and for individual identity, in fact uses a range of variable genomic features. I mention SNPs in large part because they are the easiest to explain. The differences between these features are not important for the present discussion.

8. Some commercial providers of ancestry tests do claim to use mainly or entirely functional loci, and I don't wish to query (or endorse) their claims. One reason it is difficult to do so is that most of these loci are proprietary information. As noted in the text, I don't think that anything fundamental is at stake.

9. For general discussion, see Dupré (2002), especially chapters 1, 2, and 8.

REFERENCES

Bamshad, M. J., Wooding, S., Watkins, W. S., Ostler, C. T., Batzer, M. A., & Jorde, L. B. (2003). Human population genetic structure and inference of group membership. *American Journal of Human Genetics, 72*, 578–589.

Beurton, P., Falk, R., & Rheinberger, H.-J. (2000). *The concept of the gene in development and evolution: Historical and epistemological perspectives.* Cambridge: Cambridge University Press.

Bolnick, D. A. (2008 [this volume]). Individual ancestry inference and the reification of race as a biological phenomenon. In B. A. Koenig, S. S.-J. Lee, & S. S. Richardson (Eds.), *Revisiting race in a genomic age* (pp. 70–85). New Brunswick, NJ: Rutgers University Press.

Dupré, J. (2002). *Humans and other animals.* Oxford: Oxford University Press.

Dupré, J. (2003). *Darwin's legacy: What evolution means today.* Oxford: Oxford University Press.

Edwards, A.W.F. (2003). Human genetic diversity: Lewontin's fallacy. *Bioessays, 25*, 798–801.

Fullwiley, D. (2008 [this volume]). The molecularization of race: U.S. health institutions, pharmacogenetics practice, and public science after the genome. In B. A. Koenig, S. S.-J. Lee, & S. S. Richardson (Eds.), *Revisiting race in a genomic age* (pp. 149–171). New Brunswick, NJ: Rutgers University Press.

Graves, J. L. (2005, April 25). What we know and what we don't know: Human genetic variation and the social construction of race. See http://raceandgenomics.ssrc.org/Graves/pf/.

Haslanger, S. (2008 [this volume]). A social constructionist analysis of race. In B. A. Koenig, S. S.-J. Lee, & S. S. Richardson (Eds.), *Revisiting race in a genomic age* (pp. 56–69). New Brunswick, NJ: Rutgers University Press.

Kaplan, J., & Pigliucci, M. (2003). On the concept of biological race and its applicability to humans. *Philosophy of Science* (Supplement), *70* (5), 1161–1172.

Lewontin, R. C. (1972). The apportionment of human diversity. *Evolutionary Biology, 6*, 381–398.

Lewontin, R. C. (2005, April 20). Confusions about race. See http://raceandgenomics.ssrc.org/Lewontin/pf/.

Moss, L. (2003). *What genes can't do*. Cambridge: MIT Press.

Rosenberg, N. A., Pritchard, J. K., Weber, J. L., Cann, H. M., Kidd, K. K., Zhivotovsky, L. A., et al. (2002). Genetic structure of human populations. *Science, 298*, 2981–2985.

Stotz, K., & Griffiths, P. E. (2004). Genes: Philosophical analyses put to the test. *History and Philosophy of the Life Sciences, 26*, 5–28.

3

A Social Constructionist Analysis of Race

SALLY HASLANGER

In the contemporary world the term 'race'[1] is used widely both in American popular culture and in a variety of academic disciplines, and its meanings evolve in different ways in response to the pressures in each. This chapter brings philosophical analysis to bear on the debate among geneticists, humanists, and social scientists over the meaning of the term 'race' in a genomic age—a debate that extends beyond our immediate disciplines and into the public domain. What are the genuine disagreements and what are only apparent disagreements due to the use of different vocabularies? Why does it matter which of the positions we accept? What sort of evidence is relevant to adjudicating the claims? How should we go about resolving the controversy? In answering these questions, I develop a realist, social constructionist account of race. I recommend this as an account that does justice to the meanings of 'race' in many ordinary contexts and also as an account that serves widely shared antiracist goals.

I argue that in debates over the meaning of 'race' in a genomic age we are better served by shifting from the metaphysical/scientific question, Is race real? to the political question, What concept of race should we employ in order to achieve the antiracist goals we share? To answer this question, I contend that we must also look at the semantics of the term 'race' in public—specifically nonscientific—discourse, for this popular notion of race is what we use to frame our identities and political commitments. My argument is based on a view of language as a collective social practice rather than a set of terms stipulated by an authority. On this view, the issue is not whether groups of people—experts in a particular field or folk in a neighborhood—are entitled to use the term 'race' for the divisions in which they are interested. Of course they are: there are no "language police," and people can appropriate and transform language for their own purposes. Similarly, what 'race' means outside of the stipulated meaning operative in the biology lab is not up to the biologist.[2] Just as there is no "language police" to judge that the biologist is wrong to use the term 'race' in a particular way, likewise there is no "language police" or even "language legislature" to determine what a term will mean in public discourse. Language evolves in complicated and subtle ways. Thus, I argue that anyone using the term 'race' in public life should be aware of its ordinary meanings; and if we want to change or refine the concept of race, we should be aware of where we are starting from as well as the normative basis for where we want to go.

Race Eliminativism, Race Constructionism, and Race Naturalism

Questions of what the term 'race' means and whether race is *real* have become tied up with different political goals and strategies for achieving them.[3] Race *eliminativists* maintain that talk of races is no better than talk of witches or ghosts, and in order to achieve racial justice we should stop participating in a fiction that underwrites racism (Appiah, 1996; Zack, 2002). Race *constructionists* argue that races are real, but that they are social rather than natural groups; on the constructionist view, racial justice requires us to recognize the mechanisms of racial formation so that we can undo their damage (Omi and Winant, 1994; Mills, 1997; Haslanger, 2000). Present-day race *naturalists* agree with the eliminativists and constructionists that races are not what they were once thought to be—they are not groups with a common racial essence that explains a broad range of psychological and moral features of the group's members—but they disagree with both other views in maintaining that the human species can be divided on the basis of natural (biological, genetic, physical) features into a small set of groups that correspond to the ordinary racial divisions (Kitcher, 1999; Andreason, 2000; Rosenberg et al., 2002; Mountain & Risch, 2004), *and* that this natural division is socially and politically important for the purposes of achieving racial justice, for example, by enabling us to address racially divergent medical needs (Risch, Burchard, Ziv, & Tang., 2002; cf. Lee, Mountain, & Koenig, 2001).

Although the choice between these approaches to race may seem to some as "just semantics" (in the pejorative sense), the debate plays a role in framing and evaluating social policy. For example, consider the FDA approval of BiDil, a drug to treat heart failure, for Black[4] patients. Eliminativists, naturalists, and constructionists will have very different approaches to this decision. For example, if, as the eliminativist argues, race is not real, then the approval of BiDil for Blacks is as (un)justified as the approval of BiDil for witches. The category *Black*, on the eliminativist view, is a fiction projected onto the world, and the FDA has done social harm by reinforcing the illusion that the category is scientifically grounded. In contrast, a race naturalist could support the FDA's action—or if not in the particular case of BiDil, in a similar sort of case—arguing that racial categories map biological categories that may have significant health consequences and should not be ignored in developing new medicines. On the naturalist's view, it is as politically important for the FDA to address the biological implications of race differences as it is to address the biological implications of any other genetic differences that have medical implications; in fact, to ignore the real differences between the races would be a form of injustice. The constructionist would disagree with the naturalist that there are natural differences between the races that warrant different medical treatment, but could allow that the social differences race makes must be taken into account in deciding a course of treatment or the approval of a drug. Although disagreeing with the eliminativist's rejection of race, the constructionist would be sympathetic with the eliminativist's worry that the FDA has reinforced a pernicious belief in the natural basis for racial categories. But how should we adjudicate these different positions?

Natural and Social Kinds

Some are tempted to view the debate between eliminativists, constructionists, and naturalists as (primarily) a metaphysical/scientific debate about the reality of race. On this construal, the question is whether races are natural kinds. Eliminativists and naturalists agree that races, *if they exist*, are natural kinds. Naturalists hold that races are a natural division of human beings, i.e., a division which rests entirely on natural properties of things; eliminativists deny it. Constructionists reject the claim that races are natural kinds, i.e., they allow that races are kinds, but hold that the division rests at least partly on social properties (being viewed and treated in a certain way, functioning in a certain social role, etc.) of the things in question. This requires understanding social kinds as just as fully real as natural kinds (see table 3.1). There are semantic issues: What does 'race' mean? Is it part of the *meaning* of 'race' that races are natural kinds? There are scientific/metaphysical issues: Is race real? Do races exist? And there are moral/political issues: How should we, as a nation, address the problem of racial injustice?

Following Aristotle, the term "kind" is sometimes used to capture the classification of objects in terms of their *essence*. On this view, objects—genuine objects as opposed to heaps or weird scattered bits and parts of things—are distinctive because they have an essence. The rose bush in my garden is an object because of its rose-essence; the scattering of petals, leaves, dirt, pebbles, gum wrappers, and fertilizer under it is not an object because it has no essence. The essence of the individual is (roughly) that set of properties

TABLE 3.1

Sources of Disagreement

	Eliminativism	Constructivism	Naturalism
Is race a natural category?	Yes	No	Yes
Is race real?	No	Yes	Yes

without which the object cannot exist and which serves in some important way in explanations of the object's characteristic behavior.

Are *races* Aristotelian *kinds*? Traditional racialists would probably think they are (Appiah, 1993, chap. 2): Whites and Blacks have different natures that explain their characteristic behaviors, and this nature is essential to who they are. However, this view is not credible at this point in time. It would be implausible to claim that an individual could not have existed as a member of a different race. In fact, people can travel from the United States to Brazil and function socially as a member of a different race; and features as superficial as skin color, hair texture, and eye shape are clearly not essential (they, too, can be changed with chemicals and surgery). If one thinks that one has one's entire genetic makeup necessarily (something with even a slight difference from your genetic makeup wouldn't be you), then there might be a case to be made for the claim that one could not have been a member of a different race. But essences are supposed to be rich explanatory resources for explaining the characteristic behavior of the individual, and there is no support for the idea that there are racial essences of this sort.

Locke has a different account of kinds than Aristotle. For Locke, kinds are highly unified, but not by virtue of the essences of their members. So, e.g., red things constitute a kind (their unity consists in their all being red), even though redness is seldom an essential property of the things that have it. On a Lockean view, the main contrast to consider is between "real" kinds and "nominal" kinds. Real kinds are those types unified by properties that play a fundamental role in the causal structure of the world and, ideally, in our explanations. Nominal kinds are types unified by properties that happen to be useful or interesting to us. Whether there are real kinds corresponding to (and underlying) the nominal kinds we pick out is an open question. On this view *concepts* or *properties* (and, contra Aristotle, not individuals) have essences.

Are races Lockean kinds? Can we give necessary and sufficient conditions for being a member of a particular race? This question actually opens a long debate between realists and nominalists that (fortunately!) we don't need to get into about whether one can *ever* give necessary and sufficient

conditions for membership in a kind. If our goal is to do justice to our pre-
theoretical judgments about membership in a given race, then there are reasons to doubt whether races are definable in the sense required. However, if
we stipulate a definition, either as a nominal essence to pick out a group of
things we are interested in, or in postulating explanatory categories as part
of a theoretical project, then the definition will give the Lockean essence of
the kind.

Note that on both the Lockean and Aristotelian accounts, kinds or
types may be either social or natural. Types are *natural* if the properties that
constitute their unity are natural, and *social* if the properties are social. It
is notoriously difficult to characterize the distinction between natural and
social properties (and relations), but for our purposes we could take natural properties of things to be those studied by the natural sciences and the
social properties to be those studied by the social sciences. So the set of
quarks is a natural type; the set of adoptive families is a social type. Plausibly, there is *some* degree of unity in the members of a race, e.g., one could list
a cluster of physical, historical, and sociological properties associated with
each race such that members of the race share a weighted subset of those
properties. If for a category to be real is just for it to pick out a set with some
loose connection amongst the members, then there is a sense in which, on
any non-empty construal of race, races are real. It takes very little to be an
objective type in this sense.

Can "Facts" Settle the Matter?

Some may find it tempting to respond that to resolve this issue, we just need
to look at the facts: either there are races or there aren't; either races are
social or they aren't. One significant problem with this approach is that we
can determine whether there "really are" races only if the term 'race' has a
specified meaning; and what it means—at least for the purposes at hand—is
part of the question. Consider a different example. Suppose we ask, What
percentage of the U.S. population is on welfare? Well, it depends on what you
mean by 'welfare.' Do we include only those who receive TANF (Temporary
Assistance for Needy Families, the successor to "welfare as we know it")? Or
do we include those who receive social security benefits? What about "corporate welfare" in the form of tax breaks? We ask, Is race real? Well, it depends
on what you mean by 'race.'

This is not to say that the controversy will dissipate if we only would
make clear our stipulated definitions. If I maintain that 99% of the U.S.
population is on welfare, then presumably I am using a non-mainstream
definition of 'welfare.' For me to justify my claim it would not be sufficient to

say that given *my* meanings, I've uttered a truth, if *my* meaning of 'welfare' is idiosyncratic and beside the point. But it may be that what I say is true and especially useful in the context of the debate in which I engage. In such a case the task of justification would be to show that my definition of 'welfare' better tracks what is important for the purposes at hand (Anderson, 1995).

The reason why the facts don't settle the issue is that simply establishing that there is a fact of the matter about something doesn't establish that it is a significant or relevant fact for the purposes at hand. Suppose I say that I'm going to use the term 'White' for all and only those who have blonde hair. Whites, then, are a natural kind. Turn now to the public context in which we are discussing, say, affirmative action. If I argue that non-Whites should be given preferential treatment because of historical injustice, my claim sounds familiar, but the category I am using is not the most apt for considering the justice of affirmative action. The fact that 'White,' as I defined the term, captures a real kind, even combined with the truth that (some) non-Whites have been treated unjustly, does not usefully further the debate because I have chosen categories for addressing the problem that are ill-suited to the task (see Anderson, 1995). Truth alone does not sct us free; there are too many irrelevant and misleading truths. The choice of truths must—at the very least—be insightful and judicious.

Lessons from Philosophy of Language

So it would seem that the next step in our inquiry should be to adjudicate what the term 'race' means. As I mentioned before, there need not be only one meaning for the term. But for the purposes of engaging in discussion concerning matters of biological research on race, it would be useful to have a shared understanding of race. And to achieve this, we should have a sense of what the folk concept of race is. This is not because I believe that we should honor the folk concept as the *true meaning*, but because in any context where communication is fraught, it is useful to understand the competing meanings at issue. If there is a socially dominant understanding of race, then even if we want to recommend a change in the concept, we should know what it is.

This suggests that we must not simply resolve semantical disagreements in order to make headway in the debate. We must look more closely at our purposes and how we might achieve them: should we as biologists, social scientists, scholars, citizens, and as people who care about social justice frame our dialogue—our narratives of explanation, justification, and justice—in terms of race? And if so, then what concept of race should we employ? These questions can be broken down further:

- Is there currently a single or dominant public meaning (or folk concept) of 'race'? If so, what is it (or what are the contenders)?
- In the quest for social justice, e.g., in debating health policy, do we need the concept of race? For what purposes? If so, can we make do with the folk concept or should we modify the concept?
- If the folk concept of race is not an adequate tool to help achieve social justice (if, perhaps, it is even a barrier), then how should we proceed?

In what follows, I will suggest that an answer to the first question, in particular, is not straightforward; and yet if we are going to speak meaningfully in a public context, then we need to recognize the force and implications of our words in that context. In science it is commonplace to define or redefine terms in whatever way suits the theory at hand (e.g., 'atom,' 'mass,' 'energy,' 'cell,'), without much concern with the ordinary meanings these terms have or the political import of stipulating new meanings. But semantic authority cannot be granted to the biologist in considering a term like 'race' that plays such a major role in our self-understandings and political life.

In undertaking conceptual analysis of, say, *Fness* (in our case, *Fness* might be 'Blackness,' 'Whiteness,' 'Asianness,' or the broader category, 'race'), it is typically assumed that it is enough to ask competent users of English under what conditions someone *is F*. After all, if competent speakers know the meaning of their terms, then all that is needed is linguistic competence to analyze them. However, this stance is not plausible if one takes into account arguments in philosophy of language over the past 30 years that call into question the assumption that competent users of a term have full knowledge of what the term means. This assumption in particular is challenged by the tradition of semantic externalism. Externalists maintain that the content of what we think and mean is determined not simply by what we think or intend, but at least in part by facts about our social and natural environment. For example, one can be competent in using the term 'water' without knowing that water is H_2O; one can use the term 'elm' meaningfully even if one cannot tell the difference between a beech and an elm. When I say, 'Elm trees are deciduous' I say something meaningful and true, even though I couldn't identify an elm or give any clear description of one. The externalist holds that these sorts of cases point to two features of language that the traditional picture ignored: *reference magnetism* and *the division of linguistic labor*. These ideas can be expressed *very* roughly as follows:

> *Reference magnetism* (Putnam, 1973, 1975; Kripke, 1980): type-terms (such as general nouns) pick out a type, whether or not we can state the essence of the type, by virtue of the fact that their meaning is determined by a selection of paradigms together

with an implicit extension of one's reference to things of the same type as the paradigms. For example, the marketing department and the R&D department of a toy manufacturer have a meeting. R&D has produced a new "squishy, stretchy substance that can transform into almost anything," and they present a sample. The marketing director points to it and says, "Let's call the stuff 'Floam.'" Bingo. 'Floam' now refers to a whole kind of stuff, some of which has not yet been produced, and the ingredients of which are totally mysterious. Which stuff? Presumably, 'floam' refers to the most unified objective type of which the sample is a paradigm instance. This example is artificial, but the phenomenon of reference magnetism is ubiquitous.

Division of linguistic labor (Putnam, 1975, Burge, 1979): the meaning of a term used by a speaker is determined at least in part by the linguistic usage in his or her community, including, if necessary, expert usage. For example, before the invention of chemistry, people used the term 'water' to refer to H_2O because the kind H_2O was a "reference magnet" for their term. However, in cases where one cannot even produce a paradigm, e.g., when I can't tell the difference between a beech and an elm, my use of the term 'elm' gets its meaning not from *my* paradigms, but from the linguistic labor of others in my community, including botanists. The division of linguistic labor may also play an important role if I have idiosyncratic paradigms. The idea is that what I mean in using a term such as 'elm' or 'arthritis' is not just a matter of what is in my head, but is determined by a process that involves others in my language community.

Most commonly, externalist analyses have been employed to provide *naturalistic* accounts of knowledge, mind, etc.; these seek to discover the *natural* (non-*social*) kind within which the selected paradigms fall. But it is possible to pursue an externalist approach within a social domain as long as one allows that there are social kinds or types, such as 'democracy' and 'genocide,' or ethical terms such as 'responsibility' and 'autonomy.'

Of course, an externalist analysis of a social term cannot be done in a mechanical way and may require sophisticated social theory both to select the paradigms and analyze their commonality. It may take sophisticated social theory to determine what 'parent' or 'Black' means. In an externalist project, intuitions about the conditions for applying the concept should be considered secondary to what the cases in fact have in common: as we learn more about the paradigms, we learn more about our concepts.

Is Race a Fiction?

If we are externalists about meaning, which is the approach I am recommending, then the eliminativist about race is in a very weak position. We can all confidently identify members of different races. Martin Luther King, Nelson Mandela, Malcolm X, Toni Morrison, Oprah Winfrey, W.E.B. DuBois, Kofi Annan, Thabo Mbeki (insert here your choice of various friends and relatives) are Black. George Bush, Arnold Schwarzenegger, Margaret Thatcher, Golda Meir, Bertrand Russell, Vincent Van Gogh (insert here your choice of various friends and relatives) are White. Similar lists can be constructed for Asians, Latino/as, and other groups usually considered races. But if this is the case, then the terms 'Black' and 'White' pick out the best fitting and most unified objective type of which the members of the list are paradigms—even if I can't describe the type or my beliefs about what the paradigms have in common are false. What that type is is not yet clear. But given how weak the constraints on an objective type are, undoubtedly there is one. The term 'race' then, picks out the more generic type or category of which 'Black,' 'White,' etc. are subtypes.

I believe that these considerations about meaning show that eliminativism is the wrong approach to understand the public or folk meaning of 'race.' It is compatible with this that we should work to change the public meaning of 'race' in keeping with the eliminativist strategy so that it becomes clear that the racial terms are vacuous. In other words, eliminativism may still be a goal for which to aim. But as things stand now, race is something we *see* in the faces and bodies of others; we are surrounded by cases that function to us as paradigms and ground our meanings. The eliminativist's suggestion that "our" concept of race is vacuous is not supported by the observation that we tend to think of races as natural kinds because the meaning of 'race' isn't determined simply by what we think races are. So the eliminativist project needs to be rethought.

Race as a Social Kind

Recent work in race genetics and biology leads me to believe that there are no very unified natural types that are good candidates for the reference of race terms, where the reference of these terms is fixed by generally acceptable paradigms of each race (see Feldman and Lewontin, this volume; Bolnick, this volume). What "we" in public discourse call race is not a natural or genetic category. Rather, the ordinary term 'race' picks out a social type, i.e., the objective type that attracts our reference is unified by social features rather than natural ones.[5] Let me sketch one suggestion along these lines.

Feminists define 'man' and 'woman' as *genders* rather than sexes (male and female). The slogan for understanding gender is this: gender is the social meaning of sex. It is a virtue, I believe, of this account of gender that, depending on context, one's sex may have a very different meaning and it may position one in very different kinds of hierarchies. The variation will clearly occur from culture to culture (and subculture to subculture); so, e.g., to be a Chinese woman of the 1790s, a Brazilian woman of the 1890s, or an American woman of the 1990s may involve very different social relations and very different kinds of oppression. Yet on the analysis suggested, these groups count as women insofar as their subordinate positions are marked and justified by reference to (female) sex.

With this strategy of defining gender in mind, let's consider whether it will help in giving some content to the social category of race. The feminist approach recommends this: don't look for an analysis that assumes that the category's meaning is always and everywhere the same; rather, consider how members of the group are *socially positioned* and what *physical markers* serve as a supposed basis for such treatment.[6]

I use the term 'color' to refer to the (contextually variable) physical markers of race, just as the term 'sex' to refers to the (contextually variable) physical markers of gender. "Color" is more than just skin tone: racial markers may include eye, nose, and lip shape, hair texture, physique, etc. Virtually any cluster of physical traits that are assumed to be inherited from those who occupy a specific geographical region or regions can count as "color." (Although the term 'people of color' is used to refer to non-Whites, the markers of "Whiteness" also count as "color.") Borrowing the slogan used before, we can say then that race is the social meaning of the "colored," i.e., geographically marked, body (see fig. 3-1).

To develop this briefly, consider the following account.[7] A group is *racialized* (in context C) if and only if (by definition) its members are (or would be) socially positioned as subordinate or privileged along some dimension (economic, political, legal, social, etc.) (in C), and the group is "marked" as a target for this treatment by observed or imagined bodily features presumed to be evidence of ancestral links to a certain geographical region.

In other words, races are those groups demarcated by the geographical associations accompanying perceived body type when those associations take on evaluative significance concerning how members of the group should be viewed and treated. Given this definition, we can say that S is of the White (Black, Asian, etc.) race (in C) if and only if (by definition) Whites (Blacks, Asians, etc.) are a racialized group (in C) and S is a member.[8]

Note that on this view, whether a group is racialized, and so how and whether an individual is raced, will depend on context. For example, Blacks, Whites, Asians, and Native Americans are currently racialized in the United

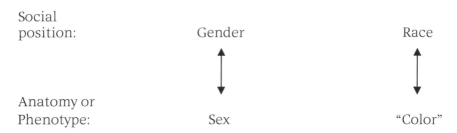

FIGURE 3-1. Meanings given to the body generate social positions, which, in turn, produce new interpretations of (and sometimes modifications of) the body.

States insofar as these are all groups defined in terms of physical features associated with places of origin and membership in the group functions as a basis for evaluation. However, some groups are not currently racialized in the United States but have been so in the past and possibly could be again (and in other contexts are), e.g., the Italians, the Germans, the Irish.

I offer the constructionist analysis of 'race' just sketched as one that captures our ordinary use of the term. The social constructionist analysis of race presents the strongest conceptual framework and consensus point for cross-disciplinary and public discussions around race and genetics research. I believe it also provides important resources in politically addressing the problem of racial injustice; specifically, it gives us a way of capturing those groups that have suffered injustice due to assumptions about "color." These are groups that matter if we are going to achieve social justice. Moreover, we already use racial terms in ways that seem to track these groups (or groups very close to them). So by adopting the constructionist account we can proceed politically without recommending a semantic revolution as well.

Conclusion

I have argued that the debate between eliminativists, constructionists, and naturalists about race should be understood as not simply about whether races are real or whether they are natural kinds, but about how we should understand race and employ racial concepts in our public discourse. I have argued that the debate cannot be settled simply by considering "the facts" of genetics, but requires close attention to the language of 'race' and 'kind' as well as contemporary racial politics. With this reframing of the question, I have argued that our ordinary concept of race is of a social kind and for a particular analysis of race that highlights social hierarchy. Given the history of racial injustice and the need to address this history, it is important for us to attend publicly to those who have suffered from what we might call *color hierarchy*. Since we have reason to track racial injustice, and since the naturalist and eliminativist accounts do not come close to matching our ordinary term for 'race,' constructionism about race is currently the best candidate of the three views considered. My conclusions are qualified, however. I do not argue that my account of race captures *the meaning of 'race'* (or what we should mean by 'race') for all time and in all contexts; it would be foolhardy for anyone to attempt that. More specifically, it would reveal a misunderstanding of how language, as a collective social practice, works.

NOTES

1. In this chapter, I follow the philosopher's convention of distinguishing between use of an expression and mention of it. When a word is mentioned, i.e., when the subject matter is the word or term and not what the word or term usually means, it is enclosed in single quotes. 'Race' in single quotes refers to the word itself; without the quotes it has the conventional meaning. Double quotes are used for quotation of another's text or as scare quotes. Scare quotes indicate that the author is distancing himself or herself from the choice of term and is relying on a known, potentially problematic, usage.

2. Note that the term 'race' did not originate as a biological term but plausibly has religious/metaphysical origins (Stocking, 1994).

3. I sometimes frame the question as whether race is real as opposed to whether races exist because sometimes the debate is muddled by those who want to allow that races *exist* (e.g., "in the head" or "in society") but that they aren't *real*. As I see it, if races exist only in the head, then they don't exist (just as people may believe in unicorns, but this is not to say that they exist); and if races exist in society, then they do exist, since social categories are real. But to avoid potential disagreements over what it means to say that something exists, I've framed the question instead as whether races are real.

4. In this chapter, upper case is used for names of races, i.e., Black and White; lower case is used for color terms.

5. It is a controversial issue as to what counts as a "social fact" and in what sense the social is "constructed." In my discussion I assume very roughly that social facts are "interpersonal" facts or facts that supervene on such facts. So, simplifying considerably, *I am Deb's friend* is a social fact because it supervenes on a certain base set of interpersonal actions and attitudes. Others, such as John Searle (1995), have much higher demands on what counts as a social fact, including controversial "we-intentions," assignment of function, and the generation of constitutive rules. These elements are more plausibly required in creating institutional facts or conventional facts (his standard example is the social constitution of money); it is too demanding to capture much of ordinary, informal social life. E.g., we can have coordinated intentions without them being "we-intentions"; things can have a social function even if they aren't assigned it; and social kind membership isn't always governed by rules. Searle's analysis is not well-suited to the project of analyzing gender and race, which are the heart and soul (so to speak!) of ordinary, informal social life.

6. This analysis is part of a larger project aiming to identify sites of structural subordination; other projects, such as those undertaking to define a basis for racial or

ethnic identity (McPherson and Shelby, 2004) or those offering reconstructions of the notion of race (Gooding-Williams, 1998; Alcoff, 2000), are not incompatible with this.

7. On this I am deeply indebted to Stevens (1999, chap. 4) and Omi and Winant (1994, esp. pp. 53–61). I develop this definition more fully in Haslanger, 2000. Note that if this definition is adequate, then races are not only objective types but are Lockean (social) kinds.

8. As in the case of gender, I recommend that we view membership in a racial/ethnic group in terms of how one is viewed and treated *regularly and for the most part* in the context in question; one could distinguish *being* a member of a given race from *functioning as* one in terms of the degree of entrenchment in the racialized social position.

REFERENCES

Alcoff, L. M. (2000). Is latino/a identity a racial identity? In J. E. Gracia and P. De Grieff (Eds.), *Hispanics and Latinos in the United States: Ethnicity, race, and rights* (pp. 23–44). New York: Routledge.

Anderson, E. S. (1995). Knowledge, human interests, and objectivity in feminist epistemology. *Philosophical Topics, 23,* 27–58.

Andreason, R. (2000). Race: Biological reality or social construct? *Philosophy of Science, 67* (supplementary volume), S653–666.

Appiah, K. A. (1993). *In my father's house.* New York: Oxford University Press.

Appiah, K. A. (1996). Race, culture, identity: Misunderstood connections. In K. A. Appiah and A. Gutmann (Eds.), *Color conscious: The political morality of race* (pp. 30–105). Princeton: Princeton University Press.

Armstrong, D. (1989). *Universals: An opinionated introduction.* Boulder, CO: Westview Press.

Bolnick, D. A. (2008 [this volume]). Individual ancestry inference and the reification of race as a biological phenomenon. In B. A. Koenig, S. S.-J. Lee, & S. S. Richardson (Eds.), *Revisiting race in a genomic age* (pp. 70–85). New Brunswick, NJ: Rutgers University Press.

Burge, T. (1979). Individualism and the mental. *Midwest Studies in Philosophy, 4,* 73–121.

Burge, T. (1986). Intellectual norms and foundations of mind. *Journal of Philosophy, 83,* 697–720.

Delphy, C. (1984/1970). *Close to home: A materialist analysis of women's oppression* (D. Leonard, Trans.). Amherst: University of Massachusetts Press.

Feldman, M. W., & Lewontin, R. C. (2008 [this volume]). Race, ancestry, and medicine. In B. A. Koenig, S. S.-J. Lee, & S. S. Richardson (Eds.), *Revisiting race in a genomic age* (pp. 89–101). New Brunswick, NJ: Rutgers University Press.

Gooding-Williams, R. (1998). Race, multiculturalism, and democracy. *Constellations, 5,* 18–41.

Hartmann, H. (1981). The unhappy marriage of Marxism and feminism: Towards a more progressive union. In Lydia Sargent (Ed.), *Women and Revolution* (pp. 1–42). Cambridge, MA: South End Press.

Haslanger, S. (1995). Ontology and social construction. *Philosophical Topics, 23* (2), 95–125.

Haslanger, S. (2000). Gender and race: (What) are they? (What) do we want them to be? *Noûs, 34,* 31–55.

Haslanger, S. (2003). Social construction: The "debunking" project. In Frederick F. Schmitt (Ed.), *Socializing metaphysics: The nature of social reality* (pp. 301–325). Lanham, MD: Rowman and Littlefield.

Kitcher, P. (1999). Race, ethnicity, biology, culture. In L. Harris (Ed.), *Racism* (pp. 87–117). New York: Humanity Books.

Kripke, S. (1980). *Naming and necessity.* Cambridge, MA: Harvard University Press.

Lee, S. S.-J., Mountain, J., & Koenig, B. (2001). The meanings of "race" in the new genomics: Implications for health disparities research. *Yale Journal of Health Policy, Law and Ethics, 1,* 33–75.

MacKinnon, C. (1987). *Feminism unmodified.* Cambridge, MA: Harvard University Press.

McPherson, L., & Shelby, T. (2004). Blackness and blood: Interpreting African-American identity. *Philosophy & Public Affairs, 32,* 171–192.

Mills, C. (1997). *The racial contract.* Ithaca, NY: Cornell University Press.

Mountain, J. L., & Risch, N. (2004). Assessing genetic contributions to phenotypic differences among "racial" and "ethnic" groups. *Nature Genetics, 36* (11 Suppl), S48–53.

Omi, M., & Winant, H. (1994). Racial formation. In M. Omi and H. Winant, *Racial formation in the United States* (pp. 53–76). New York: Routledge.

Putnam, H. (1973). Meaning and reference. *The Journal of Philosophy, 70,* 699–711.

Putnam, H. (1975). The meaning of "meaning." In H. Putnam, *Mind, language, and reality.* Vol. 2 of *Philosophical Papers* (pp. 215–271). Cambridge, MA: Cambridge University Press.

Risch, N., Burchard, E., Ziv, E., & Tang, H. (2002). Categorization of humans in biomedical research: Genes, race, and disease. *Genome Biology, 3,* 2007.1–2007.12.

Rosenberg, N. A., Pritchard, J. K., Weber, J. L., Cann, H. M., Kidd, K. K., Zhivotovsky, L. A., et al. (2002). Genetic structure of human populations. *Science, 298,* 2381–2385.

Scott, J. (1996). Gender: A useful category of historical analysis. In J. Scott (Ed.), *Feminism and History* (pp. 152–180). Oxford: Oxford University Press.

Searle, J. (1995). *The construction of social reality.* New York: The Free Press.

Stevens, J. (1999). *Reproducing the state.* Princeton, NJ: Princeton University Press.

Stocking, G. (1994). The turn-of-the-century concept of race. *Modernism/Modernity, 1* (1), 4–16.

Wittig, M. (1992). *The straight mind and other essays.* Boston: Beacon Press.

Zack, N. (2002). *Philosophy of science and race.* New York: Routledge.

4

Individual Ancestry Inference and the Reification of Race as a Biological Phenomenon

DEBORAH A. BOLNICK

Anthropological ideas about the pattern of human diversity shifted drastically during the 20th century. Prior to World War II, *Homo sapiens* was generally perceived as a polytypic species with biologically distinct subgroups, or races (Stepan, 1982; Marks, 1995). This biological differentiation was thought to be the result of long periods of independent evolution when each race was largely isolated from the others. Anthropologists gradually moved away from such typological thinking during the latter half of the 20th century, in part because new genetic data did not support this paradigm. Instead, genetic research suggested that humans could not be neatly divided into a few discrete, isolated races (Brown & Armelagos, 2001; Kittles & Weiss, 2003). Studies of human biological diversity therefore began to focus less on classification and more on the actual patterns of variation among populations, as well as on the evolutionary processes that shaped those patterns.

With this shift away from typological thought has come an increased interest in *individuals* and what genetics can tell us about the unique identity and history of each person. As part of this trend, anthropologists and geneticists have recently begun to explore how genomic data can be used to infer an individual's "ancestry." I will consider the meaning of this term in more detail later in this chapter, but "ancestry" is generally used to refer to the geographic region or regions where one's biological ancestors lived (Jorde & Wooding, 2004; Race, Ethnicity, and Genetics Working Group, 2005).

Several methods have been developed for inferring an individual's ancestry from genetic data (Rannala & Mountain, 1997; McKeigue, Carpenter, Parra, & Shriver, 2000; Pritchard, Stephens, & Donnelly, 2000), and these methods are starting to be used in a variety of contexts. For example, individual ancestry inference has important biomedical applications because ancestry may influence disease susceptibility and drug response (Wilson et al., 2001; Risch, Burchard, Ziv, & Tang, 2002; Helgadottir et al., 2005; Tate & Goldstein, this volume). Individual ancestry inference can also aid forensic investigations by determining the genetic heritage of DNA left at a crime scene, which can then be used to narrow the pool of potential suspects (Frudakis et al., 2003; Shriver, Frudakis, & Budowle, 2005). Finally, these methods are also of great interest to members of the general public who want to reconstruct their personal genealogical histories (Elliott & Brodwin, 2002; Bolnick, 2003; TallBear, 2005; Greely, this volume; Shriver & Kittles, this volume; TallBear, this volume).

Although this body of work emphasizes the *individual* as the crucial unit of analysis, individual ancestry inference is closely tied to our understanding of human *groups* and the distribution of genetic variation among them. Inferring an individual's genetic ancestry entails deciding that his or her DNA was inherited from a certain group or groups, and that cannot be accomplished unless one first distinguishes groups that differ genetically in some way. Thus, even such individually oriented genetic research has implications for our understanding of race and the pattern of human biological diversity.

In this chapter, I begin with an overview of our current understanding of the pattern of human biodiversity. I then examine two widely cited studies that use the *structure* program (Pritchard et al., 2000) to infer individual ancestry (Rosenberg et al., 2002; Bamshad et al., 2003) and discuss what these studies imply about the relationship between human genetic structure and traditional notions of race.

The Distribution of Human Genetic Variation

Our current understanding of human genetic structure is based on hundreds of studies that have been conducted over the past few decades. Both mitochondrial DNA and nuclear loci have been surveyed using many different types of markers (Tishkoff & Verrelli, 2003). While the specific findings of each study have varied, two general patterns have consistently emerged.

First, African populations exhibit greater genetic diversity and less linkage disequilibrium than non-African populations (Tishkoff & Williams, 2002; Kittles & Weiss, 2003).[1] This pattern reflects the evolutionary and demographic history of our species. Because *Homo sapiens* evolved in Africa before dispersing throughout the rest of the world (Klein, 1999), African populations are older and have had more time to accumulate genetic differences through mutation. Similarly, the greater age of African populations helps to explain the lower levels of linkage disequilibrium in Africa since linkage disequilibrium decreases over time due to recombination (Kittles & Weiss,

2003). Differences between African and non-African populations also reflect a genetic bottleneck that occurred when humans dispersed out of Africa. The individuals who left Africa carried only a subset of the genetic variants found in the ancestral African population. Consequently, non-Africans are less genetically diverse and exhibit increased linkage disequilibrium compared to Africans (Bamshad, Wooding, Salisbury, & Stephens, 2004; Tishkoff & Kidd, 2004).

The second pattern that has emerged from many genetic studies is that human variation is clinally distributed (see fig. 4-1). Allele frequencies change gradually across geographic space, with few sharp discontinuities (Barbujani, 2005). Populations are most genetically similar to others that are found nearby, and genetic similarity is inversely correlated with geographic distance (Relethford, 2004; Ramachandran et al., 2005).

There are several reasons for this pattern. First, it reflects localized gene flow and isolation by distance (Cavalli-Sforza, Menozzi, & Piazza, 1994; Relethford, 2004). In other words, because geographic distance limits migration, individuals tend to mate with those who live nearby and geographically close populations tend to exchange more genes than geographically distant ones (Wright, 1943; Malécot, 1969). Restricted gene flow therefore

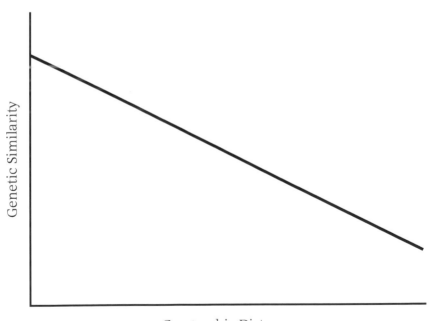

FIGURE 4-1. The relationship between genetic similarity and geographic distance under a pattern of clinal variation.

contributes to the observed pattern of decreasing genetic similarity with increasing geographic distance.

The clinal pattern of human genetic variation also reflects successive founder effects that occurred as humans migrated out of Africa to populate the rest of the world (Relethford, 2004; Prugnolle, Manica, & Balloux, 2005; Ramachandran et al., 2005). Prugnolle et al. (2005) and Ramachandran et al. (2005) suggest that this form of genetic drift played a particularly important role in shaping the patterns of variation among human populations. According to their analyses, serial founder effects explain 76–85% of the observed variation (Prugnolle et al., 2005; Ramachandran et al., 2005).

Finally, clinal variation at some loci reflects selection in response to environmental gradients. Clines due to selection vary from locus to locus (i.e., allele frequencies at one locus change faster than those at another locus over the same geographic distance). Since many loci show similar patterns of allele frequency change across human populations, the overall pattern of genetic variation in our species reflects selection less than serial founder effects and restricted gene flow with isolation by distance (Relethford, 2004).

Because of the patterns of human genetic variation described here, many anthropologists argue that traditional notions of race misrepresent human biological diversity and the evolutionary history of our species. While traditional notions of race are extremely variable—no consensus has ever been reached regarding the number or composition of human races, for example—most describe racial groups as equivalent, biologically distinct units (Barbujani, 2005). However, the patterns described above suggest that this is not the case. From a genetic perspective, non-Africans are essentially a subset of Africans (Quintana-Murci et al., 1999; Underhill et al., 2000; Kidd, Pakstis, Speed, & Kidd, 2004). No discrete boundaries separate humans into a few genetically distinct groups, and the members of each racial group are highly variable (Brown & Armelagos, 2001). Consequently, human racial groups do not appear to be distinct genetic groups.

Individual Ancestry Inference, Race, and Genetic Structure

Several recent studies of individual ancestry seem to challenge this understanding of the distribution of human genetic variation. These new studies instead suggest genetic differentiation among what are essentially races based on continental ancestry. For example, Rosenberg et al. "identified six main genetic clusters, five of which correspond to major geographic regions" (2002, p. 2381). Since the five "major geographic regions" comprise Africa, Eurasia, East Asia, Oceania, and America (Rosenberg et al., 2002), these results have been interpreted as showing that racial divisions based on continental ancestry are biologically significant (Burchard et al., 2003; Mountain & Risch, 2004). Similarly, Bamshad et al. (2003) identified three genetic clusters that correspond to Africa, Europe, and Asia. These studies have been widely cited as verifying traditional ideas about race and the pattern of human biological diversity (Wade, 2002; Seebach, 2003).[2]

The conclusions of both the Rosenberg et al. (2002) and Bamshad et al. (2003) studies were based on the Bayesian computer program *structure* (Pritchard et al., 2000). To understand the results of these two studies and what they imply about the structure of the human gene pool, it is important to first understand how this computer program works.

The *structure* program implements a model-based clustering method to infer population structure from multilocus genotype data and then allocates individuals into populations (Pritchard et al., 2000). It can be used to estimate the number of genetic clusters or populations present in a given data set as well as the population of origin of each individual. The populations are expected to be in Hardy-Weinberg equilibrium. Pritchard et al. (2000) assume a model in which a number (K) of populations exist, each of which is characterized by a set of allele frequencies.

A data set of multilocus genotypes (X) is therefore viewed as being made up of individuals sampled from K separate populations (see fig. 4-2). When running the *structure* program, the user defines K in advance. *Structure* then assigns individuals probabilistically to K populations with the goal of maximizing Hardy-Weinberg equilibrium in each population. In other words, for any given value of K, *structure* searches for the most probable way to divide the sampled individuals into that pre-defined number of clusters based on their genotypes. If an individual's genotype suggests that he or she has ancestry from more than one population, *structure* can assign the individual jointly to two or more populations and estimate the proportion of ancestry from each. The analysis can (and should) be performed for multiple different values of K.

Thus, the fact that *structure* identifies a particular number of clusters is insignificant: it does so simply because the user told it to do so. What is more

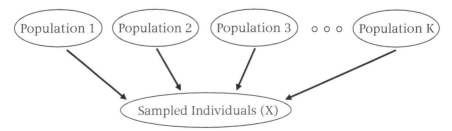

FIGURE 4-2. The population model assumed by the *structure* program.

important is that *structure* provides a way to determine the value of K that is most appropriate for the data set in question (i.e., the most likely number of clusters or populations represented). The "best" value of K is the one that maximizes the probability of observing that set of data. *Structure* calculates the probability of the data given each value of K submitted (i.e., Pr [X|K]), and the inferred value for K is the one associated with the highest Pr (X|K).

However, it is not entirely straightforward to determine the true number of genetic clusters in a given data set for three reasons. First, because it is computationally difficult to estimate Pr (X|K), *structure* provides only an approximation (Pritchard et al., 2000). Pritchard et al. note that "the assumptions underlying [this approximation] are dubious at best, and we do not claim (or believe) that our procedure provides a quantitatively accurate estimate of the posterior distribution of K. We see it merely as an *ad hoc* guide to which models are most consistent with the data, with the main justification being that it seems to give sensible answers in practice" (2000, p. 949). Thus, because *structure*'s estimates of Pr (X|K) may or may not be accurate, the value of K estimated as maximizing the probability of the data may not actually do so.

Second, if a data set is complex, different runs of *structure* may produce substantially different results. In these cases, the composition of genetic clusters varies among runs using the same pre-defined value for K. A simplistic illustration of this situation would be a case where the analysis of four individuals (A, B, C, and D) using $K = 2$ yielded clusters (A, B) and (C, D) in run 1, but clusters (A, C) and (B, D) in run 2. Pritchard and Wen (2004) suggest that this mostly occurs with data sets containing a large number of genetic clusters ($K > 5$) and is either because the program did not run long enough (i.e., *structure* did not have enough time to determine the optimal clustering of individuals) or because there are several highly probable ways to divide the sampled individuals into that number of clusters. If the latter is the case, it may not be possible to determine a single optimal clustering scheme. Furthermore, the different ways to divide individuals into a particular number of clusters may each yield a different Pr (X|K). For example, in the above illustration with $K = 2$, the clustering scheme in run 1 might be associated with a high Pr (X|K), whereas the clustering scheme in run 2 might be associated with a low Pr (X|K). Consequently, it is not always clear which of the probabilities associated with a given K should be used when determining the "best" value for K.

Third, the underlying model used in the *structure* program is not appropriate for all data sets. In particular, Pritchard and Wen (2004) note that the *structure* model is not well suited to data shaped by restricted gene flow with isolation by distance. If *structure* is used to analyze such data, they warn that "the inferred value of K . . . can be rather arbitrary" (2004, p. 14). Thus, although the *structure* program can estimate the number and composition of genetic clusters present in a given data set, such estimates must be interpreted carefully.

Rosenberg et al.'s (2002) Study of Human Genetic Structure

Rosenberg et al. (2002) used the *structure* program to analyze genotypic variation at 377 autosomal microsatellite loci in 1,056 individuals from around the world (the HGDP-CEPH Human Genome Diversity Cell Line Panel). The abstract of their article mentioned the identification of 6 main genetic clusters (Africa, Eurasia, East Asia, Oceania, America, and the Kalash of Pakistan), but Rosenberg et al. (2002) actually presented results for multiple values of K (2–6) in the body of the paper. They also analyzed the data set using values of $K > 6$ (up to $K = 20$; N. Rosenberg, personal communication), but they did not publish those results because *structure* identified multiple ways to divide the sampled individuals into K clusters when $K > 6$ (Rosenberg et al., 2002). For example, in 10 replicates, *structure* found 9 different ways to divide the sampled individuals into 14 clusters and 10 different ways to divide them into 20 clusters (N. Rosenberg, personal communication). The different clustering schemes in these replicates were fairly similar, but they often yielded very different Pr (X|K), making it difficult to interpret the results for a given value of K when $K > 6$. Rosenberg et al. (2002) therefore published the results for $K < 7$ for the worldwide sample, as well as further analyses using regional subsets of the entire data set.

Thus, the fact that *structure* identified 6 genetic clusters is not significant in and of itself—the program also identified 2, 5, 10, and 20 genetic clusters using the same set of data. As noted above, *structure* will identify as many clusters as the user tells it to identify. While it may be interesting that 5 of the 6 clusters identified with $K = 6$ correspond to major geographic regions, such clustering does not necessarily provide a better representation of human genetic differentiation than the clustering observed when K is set to 4, 9, 12, or any other number. Only by evaluating the probability of the observed data given each value of K (i.e., Pr [X|K]) is it possible to determine the number of genetic clusters *most likely* represented in this data set.

Rosenberg et al. (2002) did not report the *most likely* number of genetic clusters, nor did they publish the probabilities of the observed data given each value of K. Since some of the larger values of K were associated with several different Pr (X|K) across runs, and since Rosenberg et al. wanted to present results that could be easily replicated, they felt that it was more informative to show the robust results for multiple small K than to focus on a larger value of K that was associated with variable clustering schemes and both high and low probabilities (N. Rosenberg, personal communication).

In other words, no single value of K clearly maximized the probability of the observed data. Probabilities increased sharply from $K = 1$ to $K = 4$ but were fairly similar for values of K ranging from 4 to 20 (N. Rosenberg, personal communication). The probability of the observed data was higher for $K = 6$ than for smaller values of K, but not as high as for some replicates with larger values of K (N. Rosenberg, personal communication). The highest Pr $(X|K)$ was associated with a particular replicate of $K = 16$, but that value of K was also associated with very low probabilities when the individuals were grouped into 16 clusters in other ways (N. Rosenberg, personal communication). Consequently, it is uncertain what number of genetic clusters *best* fits this data set, but there is no clear evidence that $K = 6$ is the best estimate.

Thus, the Rosenberg et al. (2002) study does not challenge our current understanding of human genetic structure as much as some have suggested. Indeed, the fact that it was not possible to determine a single best value for K is exactly what we would expect given the clinal variation and pattern of isolation by distance found in our species. In addition, as Tishkoff and Kidd (2004) have noted, individuals from areas near the borders of the five "major geographic regions" exhibited ancestry from multiple genetic clusters. These results suggest a gradient of change between geographic regions, not discrete boundaries.[3]

So why has so much emphasis been placed on the results of the analysis using $K = 6$? Despite the fact that Rosenberg et al. (2002) presented no evidence that $K = 6$ represented the *most likely* number of genetic clusters in their data set, virtually all references to this study in both the scientific literature and the popular press mention the identification of either 5 or 6 genetic clusters (for examples, see Wade, 2002; Seebach, 2003; Bamshad et al., 2004; Tishkoff & Kidd, 2004; Barbujani, 2005; and Tate & Goldstein, this volume). I would suggest that these particular results have been emphasized simply because they fit the general notion in our society that continental groupings are biologically significant. This notion is a legacy of traditional racial thought and seems to persist even when not clearly supported by biological data.

Bamshad et al.'s (2003) Study of Population Structure and Group Membership

Bamshad et al. (2003) analyzed 100 *Alu* insertion polymorphisms in 565 individuals from sub-Saharan Africa, East Asia, Europe, and India, as well as 60 microsatellites in 206 of the individuals from sub-Saharan Africa, Europe, and East Asia. Like Rosenberg et al. (2002), they used the *structure* program to help determine the number of genetic clusters present in their data set. Bamshad et al. (2003) first analyzed only the samples from sub-Saharan

Africa, East Asia, and Europe and ran *structure* using values of K between 1 and 6. When all of the individuals from these three regions were included in the analysis, they found that $K = 4$ maximized the probability of observing that set of data (Pr $[X|K = 4] = 1$). The sampled individuals likely represented four genetic clusters, comprising (1) East Asians, (2) Europeans, (3) sub-Saharan Africans except for the Mbuti and three other individuals, and (4) the Mbuti and three other sub-Saharan Africans. This division of sub-Saharan Africans into two genetic clusters is consistent with other evidence of greater genetic diversity and greater genetic structure among Africans (Tishkoff & Williams, 2002).

Bamshad et al. (2003) also conducted this analysis excluding the Mbuti samples. In this case, *structure* found that $K = 3$ best fit the observed data (Pr $[X|K = 3] = 1$), indicating three genetic clusters of individuals (sub-Saharan Africans, Europeans, and East Asians). Bamshad et al. set K equal to 3 for most subsequent analyses even though those analyses used the complete data set (including the Mbuti), which most likely contained four genetic clusters. Given this, Bamshad et al. may have taken other (unnamed) factors into consideration when deciding upon the number of genetic clusters present in their data. They did note that "K provides only a rough guide for determining which models may be consistent with the data" because estimates of K depend on the number of individuals per population, the number of loci studied, and the amount of differentiation between populations (Bamshad et al., 2003, p. 579).

Bamshad et al. (2003) also investigated whether genetic data could be used to correctly infer an individual's ancestry. They used *structure* to assign individuals to genetic clusters and to estimate the proportion of an individual's ancestry from each genetic cluster. An individual "was considered assigned 'correctly' if the cluster with the greatest proportion of ancestry was the same as the continent of origin of the sample" (Bamshad et al., 2003, p. 579). Thus, Bamshad et al. assumed that continental groupings were important from the start, which perhaps explains why they chose $K = 3$ as the best estimate of human genetic structure.

Using the 100 *Alu* markers and 60 microsatellites, *structure* was able to identify the correct continent of origin for 99% of the individuals from sub-Saharan Africa, Europe, and East Asia. These results demonstrate that substantial genetic differentiation exists among the populations sampled, but they do not necessarily indicate substantial genetic differentiation among continental groupings. As Bamshad et al. (2003) note, these analyses included individuals from only a few widely separated regions of Africa, Asia, and Europe. The observed genetic differentiation may therefore reflect the large geographic distances between sampled populations rather than continental divisions per se.

Additional analyses using samples from areas closer to the continental borders support this hypothesis. When Bamshad et al. (2003) included samples from India in a data set with the European and East Asian samples, *structure* found that the optimal number of genetic clusters (*K*) was one. In other words, when a more representative geographic sample was analyzed, continental groupings no longer appeared to be genetically distinct. Accordingly, Bamshad et al. concluded that "the inclusion of [geographically intermediate] samples demonstrates geographic continuity in the distribution of genetic variation and thus undermines traditional concepts of race" (2003, p. 587).

The *structure* analysis in this study therefore supports our current understanding of human genetic structure. It indicates greater genetic diversity and greater genetic structure among Africans as well as little genetic differentiation among continental groupings *when a representative geographic sample is analyzed*. The results using the expanded Eurasian sample are also consistent with previous evidence that human genetic variation is clinally distributed with few sharp discontinuities.

Despite these results, Bamshad et al.'s (2003) study has been cited as showing that groups defined by continental ancestry or race are genetically differentiated (Mountain & Risch, 2004). This interpretation likely reflects the way that Bamshad et al. (2003) presented their results, rather than the results themselves. First, as noted above, they emphasized the significance of the three continental groupings even though *structure* identified four genetic clusters in the complete data set of sub-Saharan Africans, East Asians, and Europeans. Second, while Bamshad et al. (2003) recognized that the inclusion of samples from India demonstrated the continuous distribution of genetic variation in Eurasia, they excluded those samples from most analyses. As a result, many of the reported analyses implied continental discontinuities even though the more complete data set showed that such discontinuities do not exist.

Third, *structure* found that the European, Indian, and East Asian data set most likely contained a single genetic cluster, but Bamshad et al. (2003) focused primarily on an analysis of that data set using $K = 3$. In the text of their paper, Bamshad et al. wrote: "If we assumed that three clusters were present (i.e., $K = 3$), as suggested by proxy information (i.e., place of origin), three groups were distinguished. Correct assignment of samples to their place of origin was 97% for samples from East Asia, 94% for samples from Europe, and 87% for samples from southern India" (2003, p. 584). The article abstract also made no mention of the optimal clustering scheme ($K = 1$), but instead stated that "less accurate assignment (87%) to the appropriate genetic cluster was possible for a historically admixed sample from southern India" (Bamshad et al., 2003, p. 578).

Of course, as noted earlier, *structure* will identify as many groups as the program user tells it to identify, so it is not surprising—or significant—that *structure* distinguished three groups when Bamshad et al. set *K* equal to 3. Nor is it clear how to interpret the results of this analysis since it is statistically unlikely that three genetic clusters really exist in this data set. Bamshad et al.'s (2003) presentation of this analysis obscures these issues and makes it seem as if the three racial/ethnic groups (Europeans, East Asians, and Indians) are more genetically distinct than they really are.

Finally, Bamshad et al.'s description of the Indian population as "historically admixed" (2003, p. 578) reinforces traditional racial views of human variation and human evolutionary history. Previous studies have shown that the genetic makeup of the Indian population reflects gene flow from European and Asian sources (Bamshad et al., 2001; Majumder, 2001), but Bamshad et al.'s (2003) description suggests that such gene flow occurred only in historical times. Racial studies of the early 20th century presented a similar picture of Eurasian history. For example, Hooton (1931, 1939) suggested that populations resembling multiple races (such as Indians) formed only recently due to gene flow between the primary races, which were isolated from one another during prehistoric times. There is no evidence that a significant barrier to Eurasian gene flow existed in the more distant past, though, and other studies indicate migration and gene flow throughout Eurasia at many points in human history (Templeton, 2002; Basu et al., 2003). Thus, the way that Bamshad et al. (2003) describe their research reinforces traditional racial views of human variation even though the data do not necessarily support such views.

Ancestry and Race

Given the descriptions and interpretations of the studies by Rosenberg et al. (2002) and Bamshad et al. (2003), the relationship between ancestry and race should be examined more carefully. In recent years, ancestry has been widely promoted as an objective, scientific alternative to race. The term "ancestry" is often used without being clearly defined, but it generally refers to the geographic region or regions where one's biological ancestors lived (Collins, 2004; Jorde & Wooding, 2004; Shriver et al., 2004; Race, Ethnicity, and Genetics Working Group, 2005). Because of this focus, ancestry is seen as being more specific and objective than race (Bamshad, 2005), which is highly charged and encompasses geographic origins, political history, socioeconomic status, culture, skin color, and other perceived physical, behavioral, and genetic characteristics. Jorde and Wooding also argue that ancestry is "a more subtle and complex description of an individual's genetic makeup than is race" (2004, p. S30). An individual can have ancestry from multiple geographic regions, and the concept of ancestry is flexible enough that those regions could be local (e.g., southwestern Nigeria) or much broader (e.g., all of Africa).

However, other aspects of ancestry are more problematic. Just as the term is rarely defined, there has been little discussion of the size of geographic regions, how they should be defined, or why specific geographic divisions are more relevant than others for studies of human genetic variation. Nor is it always clear what time frame should be considered when determining an individual's ancestry. For example, my grandparents lived in the United States, but my great-great-grandparents lived in Eastern Europe. My more distant ancestors, like those of all humans, lived in Africa. The time depth of interest depends on the question or hypothesis being addressed, but this issue is often discussed only briefly, if at all.

Furthermore, ancestry is not that different from race in practice. Like race, ancestry is sometimes defined politically or culturally (Race, Ethnicity, and Genetics Working Group, 2005). In individual ancestry studies, the ancestral regions are almost always continents (Risch et al., 2002; Mountain & Risch, 2004; The Unexamined "Caucasian," 2004). Since the contemporary Euro-American definition of race is based on continental geography, anthropologists and human geneticists use the term "ancestry" much as the general public uses the term "race." Indeed, some scientists explicitly define ancestry as an individual's racial group or the race of his or her ancestors (Risch et al., 2002; Frudakis et al., 2003).

Because an individual can have ancestry from multiple geographic regions, ancestry does differ from conceptions of race based on the one-drop rule, which allow an individual to belong to only a single race. However, contemporary understandings of race accept the existence of "mixed-race" individuals, as evidenced by the large number of Americans who checked the box associated with the Other category on the last U.S. census. Thus, while ancestry has the potential to be a more subtle, objective, and scientific alternative to race, it currently appears to be quite similar to race in practice.

Conclusion

Recent studies of individual ancestry have been cited as verifying traditional ideas about race, but these studies do not present new data suggesting that racial groups are genetically distinct. Rather, the data and *structure* analyses reported in the Rosenberg et al. (2002) and Bamshad et al. (2003) studies are consistent with our current understanding of human genetic structure. However, the results of these studies have been described and interpreted in ways that both reflect and reinforce traditional racial views of human biological diversity and the evolutionary history of our species. The disconnect between the results and the interpretations of these studies is unfortunate since they are playing an important role in the reification of race as a biological phenomenon.

NOTES

I would like to thank Sarah Richardson, Sandra Soo-Jin Lee, Barbara Koenig, Daniel Bolnick, and Noah Rosenberg for their helpful suggestions and comments regarding an earlier draft of this chapter. I am also grateful to the participants in the authors' conference, who provided many valuable discussions.

1. Linkage disequilibrium (LD) refers to (a) the nonrandom association of alleles at different sites and (b) the length of a chromosomal segment that is inherited without recombination from a common ancestor (Kittles & Weiss, 2003; Tishkoff & Kidd, 2004).

2. Other studies have been cited as proving the same point, but I do not discuss those studies in this chapter since they are based on different methods of analysis than the Rosenberg et al. (2002) and Bamshad et al. (2003) studies. For example, Frudakis et al. (2003) used a linear classification method and Shriver et al. (2004) used a tree-based method. DNAPrint's AncestrybyDNA test, which also suggests the validity of race as a biological phenomenon (Bolnick, 2003), is based on an admixture-mapping approach (R. Malhi, personal communication).

3. Rosenberg et al. (2005) reanalyzed 1,048 individuals from the HGDP-CEPH Human Genome Diversity Panel after expanding their data set to include 993 markers. As in the 2002 article, Rosenberg et al. (2005) presented the results of *structure* analyses using $K = 2$–6 without specifying the *most likely* number of genetic clusters represented in the data set. Rosenberg et al. (2005) also examined the effects of several variables on the "clusteredness" of individuals, or the extent to which an individual was estimated as belonging to a single cluster. Although the HGDP-CEPH Human Genome Diversity Cell Line Panel does not represent a comprehensive sample of all regions occupied by humans, they found that more continuous geographic sampling would have little impact on the observed degree of clustering. Finally, Rosenberg et al. (2005) suggested that the clustering observed with $K = 5$ reflects slightly greater genetic differences across geographic barriers like oceans or the Sahara desert. Since $K = 5$ may or may not be the best estimate of the number of genetic clusters present in this data set, the biological significance of this finding is unclear.

REFERENCES

Bamshad, M. (2005). Genetic influences on health: Does race matter? *Journal of the American Medical Association, 294*, 937–946.

Bamshad, M., Kivisild, T., Watkins, W. S., Dixon, M. E., Ricker, C. E., Rao, B. B., et al. (2001). Genetic evidence on the origins of Indian caste populations. *Genome Research, 11*, 994–1004.

Bamshad, M., Wooding, S., Salisbury, B. A., & Stephens, J. C. (2004). Deconstructing the relationship between genetics and race. *Nature Reviews Genetics, 5*, 598–609.

Bamshad, M. J., Wooding, S., Watkins, W. S., Ostler, C. T., Batzer, M. A., & Jorde, L. B. (2003). Human population structure and inference of group membership. *American Journal of Human Genetics, 72*, 578–589.

Barbujani, G. (2005). Human races: Classifying people vs. understanding diversity. *Current Genomics, 6,* 215–226.

Basu, A., Mukherjee, N., Roy, S., Sengupta, S., Banerjee, S., Chakraborty, M., et al. (2003). Ethnic India: A genomic view, with special reference to peopling and structure. *Genome Research, 13,* 2277–2290.

Bolnick, D. A. (2003). "Showing who they really are": Commercial ventures in genetic genealogy. Paper presented at the American Anthropological Association Annual Meeting. November, Chicago, IL.

Brown, R., & Armelagos, G. J. (2001). Apportionment of racial diversity: A review. *Evolutionary Anthropology, 10,* 34–40.

Burchard, E. G., Ziv, E., Coyle, N., Gomez, S. L., Tang, H., Karter, A. J., et al. (2003). The importance of race and ethnic background in biomedical research and clinical practice. *New England Journal of Medicine, 348,* 1170–1175.

Cavalli-Sforza, L. L., Menozzi, P., & Piazza, A. (1994). *The history and geography of human genes.* Princeton: Princeton University Press.

Collins, F. S. (2004). What we do and don't know about "race," "ethnicity," genetics, and health at the dawn of the genome era. *Nature Genetics, 36,* S13–S15.

Elliott, C., & Brodwin, P. (2002). Identity and genetic ancestry tracing. *British Medical Journal, 325,* 1469–1471.

Frudakis, T., Venkateswarlu, K., Thomas, M. J., Gaskin, Z., Ginjupalli, S., Guntari, S., et al. (2003). A classifier for the SNP-based inference of ancestry. *Journal of Forensic Sciences, 48,* 771–778.

Greely, H. T. (2008 [this volume]). Genetic genealogy: Genetics meets the marketplace. In B. A. Koenig, S. S.-J. Lee, & S. S. Richardson (Eds.), *Revisiting race in a genomic age* (pp. 215–234). New Brunswick, NJ: Rutgers University Press.

Helgadottir, A., Manolescu, A., Helgason, A., Thorleifsson, G., Thorsteindottir, U., Gudbjartsson, D. F., et al. (2005). A variant of the gene encoding leukotriene A4 hydrolase confers ethnicity-specific risk of myocardial infarction. *Nature Genetics, 38,* 68–74.

Hooton, E. (1931). *Up from the ape.* New York: Macmillan Company.

Hooton, E. (1939). *Twilight of man.* New York: G. P. Putnam's Sons.

Jorde, L. B., & Wooding, S. P. (2004). Genetic variation, classification, and "race." *Nature Genetics, 36,* S28–S33.

Kidd, K. K., Pakstis, A. J., Speed, W. C., & Kidd, J. R. (2004). Understanding human DNA sequence variation. *Journal of Heredity, 95,* 406–420.

Kittles, R. A., & Weiss, K. M. (2003). Race, ancestry, and genes: Implications for defining disease risk. *Annual Review of Genomics and Human Genetics, 4,* 33–67.

Klein, R. G. (1999). *The human career: Human biological and cultural origins* (2nd ed.). Chicago: University of Chicago Press.

Majumder, P. P. (2001). Ethnic populations of India as seen from an evolutionary perspective. *Journal of Biosciences, 26,* 533–545.

Malécot, G. (1969). *The mathematics of heredity.* San Francisco: W. H. Freeman.

Marks, J. (1995). *Human biodiversity: Genes, race, and history.* New York: Aldine de Gruyter.

McKeigue, P. M., Carpenter, J., Parra, E. J., & Shriver, M. D. (2000). Estimation of admixture and detection of linkage in admixed populations by a Bayesian approach: Application to African-American populations. *Annals of Human Genetics, 64,* 171–186.

Mountain, J., & Cavalli-Sforza, L. L. (1997). Multilocus genotypes, a tree of individuals, and human evolutionary history. *American Journal of Human Genetics, 61,* 705–718.

Mountain, J. L., & Risch, N. (2004). Assessing genetic contributions to phenotypic differences among "racial" and "ethnic" groups. *Nature Genetics, 36*, S48–S53.

Pritchard, J. K., Stephens, M., & Donnelly, P. (2000). Inference of population structure
using multilocus genotype data. *Genetics, 155*, 945–959.

Pritchard, J. K., & Wen, W. (2004). *Documentation for Structure software: Version 2*. Chicago, IL.

Prugnolle, F., Manica, A., & Balloux, F. (2005). Geography predicts neutral genetic
diversity of human populations. *Current Biology, 15*, R159–R160.

Quintana-Murci, L., Semino, O., Bandelt, H.-J., Passarino, G., McElreavey, K., & Santachiara-Benerecetti, A. S. (1999). Genetic evidence of an early exit of *Homo sapiens
sapiens* from Africa through Eastern Africa. *Nature Genetics, 23*, 437–441.

Race, Ethnicity, and Genetics Working Group. (2005). The use of racial, ethnic, and
ancestral categories in human genetics research. *American Journal of Human Genetics, 77*, 519–532.

Ramachandran, S., Deshpande, O., Roseman, C. C., Rosenberg, N. A., Feldman, M. W.,
& Cavalli-Sforza, L. L. (2005). Support from the relationship of genetic and geo-
graphic distance in human populations for a serial founder effect originating in
Africa. *Proceedings of the National Academy of the Sciences USA, 102*, 15942–15947.

Rannala, B., & Mountain, J. L. (1997). Detecting immigration by using multilocus genotypes. *Proceedings of the National Academy of the Sciences USA, 94*, 9197–9201.

Relethford, J. H. (2004). Global patterns of isolation by distance based on genetic and
morphological data. *Human Biology, 76*, 499–513.

Risch, N., Burchard, E., Ziv, E., & Tang, H. (2002). Categorization of humans in biomedical research: Genes, race, and disease. *Genome Biology, 3*, 1–12.

Rosenberg, N. A., Mahajan, S., Ramachandran, S., Zhao, C., Pritchard, J. K., & Feldman,
M. W. (2005). Clines, clusters, and the effect of study design on the inference of
human population structure. *PLoS Genetics, 1*, 660–671.

Rosenberg, N. A., Pritchard, J. K., Weber, J. L., Cann, H. M., Kidd, K. K., Zhivotovsky, L. A.,
et al. (2002). Genetic structure of human populations. *Science, 298*, 2381–2385.

Seebach, L. (2003, May 8). Biology and race: A clearer link; new genetic research establishes firmer basis for connection. *Rocky Mountain News*, p. 58A.

Shriver, M., Frudakis, T., & Budowle, B. (2005). Getting the science and the ethics right
in forensic genetics. *Nature Genetics, 37*, 449–450.

Shriver, M. D., Kennedy, G. C., Parra, E. J., Lawson, H. A., Sonpar, V., Huang, J., et al.
(2004). The genomic distribution of population substructure in four populations
using 8,525 autosomal SNPs. *Human Genomics, 1*, 274–286.

Shriver, M. D., & Kittles, R. A. (2008 [this volume]). Genetic ancestry and the search
for personalized genetic histories. In B. A. Koenig, S. S.-J. Lee, & S. S. Richardson
(Eds.), *Revisiting race in a genomic age* (pp. 201–214). New Brunswick, NJ: Rutgers
University Press.

Stepan, N. L. (1982). *The idea of race in science*. Hamden, CT: Archon Books.

TallBear, K. (2005). Native American DNA: Narratives of origin and race. Ph.D. Dissertation. University of California at Santa Cruz.

TallBear, K. (2008 [this volume]). Native-American-DNA.com: In search of Native American race and tribe. In B. A. Koenig, S. S.-J. Lee, & S. S. Richardson (Eds.), *Revisiting
race in a genomic age* (pp. 235–252). New Brunswick, NJ: Rutgers University Press.

Tate, S. K., & Goldstein, D. B. (2008 [this volume]). Will tomorrow's medicines work for
everyone? In B. A. Koenig, S. S.-J. Lee, & S. S. Richardson (Eds.), *Revisiting race in a
genomic age* (pp. 102–128). New Brunswick, NJ: Rutgers University Press.

Templeton, A. (2002). Out of Africa again and again. *Nature, 416*, 45–51.

Tishkoff, S. A., & Kidd, K. K. (2004). Implications of biogeography of human populations for "race" and medicine. *Nature Genetics, 36*, S21–S27.

Tishkoff, S. A., & Verrelli, B. C. (2003). Patterns of human genetic diversity: Implications for human evolutionary history and disease. *Annual Review of Genomics and Human Genetics, 4*, 293–340.

Tishkoff, S. A., & Williams, S. M. (2002). Genetic analysis of African populations: Human evolution and complex disease. *Nature Reviews Genetics, 3*, 611–621.

Underhill, P. A., Shen, P., Lin, A. A., Jin, L., Passarino, G., Yang, W. H., et al. (2000). Y chromosome sequence variation and the history of human populations. *Nature Genetics, 26*, 358–361.

The Unexamined "Caucasian." (2004). *Nature Genetics, 36*, 541.

Wade, N. (2002, December 20). Gene study identifies five main human populations, linking them to geography. *New York Times*, p. A37.

Wilson, J. F., Weale, M. E., Smith, A. C., Gratrix, F., Fletcher, B., Thomas, M. G., et al. (2001). Population genetic structure of variable drug response. *Nature Genetics, 29*, 265–269.

Wright, S. (1943). Isolation by distance. *Genetics, 28*, 114–138.

PART TWO

Race-Targeted Research and Therapeutics

5

Race, Ancestry, and Medicine

MARCUS W. FELDMAN AND RICHARD C. LEWONTIN

The cost of assessing DNA-level variation in large numbers of people has steadily declined. There are now large sets of data on variation of different kinds of DNA markers among geographically diverse people. Several technical methods of studying human genetic variation are used, including analysis of single nucleotide polymorphisms (SNPs) (e.g., Voight, Kudaravalli, Wen, & Pritchard, 2006); short tandem repeats (STRs or microsatellites) (Rosenberg et al, 2002; Rosenberg et al., 2005); and Alu sequences (Bamshad et al., 2003). There are now many millions of polymorphic sites revealed in hundreds (and soon, perhaps thousands) of people available for analysis. At the same time, statistical tools for describing and interpreting observed patterns of variation have increased in sophistication.

Recent analyses of hundreds of microsatellite[1] DNA markers and a few thousand SNPs from human populations have shown that it is possible with a high degree of accuracy to assign the major geographical region (or regions) of origin of individual human beings by using a combination of a number of these polymorphic genes (Rosenberg et al., 2002; Rosenberg et al., 2005; Conrad et al., 2006). In addition, using more markers, it is possible in some cases to narrow down the population of origin to local national populations within major geographic regions. The greatly increased facility with which human genetic variation can be studied and the suggestion that this variation may be exploited to individualize medicine have fueled a growing controversy about whether race is indeed a biologically useful and meaningful concept when applied to humans, especially in a medical and pharmacological context. Four commentaries in the *New England Journal of Medicine* (Burchard et al., 2003; Cooper, Kaufman, & Ward, 2003; Phimister, 2003; Wood, 2001), one in *Science* (Sankar & Cho, 2002), an editorial ("Genes, drugs, and race," 2001) and commentary (Calafell, 2003) in *Nature Genetics*, and four reports in the *New York Times* (Satel, 2002; Wade, 2002a, 2002b, 2003) all raised the issue of the status of racial categorization as a biological concept. In particular it is claimed that these recent data are in contradiction to the widely accepted and confirmed observations that a very large proportion of human genetic diversity lies within geographical regions, observations that have led biologists and anthropologists to abandon the notion of human races over the last 30 years. What we wish to do is to explain that there is no contradiction between these two well-substantiated bodies of data because they speak to two quite different questions that have been confused.

The microsatellite data and data on other DNA polymorphisms (Bamshad et al., 2003) are relevant, among other things, to the problem of the assignment of individuals to lines of geographical ancestry. The question asked is whether it is possible to find genes that are polymorphic in the human species and whose frequencies of alternate alleles are sufficiently different in the different major geographical regions to allow a correct assignment of geographical origin with high probability. The answer to this question is "yes," and that answer has been known for 50 years from studies of genetic polymorphisms. This is a problem in biological systematics.

The data on general genetic polymorphism for proteins and nucleotide substitutions, also addressed by the study of microsatellites and SNPs, can also be employed to ask a quite different question, which is, What *fraction* of all human genetic variation, whether based on protein coding genes, microsatellites, or any other polymorphic DNA sequences, lies within geographically separated populations and what fraction lies between these populations? This is not an assignment problem, but a question of the average *amount* of genetic diversification between and within geographical groups. The two problems can be related to each other by posing the question, Are the genes that are geographically highly differentiated in their allelic frequencies typical of the human genome in general? The answer to that question turns out to be "no." While there are indeed genes whose allelic frequencies differ markedly between geographical regions and can be used for taxonomic purposes, these are not typical of the human genome in general.

The Problem of Ancestry

It has long been known that some loci are highly differentiated between geographical populations. Indeed, one does not need to be a geneticist to solve the taxonomic problem at first sight. Using skin color, facial shape, and hair form, all obviously largely genetically determined (although the genes influencing these characters have only begun to be localized), no one has any difficulty in differentiating between a random person taken from

West Africa, from China, from Norway, and from the tropical rainforest of the Orinoco basin. With only a little more subtlety one can differentiate Amharic-speaking natives of Ethiopia from Zulus, Chinese from Japanese, and villagers of Andhra Pradesh from Afghanis by external morphology. Some classic blood-group polymorphisms are highly differentiated geographically, although most are not (Cavalli-Sforza, Menozzi, & Piazza, 1994). Thus, it is not surprising that DNA sequences such as microsatellite markers can be used to infer the major region of origin of individuals. In the case of the microsatellites, the differences in allelic frequencies between groups are, in fact, small so that data from a very large number of markers had to be subjected to a sophisticated statistical clustering technique to make reliable inferences about geographical origin.

Let us first make clear what has been discovered about ancestry from two recent studies of microsatellites and insertion-deletion polymorphisms from the Human Genome Diversity Cell Line Panel (HGDP-CEPH) collection. In the first study (Rosenberg et al., 2002), 377 microsatellite markers were studied in 1,056 individuals from 52 sites representing native populations from all continents. The second study (Rosenberg et al., 2005) included 783 microsatellite markers and 210 insertion-deletion polymorphisms. There was a slight difference between the samples of individuals used in the two studies, with 1,048 individuals representing 53 populations in the second study. The reasons for this difference have to do with possible duplication of a small number of samples and the inclusion of a small group of Bantu samples into a single group.

For the two studies, the conclusions about continental ancestry are remarkably similar despite the difference in the size of the data sets. The essential finding is that these highly variable markers can be used to form affinity clusters on the basis of similarities between individuals in their genotypes. The statistical technique used (Pritchard, Stevens, & Donnelly, 2000) finds the most probable assignment of individuals to clusters, and this is done blind to knowledge of the actual geographic origin of the individuals.[2] After these clusters are formed, they can then be compared to the actual geographic origins of all individuals. For one of the clustering schemes, with five clusters the result was a close fit of the clusters to continents or subcontinents. Many individuals had ancestry from two or more of the clusters, and some clusters showed a great deal of multiple ancestry, a signature of past migrations or conquests or of the continuity of genetic variation in space (King & Motulsky, 2002). Almost all of the sample of Mozabites from Algeria, for example, belonged both to clusters that corresponded to Eurasia and Africa. And the Altaic-speaking Uygurs of northwestern China showed strong ancestry from East Asia and Eurasia. Europe, West Asia, and South/ Central Asia are regions of especially mixed ancestry; the separate linguistic groups in these areas are difficult to separate genetically, even with 783 markers.

The continental clustering in these large sets of data derives mainly from small differences in allele frequencies at large numbers of markers, not from diagnostic genotypes. This clustering reflects the history of human migrations that began when modern humans left Africa 50,000–100,000 years ago (King & Motulsky, 2002; Excoffier, 2003; Cavalli-Sforza & Feldman, 2003). For those geographical regions such as Europe, West Asia, and South/Central Asia that have a long history of migration and colonization, finer resolution of the clusters is very difficult and will probably require more samples and many more polymorphic markers.

It takes a lot of polymorphic microsatellite markers to produce reliable genetic clusters, 50–150 polymorphisms for reliable assignment of continental ancestry. Even with 993 polymorphisms, it remains difficult to resolve finer subdivisions within continents, especially in Europe and Asia. We can conclude, however, that in most cases, self-reported ancestry coincides with the broad continental clustering seen from the genetic markers that were used.

It has been claimed (Serre & Pääbo, 2004) that the geographical clustering seen in the studies referred to above (Rosenberg et al., 2002; Rosenberg et al., 2005) is an artifact of the geographic pattern of samples in the particular data set studied, the HGDP-CEPH (CEPH, n.d.). We had remarked on the existence of "continuous gradients across regions or admixture of neighboring groups" (Rosenberg et al., 2002, p. 2382). In fact, geographic clustering and spatial gradients are both features of these large data sets (Rosenberg et al., 2005). For population pairs from the same cluster, as geographic distance increases, genetic distance increases linearly, consistent with a clinal[3] structure. But for pairs of populations from different clusters, genetic distance is generally larger than between pairs of populations from the same clusters that have the same genetic distance. This suggests that the clusters are formed by the small discontinuous jumps in genetic distance caused by major geographic barriers: oceans, mountain ranges, or deserts. Indeed, the history of migration is important (Ramachandran et al., 2005), but migration is not geographically uniform.

While the great majority of the DNA markers used to define these continental clusters show only small allelic frequency differences between populations, some genes do have greater frequency differences among populations or continents (Bamshad & Wooding, 2003). Duffy and Rh, two of the genes in the original survey of within- and between-population diversity (Lewontin, 1972), show more variation among populations than most blood group or protein genes, microsatellites, or SNPs. The presence of hemoglobin S (causing sickle cell disease) or G6PD (causing favism) in an individual markedly increases the likelihood that that person has ancestors from a geographic region where malaria was present, while an individual carrying the Tay-Sachs allele is most likely to have Ashkenazi Jewish or French Canadian ancestry. These are cases where populational rather than continental ancestry is the relevant dimension for the allelic differences.

Finally, it must be borne in mind that the taxonomic problem cannot be inverted. That is, while clustering methods are capable of assigning an individual to a geographic population with a high degree of certainty, given that individual's genotype, it is not possible to predict accurately the genotype of an individual given his or her geographical origin. Thus, knowing an individual's ancestry only slightly improves the ability to predict his or her genotype. The more polymorphic the markers, the more difficult this is. This is illustrated in figure 5-1. There are gene alleles that appear only in one group, as for example the Fy^b which is present only in individuals with some European ancestry, but there does not exist any gene for which one major geographical cluster includes 100% of one genotype while another major geographical cluster has 100% of another genotype. Even when the explicit purpose of studies has been to identify markers that show strong differentiation between groups, none that show a complete difference between major groups has been found. In the microsatellite study mentioned earlier (Rosenberg et al., 2002), the most geographically informative loci in the data set have some striking differences, as shown in figure 5-1, but nowhere near 100%. In figure 5-1, the size of the pie slice with a given degree of grayness represents its frequency in the region specified by the column. In some regions, some alleles are rare and do not occupy enough area to be seen. The top marker (D12S2070) shows very different allele frequencies in the different regions, the middle one (D10S1425) shows moderate frequency differences, and the bottom marker (D6S474) shows very small differences. All three loci have eight alleles, which are shown in increasing order of allele size (i.e., number of species) in a counterclockwise manner, starting from the top of each circle (for each locus, the smallest allele is shown in white, the largest is shown in black).

The Problem of Allocation of Variation

When we turn from the problem of finding genes that will discriminate ancestry to the problem of the relative amount of human genetic diversity that lies within and between populations, there is no controversy. The first survey, in 1972, of genetic diversity over a very large sample of local human populations from major geographical regions used all the available data for blood groups and enzyme proteins for every local human population that had been studied up to that time (Lewontin, 1972). The result was that 85%

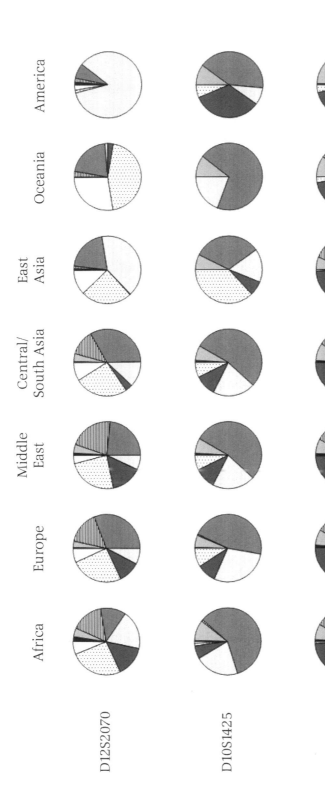

FIGURE 5-1. Distributions of allele frequencies across seven regions for three microsatellite markers.
From Rosenberg et al. (2002).

of all human genetic diversity, measured by the Shannon-Weaver information measure (or its close equivalent, heterozygosity) is present within local national groups, that is, averaging within Swedes, within Kikkuyu, within Japanese, etc. An additional 8% is present between local groups within what were designated classically as races, between Swedes, Italians, and Greeks, or between Kikkuyu, Zulu, and Hutu. The remaining 7% lies between the classical major races, between sub-Saharan Africans, East Asians, Australian Aborigines, Europeans, etc. Several similar studies were subsequently carried out on smaller geographical samples with similar results for the within-population variation, but with roughly 5% of the variation between local populations and 10% among the major "races." When these studies were repeated using a limited amount of DNA sequence variation (Barbujani, Magagni, Minch, & Cavalli-Sforza, 1997), again 85% of the variation was found within local populations, with about 5% between local populations and 10% among major classical races. The studies of the DNA markers discussed in the previous section (Rosenberg et al., 2002; Rosenberg et al., 2005) in the context of the taxonomic problem also partitioned total variation using a different measure of diversity. With this measure, 86% to 95% of the diversity was assigned within local populations, between 2% and 6% among populations within major geographical regions and between 3% and 10% among major regions (classical races).

The higher than usual estimate of between 93% and 95% for the within-population component of genetic variation arises from several sources. The previous studies used samples from isolated and geographically well-separated populations. On restriction of the microsatellite analysis to populations chosen to mimic the pattern in these previous studies, the within-population component in the microsatellite study was reduced to 89.8%. A further contribution to the difference is from the greater heterozygosity of microsatellites, of which only 30 were included in the 109 genes studied by Barbujani et al. Differential selection on protein variants across geographical regions might also augment the between-population component as compared to the microsatellite study. The different studies differ on the exact partition of variance. Nevertheless, the strong overall conclusion is that although it is possible to use genetic divergences to assign individuals to regions of origin with high confidence, there is very little average genetic difference among geographical regions as compared to the variation observed within any local population.

The Problem of Race

Race as a biological concept has had a variety of disparate meanings, even within the last 50 years. In the classical taxonomic literature a race was any distinguishable type within a species, as, for example, dark-bellied and light-bellied races of small mammals. While such races sometimes corresponded to geographically separated populations, phenotypic differences on which racial classifications were based were often a consequence of single gene differences, so that two siblings could be of different "race." In reaction against this typological notion of race, Dobzhansky introduced the notion of "geographical races" which were defined as "populations of species that differ in the frequencies of one or more genetic variants, gene alleles or chromosomal structures" (1937, p. 138). The problem with this definition is that every geographical population of every species in the world is a "geographical race" because no two populations have identical allelic frequencies for polymorphic genes, so geographical race becomes synonymous with population.

The typological and the geographic notions of race are combined in the classical division of human races because it is observed that the native inhabitants of different major regions of the world are characterized by clear phenotypic differences of color, facial features, and hair form. Variation in these phenotypes is also observed among individuals within races to the extent that even categorization according to such traits can be difficult (Brown & Armelagos, 2001). An underlying assumption of human race classification, a classification based on a small number of obvious phenotypic differences, was that these differences were characteristic of the genome in general. Just as there were large differences in genes for color, so there would be large differences in genes influencing cognitive and most physiological traits. Indeed, in the absence of any evidence to the contrary, this is not an absurd assumption, but it turns out to be wrong. The repeated and consistent results on the apportionment of genetic diversity reviewed in the previous section show that the genes underlying the phenotypic differences used to assign race categories are atypical of the genome in general and are not a reliable index to the amount of genetic differentiation between groups. Thus, racial assignment loses any general biological interest. For the human species, race assignment of individuals does not carry with it any general implication about genetic differentiation.

With the advent of large movements of populations between continents, especially with European colonial expansion and the commercial slave trade, new populations have arisen which are mixtures of the major continental groups, especially in the Western Hemisphere and Oceania. Large numbers of people, then, have ancestry from more than one major geographical region so that the association of phenotype and geography breaks down and race again becomes a typology with even less power to distinguish the genomes of those involved. What is the race of a dark-skinned person, half of whose ancestry is of sub-Saharan African origin and half of Northern European origin? Social practice in the United States makes an asymmetrical nominal assignment of race such that any detectable African ancestry makes a person "black" or "African American," but this obfuscates the biological reality.

What do the clusters constructed, for example, from the data in the microsatellite studies have to do with our common understanding of race? We must remember that the clusters are defined by markers that have no influence on obvious phenotypes. Nevertheless, there are some phenotypes that correlate well with continental origin, usually those that we can see. But even here we can be misled—dark skin is a feature of sub-Saharan Africans but also of southern Indians and Australian aborigines. Thus, if we were to use skin color alone, continental clustering, i.e., common racial classification, fails.

The Use of Race and Ancestry in Medicine

It is often claimed that racial categorization is of considerable importance in medicine because there are a number of loci of medical relevance that are highly differentiated between geographical populations.

A focus of recent studies on the medical relevance of race and/or ethnicity concerns variation in drug metabolizing enzymes, DMEs (Yancy et al., 2001; Exner, Dries, Domanski, & Cohen, 2001; Xie, Kim, Wood, & Stein, 2001; Schwartz, 2001; Wilson et al., 2001; Risch, Burchard, Ziv, & Tang, 2002). Some of these enzymes appear to differ in allele frequencies among Americans of different ethno-cultural backgrounds (Yancy et al., 2001; Exner et al., 2001; Xie et al., 2001). Wilson and his group compared differences in DME frequencies among genetically estimated clusters obtained using 39 microsatellites assayed on individuals from eight populations with corresponding differences among the same individuals classified by ethnicity. They claim that the ethnic labels are "insufficient and inaccurate" surrogates for the genetic clusters and are less valuable than the latter in resolving group-specific profiles of DMEs. This finding is contested by Risch and colleagues (2002), who go further to claim that the use of genetic clusters instead of a racial classification may cause the effects of socioeconomic, environmental, and lifestyle variation on a disease to be underestimated.

For DMEs, as we know for blood groups and other enzymes, it is reasonable to predict that variation within will far outweigh that between continental groups. For these genes, too, only variation at a much finer level than continents or races may provide information about ancestry that is phenotypically relevant.

The situation is even more complicated when we examine diseases that appear to aggregate in the classically defined races. Sickle cell disease is one that is often thought to be an African trait. But it exists in a number of

Mediterranean and Indian populations as well. Sickle cell is not a marker of skin color or race, but more properly a marker of ancestry in a geographic location where malaria is or was prevalent. And, of course, not all Africans or Sardinians carry the gene responsible for sickle cell disease. Thus, classical race is not diagnostic of the disease, and the disease is not diagnostic of race. Rosenberg et al.'s clusters (Rosenberg et al., 2002; Rosenberg et al., 2005) don't tell us very much about traits that are determined by genes that have been under selection. It is nevertheless the case that a knowledge of ancestry can play an important role in medical diagnosis and drug therapy. Thus, we might make a classification based only on the sickle cell phenotype, or the Tay-Sachs phenotype, or lactose intolerance. From a medical point of view, a breakdown of humans into the hundreds if not thousands of such groups might say more about the biology of disease than mere continental ancestry. Knowledge of ancestry with respect to these subpopulations may then be informative about the risks of disease.

Both the regional heterogeneity within major geographical regions and the widespread mixture of formerly relatively isolated populations result in a confusion between race and ancestry that is critical and must be accounted for in medical practice. The assignment of racial classification to an individual hides the biological information that is needed for intelligent therapeutic and diagnostic decisions (see also Tate & Goldstein, this volume). A person classified as "black" or "Hispanic" by social convention may have any mixture of European, African, Native American, and, more rarely, Asian ancestry. Moreover, there is genetic heterogeneity among regions within these major geographical groups. If we have a serious interest in making diagnostic and therapeutic decisions based on genotype, then it is not typological race assignment that is relevant but the various contributions to a person's ancestry that are informative. The kind of questions to be asked are these: Do you have any African ancestors? If so, do you know from what part of Africa they came? Do you have any European ancestry? If so, from what part of Europe did they come? Were there any Ashkenazi Jews among your ancestors? And so on. The detailed information about local geographical origins will often be unavailable, but categorical racial assignments are not a substitute for some kind of more informative ancestry history.

We agree entirely with Risch et al. (2002) that conventional socially defined race which, for example, classifies all persons with visually detectable African ancestry as "black" or "African American" is of use in a medical context to the extent that it provides information about social circumstances and lifestyle conditions of patients, particularly discrimination. But these socially defined categories should not be confounded with genetically defined races. The actual distribution of human genetic variation, including the distribution of genotypes that are directly relevant to the diagnosis and treatment of disease, is such that race is not a useful biological concept when applied to humans. It is nevertheless true that data about the various lines of ancestry of an individual can provide information on the likelihood that the person carries certain gene alleles. Lines of ancestry, rather than genetically arbitrary racial categories, can provide much accurate, biologically interesting, and potentially medically useful information. For diagnosis and treatment, however, individual genotypes will, in the long run, provide the most useful information.

NOTES

1. A microsatellite is any of numerous short segments of DNA that are distributed throughout the genome, that consist of repeated sequences of usually two to five nucleotides, and that tend to vary from one individual to another.

2. See Bolnick (this volume) for a discussion of Pritchard's computer program, *structure*.

3. Clinal genetic variation refers to a gradient of change in a group of related organisms, usually along a line of environmental or geographic transition.

REFERENCES

Bamshad, M., & Wooding, S. P. (2003). Signatures of natural selection in the human genome. *Nature Reviews Genetics, 4*, 99–111.

Bamshad, M. J., Wooding, S. P., Watkins, W. S., Ostler, C. T., Batzer, M. A., & Jorde, L. B. (2003). Human population genetic structure and inference of group membership. *American Journal of Human Genetics, 72*, 578–589.

Barbujani, G., Magagni, A., Minch, E., & Cavalli-Sforza, L. L. (1997). An apportionment of human DNA diversity. *Proceedings of the National Academy of the Sciences USA, 94*, 4516–4519.

Bolnick, D. A. (2008 [this volume]). Individual ancestry inference and the reification of race as a biological phenomenon. In B. A. Koenig, S. S.-J. Lee, & S. S. Richardson (Eds.), *Revisiting race in a genomic age* (pp. 70–85). New Brunswick, NJ: Rutgers University Press.

Brown, R. A., & Armelagos, G. J. (2001). Apportionment of racial diversity: A review. *Evolutionary Anthropology, 10*, 34–40.

Burchard, E. G., Ziv, E., Coyle, N., Gomez, S. L., Tang, H., Karter, A. J., et al. (2003). The importance of race and ethnic background in biomedical research and clinical practice. *New England Journal of Medicine, 348*, 1170–1175.

Calafell, F. (2003). Classifying humans. *Nature Genetics, 33*, 435–436.

Cavalli-Sforza, L. L., & Feldman, M. W. (2003). The application of molecular genetic approaches to the study of human evolution. *Nature Genetics, 33* (Suppl.), 266–275.

Cavalli-Sforza, L. L., Menozzi, P., & Piazza, A. (1994). *The history and geography of human genes*. Princeton, NJ: Princeton University Press.

Centre d'Etude du Polymorphisme Humain (CEPH) (n.d.). HGDP-CEPH Human Genome Diversity Cell Line Panel. Retrieved August 7, 2006, from http://www.cephb.fr/HGDP-CEPH-Panel/.

Conrad, D. F., Jakobsson, M., Coop, G., Wen, X., Wall, J. D., Rosenberg, N. A., et al. (2006). A worldwide survey of haplotype variation and linkage disequilibrium in the human genome. *Nature Genetics, 38,* 1251–1260.

Cooper, R. S., Kaufman, J. S., & Ward, R. (2003). Race and genomics. *New England Journal of Medicine, 348,* 1166–1170.

Dobzhansky, T. (1937). *Genetics and the origin of species.* New York: Columbia University Press.

Excoffier, L. (2003). Human diversity: Our genes tell where we live. *Current Biology, 13,* R134–R136.

Exner, D. V., Dries, D. L., Domanski, M. J., & Cohen, J. N. (2001). Lesser response to angiotensin-converting-enzyme inhibitor therapy in black as compared with white patients with left ventricular dysfunction. *New England Journal of Medicine, 344,* 1351–1357.

Genes, drugs, and race. (2001). *Nature Genetics, 29,* 239–240.

King, M., & Motulsky, A. G. (2002). Mapping human history. *Science, 298,* 2342–2343.

Lewontin, R. C. (1972). The apportionment of human diversity. *Evolutionary Biology, 6,* 381–398.

Phimister, E. G. (2003). Medicine and the racial divide. *New England Journal of Medicine, 348,* 1081–1082.

Pritchard, J. K., Stephens, M., & Donnelly, P. (2000). Inference of population structure using multilocus genotype data. *Genetics, 155,* 945–959.

Ramachandran, S., Deshpande, O., Roseman, C. C., Rosenberg, N. A., Feldman, M. W., & Cavalli-Sforza, L. L. (2005). Support from the relationship of genetic and geographic distance in human populations for a serial founder effect originating in Africa. *Proceedings of the National Academy of the Sciences USA, 102,* 15942–15947.

Risch, N., Burchard, E., Ziv, E., & Tang, H. (2002). Categorization of humans in biomedical research: Genes, race and disease. *Genome Biology, 3,* 2007.1–2007.12.

Rosenberg, N. A., Mahajan, S., Ramachandran, S., Zhao, C., Pritchard, J. K., & Feldman, M. W. (2005). Clines, clusters, and the effect of study design on the inference of human population structure. *PloS Genetics, 1,* 660–671.

Rosenberg, N. A., Pritchard, J. K., Cann, H., Weber, J., Kidd, K. K., Zhivotovsky, L. A., et al. (2002). Genetic structure of human populations. *Science, 298,* 2381–2385.

Sankar, P., & Cho, M. K. (2002). Toward a new vocabulary of human genetic variation. *Science, 298,* 1337–1338

Satel, S. (2002, May 5). I am a racially profiling doctor. *The New York Times,* sec. 6, p. 56.

Schwartz, R. S. (2001). Racial profiling in medical research. *New England Journal of Medicine, 344,* 1392–1393.

Serre, D., & Pääbo, S. (2004). Evidence for gradients of human genetic diversity within and among continents. *Genome Research, 14,* 1679–1685.

Stephens, J. C., Schneider, J. A., Tanguay, D. A., Choi, J., Acharya, T., Stanley, S. E., et al. (2001). Haplotype variation and linkage disequilibrium in 313 human genes. *Science, 293,* 489–493.

Tate, S. K., & Goldstein, D. B. (2008 [this volume]). Will tomorrow's medicines work for everyone? In B. A. Koenig, S. S.-J. Lee, & S. S. Richardson (Eds.), *Revisiting race in a genomic age* (pp. 102–128). New Brunswick, NJ: Rutgers University Press.

Voight, B. F., Kudaravalli, S., Wen, X., & Pritchard, J. K. (2006). A map of recent positive selection in the human genome. *PLoS Biology, 4,* e72.

Wade, N. (2002a, July 30). Race is seen as real guide to track roots of disease. *New York Times*, Science sec., p. 1.

Wade, N. (2002b, December 20). Gene study identifies f main human populations, linking them to geography. *New York Times*, National sec., p. 29.

Wade, N. (2003, March 20). 2 scholarly articles diverge on role of race in medicine. *New York Times*, Late ed.–final, sec. A, p. 30.

Wilson, J. F., Weale, M. E., Smith, A. C., Gratrix, F., Fletcher, B., Thomas, M. G., et al. (2001). Population genetic structure of variable drug response. *Nature Genetics, 29*, 265–269.

Wood, A.J.J. (2001). Racial differences in the response to drugs—pointers to genetic differences. *New England Journal of Medicine, 344*, 1393–1395.

Xie, H. G., Kim, R. B., Wood, A. J., & Stein, C. M. (2001). Molecular basis of ethnic differences in drug disposition and response. *Annual Review of Pharmacology and Toxicology, 41*, 815–850.

Yancy, C. W., Fowler, M. B., Colucci, W. S., Gilbert, E. M., Bristow, M. R., Coln, J. N., et al. (2001). Race and the response to adrenergic blockade with carvedilol in patients with chronic heart failure. *New England Journal of Medicine, 344*, 1358–1365.

6

Will Tomorrow's Medicines Work for Everyone?

SARAH K. TATE AND DAVID B. GOLDSTEIN

Most current health disparities have little or perhaps nothing to do with genetics, depending instead on socioeconomic and other environmental factors (Cooper, 2001), including inequalities in the delivery of health care. But this fact should not make us reject the possibility that biological differences associated with "race" or "ethnicity" will contribute to disparities in the future. Substantial changes are occurring in how medicines are developed and how they are used, and some of these changes present risks that medicines will become less inclusive. Some of the variation in how medicines work may correlate with "racial" or "ethnic" groups, exacerbating health disparities. Here we review ways in which genetics could contribute to future health disparities and outline some general research approaches that could counter this.

Why Might Medicines Become Less Inclusive?

Advances in genetic technologies should improve our understanding of disease etiology and of the factors influencing response to treatment (Collins, Green, Guttmacher, & Guyer, 2003). Although there has been relatively little progress to date in using genetics to improve the treatment of common diseases, there are some encouraging signs of progress in basic research (Goldstein, Tate, & Sisodiya, 2003; Laitinen et al., 2004; Hugot et al., 2001). If genetics does eventually prove relevant to the treatment of common diseases, then to the extent that genetic advances are uneven among racial and ethnic groups, disparities may result. Perhaps the most immediate concerns about the clinical use of genetics involve pharmacogenetics, which seeks to identify the genetic factors that influence responses to medicines. Studying drug response might be more clinically useful than studying disease predisposition (Goldstein et al., 2003; Schmith et al., 2003), at least in the near term. Pharmacogenetics also seems poised to become a common component of the drug-development pipeline (Food and Drug Administration, 2005), with proof-of-concept studies having demonstrated its potential (Danoff et al., 2004). For these reasons, we focus this perspective on pharmacogenetics.

For convenience of presentation, we shall not attempt to define "race" or "ethnicity" here. Our primary aim is to discuss the evidence for differences among groups, and so we consider mainly how people are distinguished in the current medical literature, for example, in clinical trials; therefore, we refer to different groups as racial or ethnic groups. We do, however, briefly review debates about the extent to which human genetic structure correlates with racial or ethnic groupings.

The influence of race or ethnicity on the efficacy and safety of existing medicines may guide our expectations for the future. Therefore, we begin with a review of evidence concerning whether specific drugs work differently in different racial or ethnic groups. The concerns about how genetics might influence the inclusiveness of medicines can be broadly divided into three partially overlapping categories: (1) pharmacogenetic diagnostics, in which individual genotype is used to guide the selection of medicines; (2) molecular subclassification of disease, in which genetic features of a disease are used to guide the choice of therapy; and (3) pipeline pharmacogenetics, in which genetics is used during the evaluation of new chemical entities.

Inclusiveness of Current Medicines

At least 29 medicines (or combinations of medicines) have been claimed, in peer-reviewed scientific or medical journals, to have differences in either safety or, more commonly, efficacy among racial or ethnic groups (table 6.1). But these claims are universally controversial (Cooper, Kaufman, & Ward, 2003; Burchard et al., 2003; Schwartz, 2001), and there is no consensus on how important race or ethnicity is in determining drug response.

Nevertheless, from 1995 to 1998, the labels of 8% of new drug products (15 of 185) contained a statement about racial or ethnic differences in effectiveness (Evelyn et al., 2001). For an overview of the relationship between racial or ethnic group and drug response, we compiled a list of drugs claimed to have different effects in different racial or ethnic groups on the basis of reviews (Bjornsson et al., 2003) and PubMed literature searches using combinations of the following key words and phrases: ethnic*, race, racial, drug response, pharmacokinetic*, cytochrome P450. We also determined whether there is any evidence for a genetic or a physiological contribution to the reported racial or ethnic differences in drug response, or any other evidence

TABLE 6.1

Examples of Drugs Reported to Have Different Response in Different Racial or Ethnic Groups

Drug Class	Examples	Difference in Response	Sample Size	Is the Difference Real?	Evidence
ACE inhibitors	Enalapril	Less response in AAs than in EUs with left ventricular dysfunction.	1,196 EUs, 800 AAs	B	Probably related to lower bioactivity of endogenous nitric oxide in AAs than in EUs.
	Lisinopril	Response (lowering of blood pressure) in EUs but not in AAs.	124		
	Trandolapril	AAs with hypertension required 2–4 times greater dose to obtain similar lowering of blood pressure than EUs.	207 EUs, 91 AAs		
Combination of two vasodilators	BiDil	Greater efficacy in AAs than in EUs with congestive heart failure.	Trial 1: 180 AAs, 450 EUs; Trial 2: 215 AAs, 574 EUs	B	
Vasodilator antihypertensive	Sodium nitroprusside	Attenuated vasodilation response to methacholine and sodium nitroprusside in normotensive AAs compared with EUs.	11 AAs, 9 EUs; and 21 AAs, 19 EUs	C	Attenuated responses to multiple vasodilators commonly observed in AAs. Mechanisms not fully understood.

Beta-adreno-ceptor blocker (nonselective)	Propranolol	More effective in EUs than in AAs for initial treatment of hypertension.		A	Hypertensives and healthy volunteers carrying two copies of the allele encoding the Arg variant of the β1-adrenergic receptor have greater response to metoprolol and to atenolol, respectively. The Arg variant is more frequent in EUs than in AAs (0.723 versus 0.575). There is a higher proportion of low-renin hypertension in AAs, and nonadrenergic mechanisms contribute more to blood pressure maintenance in AAs than in EUs.
	Nadolol	More effective in EUs than in AAs for systemic hypertension.	365		
	Oxprenolol	Mean blood pressure reduction less for AAs than for EUs.			
	Bucindolol	Survival benefit only in non-AAs.	2,708		
Beta-adreno-ceptor blocker (β1-selective)	Atenolol	More effective in EUs than in AAs for hypertension.	1,105	D	Basis of ADR not known and study not replicated.

(continued)

Table 6.1. Examples of Drugs Reported to Have Different Response in Different Racial or Ethnic Groups (*continued*)

Drug Class	Examples	Difference in Response	Sample Size	Is the Difference Real?	Evidence
Vasopeptidase inhibitor	Omapatrilat	Risk of angioedemas greater in AAs than in EUs for hypertension.			
Anticoagulant	Danaparoid	Significantly more EUs than AAs had favorable outcome at 3 months (for ischemic stroke).	292 AAs, 801 EUs	D	No significant difference in other response measures. Other measures (e.g., age) are better predictors of outcome. Danaparoid is not hepatically metabolized. Results not replicated.
	Warfarin	Average warfarin dose required to maintain the INR between 2.0 and 3.0 is greater in ACs than in EUs and greater in EUs than in ASs. ACs and Indo-ASs require more warfarin to hold their INR between 3 and 4.5 than do EUs.	737 EUs, 58 ACs, 2 ASs; and 715 ACs, 18 ASs, 95 EUs	D	Functional variation in CYP2C9 affects interindividual warfarin dosing but has not been demonstrated to account for differences between populations. Differing average body weight may account for some of the differences between EUs and ASs.

Class	Drug	Effect	Sample	Grade	Comments
Alpha-adreno-ceptor blocker	Prazosin	More effective in EUs than in AAs for hypertension.	1,105	B	Nonadrenergic mechanisms contribute more to blood pressure maintenance in AAs than in EUs.
Thiazide (diuretic)	Hydrochlorothiazide	Greater systolic and diastolic blood pressure responses in AAs than in EUs.	225 AAs, 280 EUs	B	Probably related to lower bio-activity of endogenous nitric oxide in AAs than in EUs.
Calcium-channel blocker	Diltiazem	More effective in AAs than in EUs for hypertension.	1,105	B	Probably related to increased predisposition of AAs to the salt-sensitive form of essential hypertension.
Beta-adrenocep-tor gonist	Isoproterenol	Attenuated vasodilation in normotensive AAs compared with EUs. Average dose giving heart rate increase of 25 beats per min. is more than 2 times higher in AAs than in EUs.	27 AAs, 27 EUs: 18 AAs, 18 EUs; 9 AAs, 13 EUs; and 16	C	Attenuated responses to multiple vasodilators commonly observed in AAs. Mechanisms not fully understood.
Glucocorticoid	Methylprednisolone	Adverse effects (steroid-associated diabetes) more common in AAs than in EUs.	9 AAs, 9 EUs	C	Also altered pharmacokinetics between AAs and EUs.

(continued)

Table 6.1. Examples of Drugs Reported to Have Different Response in Different Racial or Ethnic Groups (*continued*)

Drug Class	Examples	Difference in Response	Sample Size	Is the Difference Real?	Evidence
Hepatitis antiviral treatment	Ribavirin (interferon alpha)	AAs have lower rate of response to treatment than do EUs.	100 AAs, 100 EUs	B	May be due to differing immune abilities. AAs produce more cytokine than do EUs, and EU responders produce less than EU nonresponders.
	Interferon	Poorer response in AAs than in EUs.	4,031 AAs, 62 EUs		
Prostaglandin analog	Travoprost	Response greater in AAs than in EUs for ocular hypertension.	1,381	D	Not replicated, no supporting evidence.
Cytotoxic agents	6-MP, methotrexate and other ALL agents	Significant difference in response by ethnicity for childhood ALL. ASs > EUs > HIs > AAs. EUs have better survival than do AAs and HIs. AA, HI, and AI children with ALL have worse survival than do EU and AS/PI children.	6,703 EUs, 1,071 HIs, 506 AAs, 167 ASs; 4,061 EUs, 518 AAs, 507 HIs; and 4,952	D	Probably due to differences in quality of health care or therapy.
	Docetaxol and Carboplatin	Greater response in ASs than in EUs with advanced non-small cell lung cancer.	68	D	Not replicated. No intermediate phenotypes to support result.

Insulin	Insulin	Insulin sensitivity significantly lower in HIs and AAs than in EUs, and greater acute insulin response in AAs than in HIs.	14 EUs, 15 AAs, 28 HIs	C	Differences remain after adjusting for body fat. Results well replicated.
Antipsychotic	Haloperidol	ASs with schizophrenia had a significantly higher rating for extrapyramidal symptoms at fixed dose and significantly lower mean required dose than did EUs. Extrapyramidal side effects higher in Chinese than non-Chinese individuals.	13 EUs, 16 ASs; and 32 Chinese, 32 non-Chinese	C	No significant difference in haloperidol serum concentrations, although EUs have greater reduced haloperidol concentrations.
	Clozapine	ASs showed a greater change than EUs in total scores of the 8-item Brief Psychiatric Rating Scale while receiving a significantly lower mean dose of Clozapine and were significantly more likely to experience anticholinergic and other side effects.	17 ASs, 17 EUs	D	Although a relevant drug target polymorphism differs in frequency between EUs and ASs (HTR2A H452Y), it has not been reliably associated with response in both populations.
	Chlorpromazine	EUs received significantly larger maintenance doses than did either ASs or HIs.		D	Probably due to different prescribing practices.
	Various	AAs were given higher doses than were EUs.	442; 173; 76 AAs, 88 EUs; and 293	D	

(continued)

Table 6.1. Examples of Drugs Reported to Have Different Response in Different Racial or Ethnic Groups (*continued*)

Drug Class	Examples	Difference in Response	Sample Size	Is the Difference Real?	Evidence
Analgesic	Morphine	Reduction in blood pressure was greater in EUs than in Chinese. AIs were more susceptible to depression of respiratory response than EUs.	8 Chinese, 8 EUs; and 22 EUs, 22 AIs	D	Conflicting results.

Source: For full references, see Tate and Goldstein, 2004, S34–S42.

Key: If the difference between racial or ethnic groups seems to be real, then it is graded according to how likely it is to have a physiological or genetic basis, as follows: A, there is an indication of genetic causation; B, the association has a reasonable underlying physiological basis; C, the difference is consistently shown but no physiological basis has been demonstrated; D, possibly false positive claim. Racial or ethnic groups: AA, African American; AC, Afro-Caribbean; AI, American Indian and Alaskan Native; AS, Asian; EU, of European ancestry; HI, Hispanic; PI, Pacific Islander. ALL, acute lymphoblastic leukemia; INR, International Normalized Ratio (for blood clotting time); NA, not applicable.

that the differences are real. In cases where the differences in response are associated with underlying physiological differences (as in the case of angiotensin I-converting enzyme [ACE] inhibitors), these physiological differences may be influenced by environment, genetics, or both. Genetic epidemiology methods, including migrant studies, studies of admixed populations, and comparisons of populations in different geographical locations (e.g., West Africans and African Americans), can be used to assess the basis of racial or ethnic differences in response to drugs (Risch, Burchard, Ziv, & Tang, 2002). The most direct way to assess differences in drug response among racial or ethnic groups is to find the causes of the variable drug response and investigate how these causes differ among the groups. In most cases, however, the genetic bases of variable drug responses are too poorly known to allow a direct assessment. The one exception is beta-blockers: a polymorphism in the drug target may contribute to the differences in responses.

Many studies of drug response are small, and study designs vary. There is also a bias toward studies comparing European Americans and African Americans, reflecting a focus on the American racial or ethnic context. Certain therapeutic areas are particularly prominent among the list of medicines, most notably those related to cardiovascular disease. For example, there is arguably a consensus in the literature that African Americans respond less well than European Americans to beta-blockers, ACE inhibitors and angiotensin-receptor blockers, some of the main agents now used to treat heart-related conditions. Conversely, African Americans may respond as well as or better than European Americans to diuretics and calcium-channel blockers.

Some have argued that these differences result from underlying differences in the causes of hypertension in individuals of (west) African ancestry and of European ancestry (Ferdinand, 2003). For example, European American and African American hypertensive individuals typically differ in characteristics such as salt sensitivity, plasma volume, and renin levels (Saunders, 1991); there could be differences in the pathogenesis of hypertension. Decreased sensitivity to beta-blockers might be associated with the higher proportion of low-renin hypertension found in African Americans, and nonadrenergic mechanisms might contribute more to blood pressure maintenance in African Americans than in European Americans (Lang et al., 1997).

Differences in response to ACE inhibitors may be related to lower bioactivity of endogenous nitric oxide in African Americans than in European Americans (Kalinowski, Dobrucki, & Malinski, 2004). The increased benefit of treatment with nitrates and hydralazine in African Americans is consistent with this hypothesis. For example, BiDil, a drug that combines isosorbide dinitrate (a nitric oxide donor) and hydralazine (a vasodilator agent), was not sufficiently effective in treating congestive heart failure in two large, ethnically mixed clinical trials (Cohn et al., 1986; Cohn et al., 1991) to win regulatory approval. A retrospective analysis of the original trials, however, indicated that the drug was more effective in treating the African American than the European American participants in the trial (Carson, Ziesche, Johnson, & Cohn, 1999). On that basis the U.S. Food and Drug Administration approved a trial of BiDil in African Americans (Kahn, 2004). The trial recently terminated early because interim analyses showed the drug to be highly effective, and the makers of BiDil will now seek approval for its use in the African American population. But any genetic basis to the difference in nitric oxide bioactivity between African Americans and European Americans remains to be elucidated. Furthermore, there have been no comparative studies investigating differences in nitric oxide bioactivity between Africans and African Americans, and any relationship between endothelial nitric oxide synthase variants and plasma nitric oxide levels or other intermediate or clinical phenotypes is not fully understood (Li et al., 2004).

Genetics probably contributes to some of these reported differences between different racial or ethnic groups. On the other hand, at least some of the differences can be attributed to confounded environmental factors.

A recent study investigated blood pressure response to quinapril, an ACE inhibitor, in 533 African Americans and 2,046 European Americans with hypertension. African Americans had a lower average response to quinapril than European Americans, but age, gender, body size, and pretreatment blood pressure significantly predicted blood pressure response (Mokwe et al., 2004). These factors correlate with race or ethnicity, and when they are accounted for, the effect attributable to race or ethnicity is reduced.

Some claims for differences among racial or ethnic groups in drug response will be false positives, and these may be more likely to be reported than a negative finding (Lohmueller, Pearce, Pike, Lander, & Hirschhorn, 2003). The blood pressure response analysis makes the point that, even for real differences, it is not clear which differences (if any) are due to genetic factors and which are due to environmental correlates with race or ethnicity. The analysis also shows that there is substantial variation within racial or ethnic groups in environmental correlates of response, as will also be the case for genetic factors.

Genetic Contributors to Differences in Drug Response

In discussing possible genetic factors, it is useful to distinguish genetic factors that are specific to the drug used from those that influence the nature of the condition itself. Drug-specific effects include both pharmacokinetics (e.g., drug-metabolizing enzymes [DMEs] and transporters) and pharmacodynamics (e.g., drug targets and target-related proteins, which are often not involved in disease etiology) of the drug.

For both the disease-associated and the drug-specific effects, there is scope for intergroup differences; many pharmacogenetic variants known to influence drug response show frequency differences among racial or ethnic groups (Goldstein et al., 2003). For example, the beta-1 adrenoceptor Arg389 variant, associated with increased response to beta-blockers (Sofowora et al., 2003; Johnson et al., 2003), is less common in African Americans than in European Americans (frequencies of 0.575 and 0.723, respectively). Many DME variants also vary in frequency among populations (Xie, Kim, et al., 2001), and for some DMEs (e.g., CYP2D6, CYP2C19, CYP2C9, and NAT2) the proportion of individuals with little or no functional enzyme varies substantially among populations. As a relatively benign example, because they carry null alleles of CYP2D6, up to 10% of individuals of Northern European ancestry experience no analgesic effect from the prodrug codeine (Bradford, 2002), whereas 98% of the inhabitants of the Arabian peninsula are able to transform codeine into the active form morphine (McLellan, Oscarson, Seidegard, Evans, & Ingelman-Sundberg, 1997).

These examples show that gene variants that causally influence how individuals respond to treatment often have important differences in frequency among racial or ethnic groups. This suggests that genetic differences among racial or ethnic groups will often lead to differences in drug response. The examples we selected are consistent with the general pattern for variants known to influence drug response. Of 42 genetic variants that have been associated with drug response in two or more studies, more than two-thirds have significant differences in frequency between people of European ancestry and people of African ancestry. The average frequency difference for all 42 variants is 0.15, and nearly one-third of the variants have frequency differences greater than 0.2 (Goldstein et al., 2003). Many of the differences in drug response among racial or ethnic groups reported in table 6.1 are therefore expected to be influenced to some degree by genetics. These differences will often be modest and may not be clinically significant, but we must expect that such differences will exist and will be partially genetic.

Overall, analyses of responses to existing medicines and new chemical entities are far from conclusive about the scope for genetic differences among groups to contribute to variable drug response. In most cases it is difficult to separate genetic from environmental factors. For example, one study found significant differences in response to treatment of childhood acute lymphoblastic leukemia among racial or ethnic groups, with greatest response in Asians, followed by Europeans, Hispanics, and African Americans (Kadan-Lottick, Ness, Bhatia, & Gurney, 2003). Other investigators, however, found no difference in outcome when African Americans and

Europeans Americans were given equal access to the most advanced therapies (Pui et al., 2003). It was suggested that the conflicting findings were
"possibly due to the specialized referral base of the unique practice of St
Jude's Hospital [where the latter study took place], which attracts patients
from an eight-state area and provides therapy at no cost to the patient's family" (Kadan-Lottick et al., 2003, p. 2012). Nevertheless, there are gene variants that influence drug response that differ among racial or ethnic groups.
It is therefore impossible to rule out the existence of such variation, as some
investigators seem to have implied. For example, Cooper, Kaufman, and
Ward stated, "Race-specific therapy draws its rationale from the presumption
that the frequencies of genetic variants influencing the efficacy of the drug
are substantially different among races. This result is hard to demonstrate
for any class of drugs, including those used to treat heart failure" (2003, p.
1166). First, therapeutic response may differ among racial or ethnic groups
because of either average genetic differences or environmental differences.
Thus, race-specific therapy does not draw its rationale from a presumption
of genetic differences. In the case of BiDil, it is not currently known whether
it works differently in African Americans and European Americans because
of genetics, environment, or both. Second, there is no shortage of gene variants known to influence drug response that have substantial differences in
frequency among racial or ethnic groups.

Despite the limited information currently available, we can draw three
general conclusions about genetic contributions to differences in drug
response. First, genetic differences among groups are graded, as opposed
to dichotomous. Second, when genetic factors have a role, identifying the
genetic factors themselves so that they can be considered directly will reduce
the need to consider race or ethnicity as a loose proxy for predicting drug
response. For example, we noted that most individuals from the Arabian
peninsula would respond to codeine, whereas up to 10% of Northern Europeans would not. A simple genetic test would indicate who would respond,
regardless of geographic ancestry. Third, many differences in drug response
associated with race or ethnicity are due to environmental correlates rather
than population genetic differences. This implies that even when the genetic
structure of a test population is taken into account, it may still be appropriate to consider race or ethnicity as a variable in order to take account of the
environmental correlations (Risch et al., 2002).

Pharmacogenetic Diagnostics

One common view of the role of pharmacogenetics is that it should provide diagnostic analyses that allow matching of medicines with the genetic
makeup of an individual, to ensure use of medicines most likely to be

effective and least likely to produce adverse drug reactions (ADRs). This role of pharmacogenetics is best exemplified by a diagnostic test for low-activity variants of the gene TPMT, which encodes an enzyme that metabolizes thiopurines, which are used to treat acute lymphoblastic leukemia, rheumatoid arthritis, and inflammatory bowel disease (McLeod & Siva, 2002). Individuals with intermediate or deficient activity (partially predictable from TPMT genotype) risk toxicity at standard doses and must take smaller doses. Another example is the ALOX5 promoter polymorphism. Here, genotype predicts response to leukotriene receptor antagonists (Drazen et al., 1999). Individuals without leukotriene activity will not respond as well. Unlike the TPMT diagnostic test, this analysis is not currently used in the clinic, probably because nonresponders do not suffer adverse effects from an ineffective drug.

For pharmacogenetic diagnostics to be useful, they must be sufficiently specific and sensitive. This will be a difficult challenge. The predictive values of pharmacogenetic tests will often be low (Holtzman, 2003), whereas prediction of an ADR, for example, may have very stringent requirements. Ideally, a test for an ADR will have high positive and negative predictive values (the proportion of people with a positive test who will experience an ADR when the drug is administered and the proportion of people with a negative test who will not have an ADR, respectively). For a severe ADR, a high negative predictive value is essential. In some cases, however, positive predictive value will also be important, for example, when alternative medicines are available. Unless there is a high positive predictive value, many individuals will be misclassified as likely to have the ADR and therefore may not get the best treatment.

Will pharmacogenetic tests be equally predictive across different ethnic groups? There are two different reasons why a diagnostic test might perform differently among racial or ethnic groups. The first and most fundamental reason is that underlying physiology may differ among racial or ethnic groups. For example, the variant that influences response to a drug in one racial or ethnic group might not have the same effect, on average, in another group because of different gene-gene or gene-environment interactions. For example, the 3435C → T polymorphism in ABCB1 has been variably associated with altered drug response (Siddiqui et al., 2003; Fellay et al., 2002), pharmacokinetics, and P-glycoprotein expression (Hoffmeyer et al., 2000), but the correlation between the polymorphism and P-glycoprotein levels is not consistent across ethnic groups (Oselin, Gerloff, Mrozikiewicz, Pahkla, & Roots, 2003; Sakaeda, Nakamura, & Oklumura, 2003). If 3435C → T is the causal polymorphism (and this has not been proven), then this could be an example of physiological differences across ethnic groups. The second, and perhaps more likely, reason is that the diagnostic test uses a proxy, for example, if it is based on linkage disequilibrium (LD). LD is the nonrandom association of alleles at different polymorphic sites in the genome, and levels are often high across long genomic stretches, making it hard to know whether an associated variant is causal or just a marker for (i.e., usually inherited with) the causal variant. The predictive value of an LD diagnostic test will depend on the degree of association between the markers and the underlying causal variants, and there is considerable variation in the pattern of LD among populations (Reich et al., 2001). Unfortunately, diagnostic tests that are based on markers serving as proxies for the causal variants will generally have different predictive properties among different racial or ethnic groups.

One recent example provides a stark warning about the applicability of diagnostic tests across racial or ethnic groups. Abacavir is an effective antiretroviral drug used to treat HIV-1 infection. Approximately 5% of people treated with Abacavir develop a hypersensitivity reaction that requires them to stop taking the drug. Pharmacogenetic studies have identified multiple markers in the human leukocyte antigen (HLA)-B chromosomal region, including HLA-B*5701, that are associated with hypersensitivity to Abacavir in European Americans (Hetherington et al., 2002; Hughes et al., 2004; Mallal et al., 2002) but not in African Americans (Hughes et al., 2004). Therefore, a pharmacogenetic diagnostic test using this allele would have no predictive value in African Americans. In this case, it is not clear whether Abacavir hypersensitivity has different underlying causes in African Americans versus European Americans. The HLA-B alleles are situated in a tract of approximately 200 kb with extensive LD, and it is not clear which variants within this tract are responsible for the association in Europeans and Hispanics. Considerably larger association studies or functional studies will be required to determine whether the HLA-B*5701 allele or an associated variant is responsible for the ADR.

These considerations also suggest that in the near term at least, pharmacogenetic diagnostics should rely on validated causal variants rather than either single markers as proxies or genome-wide SNP profiles, both of which will usually rely on LD to generate associations between the markers and the drug responses. Because the degree and pattern of LD typically vary among populations, LD marker-based tests will often need to be adjusted for different racial or ethnic groups. This requires large, expensive studies of many populations.

Genetic Subclassification of Disease

Common diseases result from complex interactions between genetic and environmental factors. As more is learned about the genetic bases of common diseases, such diseases may be divided into distinct subclasses with similar phenotypes but different underlying genetic bases. In many cases, specific drugs are indicated for specific subtypes of a disease: for example, Herceptin is indicated for the subpopulation of individuals with breast cancer who express ERBB2, and Gleevec is indicated for individuals with chronic myeloid leukemia resulting from the BCR-ABL1 gene fusion. Similarly, two recent studies identified mutations in EGFR in lung cancers, which predict response to the tyrosine kinase inhibitor gefitinib (also known by the trade name Iressa) (Paez et al., 2004; Lynch et al., 2004).

It is unclear how often the underlying genetic bases of disease differ among racial or ethnic groups, but this may occur for some diseases. For example, susceptibility to Crohn disease is associated with three polymorphisms in CARD15 in European Americans (Hugot et al., 2001), but none of these variants was present in a sample of Japanese individuals with Crohn disease (Yamazaki, Takazoe, Tanaka, & Nakamura, 2002). Another example is a CCR5 variant that protects against HIV infection and progression. Up to 25% of European Americans are heterozygous with respect to this variant, but it is practically absent in other ethnic groups (Stephens et al., 1998). The EGFR mutations that predict response to gefitinib are more frequent in Japanese individuals, possibly explaining their increased responsiveness.

Even when a genetic variant associated with disease susceptibility is present in multiple ethnic groups, it may have different effects. For example the E4 variant of APOE is associated with a substantially increased risk of Alzheimer's disease (Corder et al., 1993), which varies among racial or ethnic groups. Homozygosity with respect to the E4 allele increases the risk of Alzheimer's disease by a factor of 33 in Japanese, 15 in European Americans, and 6 in African Americans (Farrer et al., 1997). A more recent study also found differential effects of the E4 allele between Europeans Americans and African Americans (Evans et al., 2003). It is impossible to determine from current evidence whether these differences reflect an interaction of APOE with genetic background or with environment.

To the extent that progress in understanding the genetic bases of common disease is faster in some ethnic groups than in others, the ability to genetically subclassify diseases might proceed faster in some groups than others, meaning that treatment can be made more precise in these groups. In this context, it may be a concern that the HapMap Project is not sufficiently inclusive.

The HapMap Project is an international research effort to describe the common patterns of human genetic variation in order to facilitate future studies that relate genetic variation to health and disease. The project currently includes European Americans, Africans (the Yoruba in Ibadan,

Nigeria), Japanese, and Han Chinese but excludes Native Americans and Pacific Islanders, two of the five racial categories indicated by the U.S. Food and Drug Administration and argued by some to be the principal determinants of the geographic component of human genetic variation (Risch et al., 2002). Exclusion of Native Americans in particular seems inconsistent with the policy of the U.S. National Institutes of Health to include minorities in biomedical research (according to the U.S. census, there are approximately 4.3 million Native Americans or Native Americans in combination with one or more other races in the United States, making up 1.5% of the total population). The decision to exclude Native Americans was made after consultation with some representatives of the Native American health research community, who cited concerns about HapMap data being used to facilitate population history studies and comparisons among populations ("Integrating Ethics and Science," 2004). In a broader sense, given the genetic diversity among African populations, it would almost certainly have been more informative to choose a second African population as opposed to a second East Asian one.

Drug Pipeline Pharmacogenetics

In some ways the use of pharmacogenetics during the development of potential new therapies generates the most serious concerns because of its potential to influence the medicines that are brought to market. In the past, the basic model in drug development was to find drugs that are as widely applicable as possible, hence the effort to prove efficacy in large, expensive phase-three trials. Now there is interest in carrying out smaller, less expensive trials using genetics (Roses, 2002). Although the idea of focusing clinical trials on subgroups of individuals is not new, as stratification by disease subtype has always been a goal of medical research (Lindpaintner, 2003), the use of genetics in this context is new. Pharmaceutical companies have long tended, when possible, not to pursue compounds known to be metabolized largely by highly polymorphic systems, such as CYP2D6, but pharmacogenetics has otherwise had a small role in drug development. Increasingly, however, there is interest in the use of systematic genetic analyses to identify the genetic causes of variable responses during the evaluation of new chemical entities. Widespread use of such reverse genetic strategies could result in important changes in drug development, including reliance on more focused clinical trials (Roses, 2002).

Early identification of a marker for drug response could lead to smaller phase-three trials involving those individuals who are more likely to respond. Efficacy pharmacogenetics might lower the cost of phase-three clinical trials if randomization could be applied to a population of individuals in whom the drug is effective selected in phase-two trials (Roses, 2002). This would result in individuals with unfavorable genetic profiles being excluded from trials, even though a proportion of them would probably respond to the drug (albeit less frequently than the target population). There is a risk of creating "orphan genotypes" that are left untreated for either scientific (difficult to treat) or economic (too small to be economically viable) reasons. Many genetic variants, including DME polymorphisms and drug-target polymorphisms, vary in frequency among populations. If a marker for efficacy has low prevalence in a certain ethnic population, that population may be excluded from research or treatment.

How to Represent Human Population Genetic Structure

It is difficult to address concerns about differences among racial or ethnic groups if there is no agreement about what constitutes a group. This is a contentious issue, and there seems to be little hope the community will soon reach a consensus (Schwartz, 2001; Risch et al., 2002). It may be helpful, therefore, to begin with some areas of general agreement. Most importantly, no matter how groups are defined, most of the genetic variation in the species is due to differences among individuals within groups, not to differences between groups. It is also agreed, however, that individuals with the same geographic ancestry are more similar, on average, than individuals with different geographic ancestries.

The main areas of dispute are how to represent that portion of our overall variability that does correlate with geography, and how important this portion of our variation is in medicine. Risch and colleagues (2002) proposed using five racial groups in biomedical research based on continental ancestry. Although this method is easy to implement, it is unclear how well it captures human population structure. A second method, called explicit genetic inference, ignores geographic, racial, or ethnic labels and instead groups similar individuals using genetic data (Wilson et al., 2001). There has been debate about how well self-identified ethnicity corresponds with explicit genetic inference and, more generally, how well we understand the global pattern of human genetic diversity (Risch et al., 2002; Wilson et al., 2001). Our view is that worldwide patterns of genetic variation are not well known. Data from Rosenberg and colleagues seem to support the scheme proposed by Risch and colleagues of a small number of groups corresponding largely to continent of origin; in a sample of 1,056 individuals from 52 populations they identified six main genetic clusters, five of which correspond to geographic regions (Rosenberg et al., 2002). Although the sample set they used is referred to as a "diversity panel," geographic sampling is far from comprehensive. The results would be far more graded between groupings, and the groupings themselves might fade or multiply, if more comprehensive samples were analyzed.

Even when there is a generally good correspondence between self-identified ethnicity and explicit genetic inference, we believe that there are contexts where it is still advisable to obtain the most precise information possible about genetic structure. For example, in evaluating new medicines, it would be straightforward to include explicit genetic inference as part of the overall analyses, with a negligible increase in cost and complexity. Groups identified by ethnic or racial labels (or genetics) may themselves be internally structured genetically. This structure would be hard to represent with ethnic labels but is straightforward to represent using genetic inference. For example, estimates of the proportion of European ancestry in African Americans average 21%, but there is a wide range of ancestry proportions among individuals (Smith et al., 2004).

The importance of group differences in medical genetics has been the subject of much debate. Cooper, Kaufman, and Ward (2003) argue that race is not an adequate proxy for choosing a drug. Although it is true that individual genotype will always be more informative than racial or ethnic labels (for genetic effects), we believe that in some cases race and ethnicity may be useful biological proxies for the underlying genetic variation. There are many examples of variants that are known to influence drug response and that differ substantially among racial or ethnic groups. Because there are many other variants that are not known, some drug response will correlate with racial or ethnic groups, some of which may be relevant to the selection of treatment alternatives. This source of genetic variation can and should be represented during our period of ignorance of the underlying causal factors. In some settings, this may be best done by explicit genetic inference (e.g., during the evaluation of new medicines), whereas in other settings self-identified racial or ethnic labels may need to suffice (e.g., routine clinical practice, where explicit genetic inference is not currently practical).

Conclusion

Our aim was to outline how advances in genetics could contribute, even if modestly, to disparities in the quality of health care among different racial or ethnic groups. To the extent that this is agreed to be a serious concern, the solution is straightforward: more and better research in those groups that have been traditionally under-represented in clinical and other biomedical studies. The U.S. National Institutes of Health guidelines instituted in 1993 specifically require that "members of minority groups and their subpopulations must be included in all NIH-supported biomedical and behavioral research projects involving human subjects." In this spirit, Francis Collins, the director of the U.S. National Human Genome Research Institute, recently outlined a case for a large-scale prospective cohort study in the United States to identify genetic and environmental influences on disease and, specifically, called for over-sampling in ethnic minority groups (Collins, 2004). Similar European efforts have, however, simply ignored the issue. The U.K. Biobank Project, for example, is sampling minorities in proportion to their representation. This effectively excludes minorities, as the numbers collected will be too small to allow identification of gene-environment interactions specific to the minority groups. We would like to see this decision reconsidered. We welcome the recent proposal from Howard University to begin genomic research in the African diaspora, which aims to identify and characterize genetic polymorphisms in African Americans and their ancestral populations.

More specifically, it is important to carry out basic pharmacogenetics research in as broad a range of ethnic groups as possible. It is nevertheless inevitable that diagnostic tests will often be identified in specific ethnic groups, and in these cases it is essential that the diagnostic be tested explicitly in other ethnic groups, as scientists from GlaxoSmithKline did with Abacavir (Hughes et al., 2004). In this context, one valuable tool may be the use of healthy volunteers in those cases where drugs can be safely administered, or where probe drugs may indicate the effect of a gene variant on transport or metabolism. The use of healthy volunteers would also facilitate efforts to ensure that functional variation at relevant genes is equally well described in all racial or ethnic groups. Finally, even greater efforts are required to expand the diversity of drug trial populations. The results of such trials could also be interpreted with more clarity, and the effects of genetic structure more systematically assessed, if genetic structure were routinely analyzed in drug trials (Wilson et al., 2001).

Overall, it is difficult to say whether advances in genomic medicine will exacerbate or attenuate health disparities. It seems that no matter how research is done, most medicines will tend to work similarly among different human populations. Because of our youth as a species, most human genetic variation comes from an ancestral source population and is present in most current racial or ethnic groups. But rough statistical similarity in how medicines work among racial or ethnic groups is not always good enough. For example, in the American context, it is both morally and politically unacceptable for the public and private sectors to follow a research agenda leading to medicines that are 5% more effective in European Americans than in African Americans. This sort of an outcome is possible unless there are more explicit efforts made to ensure that medicines are inclusive.

NOTE

This chapter was previously published as S. K. Tate and D. B. Goldstein (2004), "Will tomorrow's medicines work for everyone?" *Nature Genetics, 36*, s34–s42.

REFERENCES

Agency for Healthcare Research and Quality (2003, July). National healthcare disparities report. Rockville, MD: U.S. Department of Health and Human Services, Agency for Healthcare Research and Quality.

Aviv, A. (1996). Cellular calcium and sodium regulation, salt-sensitivity, and essential hypertension in African Americans. *Ethnicity & Health, 1*, 275–281.

Beta-Blocker Evaluation of Survival Trial Investigators. (2001). A trial of the beta-blocker bucindolol in patients with advanced chronic heart failure. *New England Journal of Medicine, 344*, 1659–1667.

Bhatia, S., Sather, H. N., Heerema, N. A., Trigg, M. E., Gaynon, P. S., & Robinson, L. L. (2002). Racial and ethnic differences in survival of children with acute lymphoblastic leukemia. *Blood, 100*, 1957–1964.

Bjornsson, T. D., Wagner, J. A., Donahue, S. R., Harper, D., Karim, A., Khouri, M. S., et al. (2003). A review and assessment of potential sources of ethnic differences in drug responsiveness. *Journal of Clinical Pharmacology, 43*, 943–967.

Blann, A., & Bareford, D. (2004). Ethnic background is a determinant of average warfarin dose required to maintain the INR between 3.0 and 4.5. *Journal of Thrombis and Haemostasis, 2*, 525–526.

Blann, A., Hewitt, J., Siddiqui, F., & Bareford, D. (1999). Racial background is a determinant of average warfarin dose required to maintain the INR between 2.0 and 3.0. *British Journal of Haematology, 107*, 207–209.

Bradford, L. D. (2002). CYP2D6 allele frequency in European Caucasians, Asians, Africans, and their descendants. *Pharmacogenomics, 3*, 229–243.

Burchard, E. G., Ziv, E., Coyle, N., Gomez, S. L., Tang, H., Karter, A. J., et al. (2003). The importance of race and ethnic background in biomedical research and clinical practice. *New England Journal of Medicine, 348*, 1170–1175.

Cardillo, C., Kilcoyne, C. M., Cannon, R. O. III, & Panza, J. A. (1999). Attenuation of cyclic nucleotide-mediated smooth muscle relaxation in blacks as a cause of racial differences in vasodilator function. *Circulation, 99*, 90–95.

Carson, P., Ziesche, S., Johnson, G., & Cohn, J. N. (1999). Racial differences in response to therapy for heart failure: Analysis of the vasodilator-heart failure trials. Vasodilator-Heart Failure Trial Study Group. *Journal of Cardiac Failure, 5*, 178–187.

Cepeda, M. S., Farrar, J. T., Roa, J. H., Boston, R., Meng, Q. C., Ruiz, F., et al. (2001). Ethnicity influences morphine pharmacokinetics and pharmacodynamics. *Clinical Pharmacology and Therapeutics, 70*, 351–361.

Chapman, A. B., Schwartz, G. L., Boerwinkle, E., & Turner, S. T. (2002). Predictors of antihypertensive response to a standard dose of hydrochlorothiazide for essential hypertension. *Kidney International, 61*, 1047–1055.

Chung, H., Mahler, J. C., & Kakuma, T. (1995). Racial differences in treatment of psychiatric inpatients. *Psychiatric Services, 46*, 586–591.

Cohn, J. N., Archibald, D. G., Ziesche, S., Franciosa, J. A., Harston, W. E., Tristani, F. E., et al. (1986). Effect of vasodilator therapy on mortality in chronic congestive heart failure. Results of a Veterans Administration Cooperative Study. *New England Journal of Medicine, 314*, 1547–1552.

Cohn, J. N., Johnson, G., Ziesche, S., Cobb, F., Francis, G., Tristani, F., et al. (1991). A comparison of enalapril with hydralazine-isosorbide dinitrate in the treatment of chronic congestive heart failure. *New England Journal of Medicine, 325*, 303–310.

Collazo, Y., Tam, R., Sramek, J., & Herrera, J. (1996). Neuroleptic dosing in Hispanic and Asian inpatients with schizophrenia. *Mount Sinai Journal of Medicine, 63*, 310–313.

Collins, F. S. (2004). The case for a US prospective cohort study of genes and environment. *Nature, 429*, 475–477.

Collins, F. S., Green, E. D., Guttmacher, A. E., & Guyer, M. S. (2003). A vision for the future of genomics research. *Nature, 422*, 835–847.

Cooper, R. S. (2001). Social inequality, ethnicity, and cardiovascular disease. *International Journal of Epidemiology, 30* (Suppl. 1), S48–S52.

Cooper, R. S., Kaufman, J. S., & Ward, R. (2003). Race and genomics. *New England Journal of Medicine, 348*, 1166–1170.

Corder, E. H., Saunders, A. M., Strittmatter, W. J., Schmechel, D. E., Gaskell, P. C., Small, G. W., et al. (1993). Gene dose of apolipoprotein E type 4 allele and the risk of Alzheimer's disease in late onset families. *Science, 261*, 921–923.

Cubeddu, L. X., Aranda, J., Singh, B., Klein, M., Brachfeld, J., Freis, E., et al. (1986). A comparison of verapamil and propranolol for the initial treatment of hypertension: Racial differences in response. *Journal of the American Medical Association, 256*, 2214–2221.

Cushman, W. C., Reda, D. J., Perry, H. M., Williams, D., Abdellatif, M., & Materson, B. J. (2000). Regional and racial differences in response to antihypertensive medication use in a randomized controlled trial of men with hypertension in the United States. Department of Veterans Affairs Cooperative Study Group on Antihypertensive Agents. *Archives of Internal Medicine, 160*, 825–831.

Danoff, T. M., Campbell, D. A., McCarthy, L. C., Lewis, K. F., Repasch, M. H., Saunders, A. M., et al. (2004). A Gilbert's syndrome UGT1A1 variant confers susceptibility to tranilast-induced hyperbilirubinemia. *Pharmacogenomics Journal, 4*, 49–53.

De Maria, N., Colantoni, A., Idilman, R., Friedlander, L., Harig, J., & Van Thiel, D. H. (2002). Impaired response to high-dose interferon treatment in African Americans with chronic hepatitis C. *Hepatogastroenterology, 49*, 788–792.

Drazen, J. M., Yandava, C. N., Dube, L., Szczerback, N., Hippensteel, R., Pillari, A., et al. (1999). Pharmacogenetic association between ALOX5 promoter genotype and the response to anti-asthma treatment. *Nature Genetics, 22*, 168–170.

Efficacy of nadolol alone and combined with bendroflumethiazide and hydralazine for systemic hypertension. (1983). *American Journal of Cardiology, 52*, 1230–1237.

Evans, D. A., Bennett, D. A., Wilson, R. S., Bienias, J. L., Morris, M. C., Scherr, P. A., et al. (2003). Incidence of Alzheimer disease in a biracial urban community: Relation to apolipoprotein E allele status. *Archives of Neurology, 60*, 185–189.

Evelyn, B., Toigo, T., Banks, D., Pohl, D., Gray, K., Robins, B., et al. (2001). Participation of racial/ethnic groups in clinical trials and race-related labeling: A review of new molecular entities approved 1995–1999. *Journal of the National Medical Association, 93*, 18S–24S.

Exner, D. V., Dries, D. L., Domanski, M. J., & Cohn, J. N. (2001). Lesser response to angiotensin-converting-enzyme inhibitor therapy in black as compared with white patients with left ventricular dysfunction. *New England Journal of Medicine, 344*, 1351–1357.

Farrer, L. A., Cupples, L. A., Haines, J. L., Hyman, B., Kukull, W. A., Mayeux, R., et al. (1997). Effect of age, sex, and ethnicity on the association between apolipoprotein

E genotype and Alzheimer disease. A meta-analysis. APOE and Alzheimer Disease Meta Analysis Consortium. *Journal of the American Medical Association, 278*, 1349–1356.

Fellay, J., Marzolini C., Meaden, E. R., Back, D. J., Buclin, T., Chave, J. P., et al. (2002). Response to antiretroviral treatment in HIV-1-infected individuals with allelic variants of the multidrug resistance transporter 1: A pharmacogenetics study. *Lancet, 359*, 30–36.

Ferdinand, K. C. (2003). Recommendations for the management of special populations: Racial and ethnic populations. *American Journal of Hypertension, 16*, 50S–54S.

Food and Drug Administration (2005). Guideline for industry pharmacogenomic data submissions. Rockville, MD: U.S. Department of Health and Human Services, Food and Drug Administration.

Friedman B., Gray, J. M., Gross, S., & Levit, S. A. (1983). United States experience with oxprenolol in hypertension. *American Journal of Cardiology, 52*, 43D–48D.

Goldstein, D. B., Tate, S. K., & Sisodiya, S. M. (2003). Pharmacogenomics goes genomic. *Nature Reviews Genetics, 4*, 937–947.

Goran, M. I., Bergman, R. N., Cruz, M. L., & Watanabe, R. (2002). Insulin resistance and associated compensatory responses in African-American and Hispanic children. *Diabetes Care, 25*, 2184–2190.

Hassaballa, H., Gorelick, P. B., West, C. P., Hansen, M. D., & Adams, H. P., Jr. (2001). Ischemic stroke outcome: Racial differences in the trial of danaparoid in acute stroke (TOAST). *Neurology, 57*, 691–697.

Hetherington, S., Hughes, A. R., Mostellaer, M., Shortino, D., Baker, K. L., Spreen, W., et al. (2002). Genetic variations in HLA-B region and hypersensitivity reactions to abacavir. *Lancet, 359*, 1121–1122.

Higashi, M. K., Veenstra, D. L., Kondo, L. M., Wittkowsky, A. K., Srinouanprachanh, S. L., Farin, F. M., et al. (2002). Association between CYP2C9 genetic variants and anticoagulation related outcomes during warfarin therapy. *Journal of the American Medical Association, 287*, 1690–1698.

Hoffmeyer, S., Burk, O., von Richter, O., Arnold, H. P., Brockmoller, J., Johne, A., et al. (2000). Functional polymorphisms of the human multidrug-resistance gene: Multiple sequence variations and correlation of one allele with P-glycoprotein expression and activity in vivo. *Proceedings of the National Academy of the Sciences USA, 97*, 3473–3478.

Holtzman, N. A. (2003). Clinical utility of pharmacogenetics and pharmacogenomics. In M. A. Rothstein (Ed.), *Pharmacogenomics: Social, ethical, and clinical dimensions* (pp. 163–185). Hoboken, NJ: John Wiley & Sons.

Hughes, A. R., Mosteller, M., Bansal, A. T., Davies, K., Haneline, S. A., Lai, E. H., et al. (2004). Association of genetic variations in HLA-B region and hypersensitivity to abacavir in some, but not all, populations. *Pharmacogenomics, 5*, 203–211.

Hugot, J. P. Chamaillard, M., Zouali, H., Lesage, S., Cezard, J. P., Belaiche, J., et al. (2001). Association of NOD2 leucine-rich repeat variants with susceptibility to Crohn's disease. *Nature, 411*, 599–603.

Integrating ethics and science in the International HapMap Project. (2004). *Nature Reviews Genetics, 5*, 467–475.

Jann, M. W., Chang, W. H., Davis, C. M., Chen, T. Y., Deng, H. C., Lung, F. W., et al. (1989). Haloperidol and reduced haloperidol plasma levels in Chinese vs. non Chinese psychiatric patients. *Psychiatry Research, 30*, 45–52.

Johnson, J. A., Burlew, B. S., & Stiles, R. N. (1995). Racial differences in beta-adrenoceptor mediated responsiveness. *Journal of Cardiovascular Pharmacology, 25*, 90–96.

Johnson, J. A., Zineh, I., Puckett, B. J., McGorray, S. P., Yarandi, H. N., & Pauly, D. F. (2003). Beta I-adrenergic receptor polymorphisms and antihypertensive response to metoprolol. *Clinical Pharmacology & Therapeutics, 74*, 44–52.

Kadan-Lottick, N. S., Ness, K. K., Bhatia, S., & Gurney, J. G. (2003). Survival variability by race and ethnicity in childhood acute lymphoblastic leukemia. *Journal of the American Medical Association, 290*, 2008–2014.

Kahn, J. (2004). How a drug becomes "ethnic": Law, commerce, and the production of racial categories in medicine. *Yale Journal of Health Policy, Law, and Ethics, 4*, 1–46.

Kalinowski, L., Dobrucki, I. T., & Malinski, T. (2004). Race specific differences in endothelial function: Predisposition of African Americans to vascular diseases. *Circulation, 109*, 2511–2517.

Kimball, P., Elswick, R. K., & Shiffman, M. (2001). Ethnicity and cytokine production gauge response of patients with hepatitis C to interferon-alpha therapy. *Journal of Medical Virology, 65*, 510–516.

Laitinen, T., Polvi, A., Rydman, P., Vendelin, J., Pulkkinen, V., Salmikangas, P., et al. (2004). Characterization of a common susceptibility locus for asthma-related traits. *Science, 304*, 300–304.

Lang, C. C., Stein, C. M., Brown, R. M., Deegan, R., Nelson, R., He, H. B., et al. (1995). Attenuation of isoproterenol-mediated vasodilatation in blacks. *New England Journal of Medicine, 333*, 155–160.

Lang, C.C., Stein, C. M., He, H. B., Belas, F. J., Blair, I. A., Wood, M., et al. (1997). Blunted blood pressure response to central sympathoinhibition in normotensive blacks: Increased importance of nonsympathetic factors in blood pressure maintenance in blacks. *Hypertension, 30*, 157–162.

Li, R., Lyn, D., Lapu-Bula, R., Oduwole, A., Igho-Pemu, P., Lankford, B., et al. (2004). Relation of endothelial nitric oxide synthase gene to plasma nitric oxide level, endothelial function, and blood pressure in African Americans. *American Journal of Hypertension, 17*, 560–567.

Lin, K. M., Poland, R. E., Nuccio, I., Matsuda, K., Hathuc, N., Su, T. P., et al. (1989). A longitudinal assessment of haloperidol doses and serum concentrations in Asian and Caucasian schizophrenic patients. *American Journal of Psychiatry, 146*, 1307–1311.

Lindpaintner, K. (2003). Pharmacogenetics and the future of medical practice. *Journal of Molecular Medicine, 81*, 141–153.

Liu, J., Liu, Z. Q., Tan, Z. R., Chen, X. P., Wang, L. S., Zhou, G., et al. (2003). Gly389Arg polymorphism of betaI-adrenergic receptor is associated with the cardiovascular response to metoprolol. *Clinical Pharmacology and Therapeutics, 74*, 372–379.

Lohmueller, K. E., Pearce, C. L., Pike, M., Lander, E. S., & Hirschhorn, J. N. (2003). Meta-analysis of genetic association studies supports a contribution of common variants to susceptibility to common disease. *Nature Genetics, 33*, 177–182.

London Health Observatory. (2004). Ethnic disparities in health and healthcare: A focused review of the evidence and selected examples of good practice. (Institution report.) *London Health Observatory*, 1–105.

Lynch, T. J., Bell, D. W., Sordella, R., Gurubhagavatula, S., Okimoto, R. A., Brannigan, B. W., et al. (2004). Activating mutations in the epidermal growth factor receptor underlying responsiveness of non-small-cell lung cancer to gefitinib. *New England Journal of Medicine, 350*, 2129–2139.

Mallal, S., Nolan, D., Witt, C., Masel, G., Martin, A. M., Moore, C., et al. (2002). Association between presence of HLA-B*5701, HLA-DR7, and HLA-DQ3 and hypersensitivity to HIV-1 reverse-transcriptase inhibitor abacavir. *Lancet, 359*, 727–732.

Matsuda, K. T., Cho, M. C., Lin, K. M., Smith, M. W., Young, A. S., & Adams, J. A. (1996). Clozapine dosage, serum levels, efficacy, and side-effect profiles: A comparison of Korean-American and Caucasian patients. *Psychopharmacolgy Bulletin, 32*, 253–257.

McLellan, R. A., Oscarson, M., Seidegard, J., Evans, D. A., & Ingelman-Sundberg, M. (1997). Frequent occurrence of CYP2D6 gene duplication in Saudi Arabians. *Pharmacogenetics, 7*, 187–191.

McLeod, H. L., & Siva, C. (2002). The thiopurine S-methyltransferase gene locus-implications for clinical pharmacogenomics. *Pharmacogenomics, 3*, 89–98.

Millward, M. J., Boyer, M. J., Lehnert, M., Clarke, S., Rischin, D., Goh, B. C., et al. (2003). Docetaxel and carboplatin is an active regimen in advanced non small-cell lung cancer: A phase II study in Caucasian and Asian patients. *Annals of Oncology, 14*, 449–454.

Mokwe, E., Ohmit, S. E., Nasser, S. A., Shafi, T., Saunders, E., Crook, E., et al. (2004). Determinants of blood pressure response to quinapril in black white hypertensive patients: The Quinapril Titration Interval Management Evaluation trial. *Hypertension, 43*, 1202–1207.

Muir, A. J., Bornstein, J. D., & Killenberg, P. G. (2004). Peginterferon alfa-2b and ribavirin for the treatment of chronic hepatitis C in blacks and non-Hispanic whites. *New England Journal of Medicine, 350*, 2265–2271.

Netland, P. A., Robertson, S. M., Sullivan, E. K., Silver, L., Bergamini, M. V., Krueger, S., et al. (2003). Response to travoprost in black and nonblack patients with open angle glaucoma or ocular hypertension. *Advances in Therapy, 20*, 149–163.

Oselin, K., Gerloff, T., Mrozikiewicz, P. M., Pahkla, R., & Roots, I. (2003). MDR1 polymorphisms G2677T in exon 21 and C3435T in exon 26 fail to affect rhodamine 123 efflux in peripheral blood lymphocytes. *Fundamental & Clinical Pharmacology, 17*, 463–469.

Paez, J. G., Janne, P. A., Lee, J. C., Tracy, S., Greulich, H., Gabriel, S., et al. (2004). EGFR mutations in lung cancer: Correlation with clinical response to gefitinib therapy. *Science, 304*, 1497–1500.

Pollock, B. H., DeBaun, M. R., Camitta, B. M., Shuster, J. J., Ravindranath, Y., Pullen, D. J., et al. (2000). Racial differences in the survival of childhood B-precursor acute lymphoblastic leukemia: A Pediatric Oncology Group study. *Journal of Clinical Oncology, 18*, 813–823.

Pui, C. H., Sandlund, J. T., Pei, D., Rivera, G. K., Howard, S. C., Ribeiro, R. C., et al. (2003). Results of therapy for acute lymphoblastic leukemia in black and white children. *Journal of the American Medical Society, 290*, 2001–2007.

Reich, D. E., Cargill, M., Bolk, S., Ireland, J., Sabeti, P. C., Richter, D. J., et al. (2001). Linkage disequilibrium in the human genome. *Nature, 411*, 199–204.

Risch, N., Burchard, E., Ziv, E., & Tang, H. (2002). Categorization of humans in biomedical research: Genes, race, and disease. *Genome Biology, 3*, 2007.

Robertson, J. A. (2003). Constitutional issues in the use of pharmacogenomic variations associated with race. In M. A. Rothstein (Ed.), *Pharmacogenomics: Social, ethical, and clinical dimensions* (pp. 291–316). Hoboken, NJ: John Wiley & Sons.

Rosenbaum, D. A., Pretorius, M., Gainer, J. V., Byrne, D., Murphy, L. J., Painter, C. A., et al. (2002). Ethnicity affects vasodilation, but not endothelial tissue plasminogen activator release, in response to bradykinin. *Arteriosclerosis, Thrombosis, and Vascular Biology, 22,* 1023–1028.

Rosenberg, N. A., Pritchard, J. K., Weber, J. L., Cann, H. M., Kidd, K. K., Zhivotovsky, L. A., et al. (2002). Genetic structure of human populations. *Science, 298,* 2381–2385.

Roses, A. D. (2002). Genome-based pharmacogenetics and the pharmaceutical industry. *Nature Reviews Drug Discovery, 1,* 541–549.

Sakaeda, T., Nakamura, T., & Okumura, K. (2003). Pharmacogenetics of MDR1 and its impact on the pharmacokinetics and pharmacodynamics of drugs. *Pharmacogenomics, 4,* 397–410.

Saunders, E. (1991). Hypertension in blacks. *Primary Care, 18,* 607–622.

Schmith, V. D., Campbell, D. A., Sehgal, S., Anderson, W. H., Burns, D. K., Middleton, L. T., et al. (2003). Pharmacogenetics and disease genetics of complex diseases. *Cellular and Molecular Life Sciences, 60,* 1636–1646.

Schwartz, R. S. (2001). Racial profiling in medical research. *New England Journal of Medicine, 344,* 1392–1393.

Segal, S. P., Bola, J. R., & Watson, M. A. (1996). Race, quality of care, and antipsychotic prescribing practices in psychiatric emergency services. *Psychiatric Services, 47,* 282–286.

Siddiqui, A., Kerb, R., Weale, M. E., Brinkmann, U., Smith, A., Goldstein, D. B., et al. (2003). Association of multidrug resistance in epilepsy with a polymorphism in the drug-transporter gene ABCB1. *New England Journal of Medicine, 348,* 1442–1448.

Smith, M. W., Patterson, N., Lautenberger, J. A., Truelove, A. L., McDonald, G. J., Waliszewska, A., et al. (2004). A high-density admixture map for disease gene discovery in African Americans. *American Journal of Human Genetics, 74,* 1001–1013.

Sofowora, G. G., Dishy, V., Muszkat, M., Xie, H. G., Kim, R. B., Harris, P. A., et al. (2003). A common beta1-adrenergic receptor polymorphism (Arg389Gly) affects blood pressure response to beta-blockade. *Clinical Pharmacology and Therapeutics, 73,* 366–371.

Stein, C. M., Lang, C. C., Nelson, R., Brown, M., & Wood, A. J. (1997). Vasodilation in black Americans: Attenuated nitric oxide-mediated responses. *Clinical Pharmacology and Therapeutics, 62,* 436–443.

Stephens, J. C., Reich, D. E., Goldstein, D. B., Shin, H. D., Smith, M. W., Carrington, M., et al. (1998). Dating the origin of the CCR5-Delta32 AIDS-resistance allele by the coalescence of haplotypes. *American Journal of Human Genetics, 62,* 1507–1515.

Strakowski, S. M., Shelton, R. C., & Kolbrener, M. L. (1993). The effects of race and comorbidity on clinical diagnosis in patients with psychosis. *Journal of Clinical Psychiatry, 54,* 96–102.

Takahashi, H., Wilkinson, G. R., Caraco, Y., Muszkat, M., Kim, R. B., Kashima, T., et al. (2003). Population differences in S-warfarin metabolism between CYP2C9 genotype-matched Caucasian and Japanese patients. *Clinical Pharmacology and Therapeutics, 73,* 253–263.

Tate, S. K., & Goldstein, D. B. (2004). Will tomorrow's medicines work for everyone? *Nature Genetics, 36,* S34–S42. http://www.nature.com/ng/journal/v36/n11s/fig_tab/ng1437_T1.html.

Taube, J., Halsall, D., & Baglin, T. (2000). Influence of cytochrome P-450 CYP2C9 polymorphisms on warfarin sensitivity and risk of over-anticoagulation in patients on long-term treatment. *Blood, 96,* 1816–1819.

Tornatore, K. M., Biocevich, D. M., Reed, K., Tousley, K., Singh, J. P., & Venuto, R. C. (1995). Methylprednisolone pharmacokinetics, cortisol response, and adverse effects in black and white renal transplant recipients. *Transplantation, 59*, 729–736.

Walkup, J. T., Alpine, D. D., Olfson, M., Labay, L. E., Boyer, C., & Hansell, S. (2000). Patients with schizophrenia at risk for excessive antipsychotic dosing. *Journal of Clinical Psychiatry, 61*, 344–348.

Watkins, L. L., Dimsdale, J. E., & Ziegler, M. G. (1995). Reduced beta 2-receptor mediated vasodilation in African Americans. *Life Sciences, 57*, 1411–1416.

Weir, M. R., Reisin, E., Falkner, B., Hutchinson, H. G., Sha, L., & Tuck, M. L. (1998). Nocturnal reduction of blood pressure and the antihypertensive response to a diuretic or angiotensin converting enzyme inhibitor in obese hypertensive patients. *American Journal of Hypertension, 11*, 914–920.

Weir, M. R., Gray, J. M., Paster, R., & Saunders, E. (1995). Differing mechanisms of action of angiotensin-converting enzyme inhibition in black and white hypertensive patients. The Trandolapril Multicenter Study Group. *Hypertension, 26*, 124–130.

Wilson, J. F., Weale, M. E., Smith, A. C., Gratix, F., Fletcher, B., Thomas, M. G., et al. (2001). Population genetic structure of variable drug response. *Nature Genetics, 29*, 265–269.

Xie, H. G., Dishy, V., Sofowora, G., Kim, R. B., Landau, R., Smiley, R. M., et al. (2001). Arg389Gly beta 1-adrenoceptor polymorphism varies in frequency among different ethnic groups but does not alter response in vivo. *Pharmacogenetics, 11*, 191–197.

Xie, H. G., Kim, R. B., Wood, A. J., & Stein, C. M. (2001). Molecular basis of ethnic differences in drug disposition and response. *Annual Review of Pharmacology and Toxicology, 41*, 815–850.

Yamazaki, K., Takazoe, M., Tanaka, T., Kazumori, T., & Nakamura, Y. (2002). Absence of mutation in the NOD2/CARD15 gene among 483 Japanese patients with Crohn's disease. *Journal of Human Genetics, 47*, 469–472.

Zanchi, A., Maillard, M., & Burnier, M. (2003). Recent clinical trials with omapatrilat: New developments. *Current Hypertension Reports, 5*, 346–352.

Zhou, H. H., Sheller, J. R., Nu, H., Wood, M., & Wood, A. J. (1993). Ethnic differences in response to morphine. *Clinical Pharmacology and Therapeutics, 54*, 507–513.

7

Patenting Race in a Genomic Age

JONATHAN KAHN

In the growing field of biotechnology research and development, a new phenomenon is emerging—the strategic use of race as a genetic category to obtain patent protection and drug approval. The imbrication of race in the field of patent law as an adjunct to biotechnological inventions is producing new racialized spaces of intellectual property that may have profound implications for broader social understandings and mobilizations of race. When the federal government grants a patent to an invention that is based on an asserted or implied genetic basis for a particular racial group, it gives the imprimatur of the federal government to the construction of race as genetic. Moreover, once granted, such patents may provide the basis for similarly race-based clinical trial designs, drug development, capital raising, and marketing strategies that carry the construction of race as genetic out to ever widening and consequential segments of society.

A dramatic rise in the use of race in biotechnology patents since the completion of the first draft of the human genome in 2000 indicates that researchers and affiliated commercial enterprises are coming to see such social categories as presenting opportunities for gaining, extending, or protecting monopoly market protection for an array of products and services. The commercial mobilization of race and ethnicity is not merely coincidental with the proliferation of new genetic knowledge. Rather, federal initiatives have played a central role in producing, classifying, and disseminating human genetic information. Once race is conceptualized in relation to biotechnology products, the patent system itself provides incentives for using race and ethnicity in order to maximize patent scope, duration, and viability. Federal initiatives, guidelines, and approvals thus provide specific, targeted incentives to see and use race and ethnicity in relation to biotechnological innovation in a manner that promotes, indeed, rewards, the reification of race as a genetic category.

Few areas of the law currently are as fully engaged in pervasive management of genetic material and information as intellectual property law. Patent law and genetics, however, while much examined, are generally explored in terms of how best to promote the efficient production and exploitation of genetic information. While the ethical implications of patenting human genetic material have been explored at length, little attention has been given to the ways in which social categories of race and ethnicity are increasingly being mobilized in the context of biotechnology patents.

A "product of nature" cannot be patented. To be rendered patentable, it must be "purified and isolated" through human interventions to produce a substance that does not otherwise exist. Historically, this involved complex chemicals such as adrenaline (see, e.g., *Parke-Davis & Co. v. H. K. Mulford Co.*, 1911). In the genomic era, however, it has come to encompass engineered complementary DNA (cDNA). DNA as found in nature contains both nucleotide sequences that code for producing proteins (exons) and sequences that do not code for proteins (introns); cDNA is synthesized *in vitro* by using an enzyme (reverse transcriptase) that produces a molecule containing only exons. While clearly scientific and technical in origin, isolation and purification are distinctively legal concepts when it comes to granting a patent (Kahn, 2003). The U.S. Patent and Trademark Office (PTO) has asserted that "the inventor's discovery of a gene can be the basis for a patent of the genetic composition *isolated from its natural state* and processed through *purifying* steps that *separate* the gene from other molecules *naturally associated* with it" (emphasis added) (*Federal Register*, 2001).

Sheila Jasanoff observes that "biotechnology . . . renders continually problematic the boundary between the natural and the unnatural" (Jasanoff, 2002, p. 895). The authoritative discourses of science and law, however, are rendered precarious by such uncertainty. Those seeking the legal recognition of patentability for biotechnological achievements work hard to resolidify and render apparently unproblematic the boundary between natural and unnatural. Thus, the PTO recognizes arguments that scientific intervention creates a patentable object by severing it from its "natural associations." The PTO constructs cDNA as isolated, not only in the sense of separating exons from introns, but, more powerfully, in the sense of separating the genetic material itself from nature. This is not a scientific process but a legal one. The scientist may create cDNA, but the PTO draws the line between nature and artifice. Similarly, purification involves stripping the genetic material of its identity as a part of nature—purifying it of its natural associations.

This chapter will explore how the rise of racial patents inverts this traditional dynamic. Patents have long been premised on a legal recognition of how human intervention may take a product out of nature and into culture. In contrast, race enters the world of biotechnology as a social construct. It serves as an admitted surrogate for presumed underlying genetic variations in particular populations. In the patent and drug approval process necessary to bring the drug to market, race is implicitly recoded as a genetic category. The patent process takes race as a social category and recodes it as "natural" by according it legal force as a component of a biotechnological invention. Law, thereby, is taking race out of culture and locating it in nature.

The chapter begins with a consideration of diverse federal mandates that structure the collection, classification, and circulation of data about both social categories of race and genetic categories of population. When brought together in the context of biotechnology research and product development, these diverse federal classificatory schemes become easily entangled and conflated, providing a structural incentive for reifying race as genetic. It proceeds to a discussion of BiDil, the first drug ever approved by the Food and Drug Administration (FDA) with a race specific indication— for the treatment of heart failure in African Americans. It argues that the dynamic whereby commercial and legal considerations drove the development of BiDil is a portent of further commercial exploitation of race in biotechnology. The chapter then moves on to a detailed analysis of the new racial patents in biotechnology, exploring the myriad ways in which the legal imperatives of patent law are appropriating the language of science and medicine to imbue products with new commodity value by reifying race as genetic. It concludes with the observation that biotechnology corporations are mining the raw material of race as a social category and using the patent process to refine it into a natural construct in order to gain patent protection and market advantage.

The Impact of Federal Schemes of Classification

The racialization of patent law can only fully be understood when viewed in relation to broader federal initiatives that shape the production and use of racial[1] categories in biomedical research. Prominent among these are a wide array of federal mandates that dictate the characterization and application of genetically based biomedical interventions, such as pharmaceuticals and diagnostic tests, in relation to socially defined categories of race. Key federal mandates include the National Institutes of Health (NIH) Revitalization Act of 1993, which directed the NIH to establish guidelines for inclusion of women and minorities in clinical research; the Food and Drug Modernization Act of 1997, which, in the context of drug development, directed that "the Secretary [of Health and Human Services] shall, in consultation with the Director of the NIH and with representatives of the drug manufacturing industry, review and develop guidance, as appropriate, on the inclusion of women and minorities in clinical trials"; and two subsequent FDA "Guidances for Industry." The first, a 1999 guidance titled "Population Pharmacokinetics," made recommendations on the use of population pharmacokinetics in the drug development process to help identify differences in drug safety and efficacy among population subgroups, including race and ethnicity (Food and Drug Modernization Act, 1997; U.S. FDA, 1999); and the second was a 2005 guidance entitled "Collection of Race and Ethnicity Data in Clinical Trials" (U.S. FDA, 2005b), which recommends a standardized approach for collecting and reporting race and ethnicity information in clinical trials that produce data for applications to the FDA for drug approval.

Underlying the standardization of data collection in all of these mandates is the Office of Management and Budget's (OMB) Revised Directive 15 on "Standards for Maintaining, Collecting, and Presenting Federal Data on Race and Ethnicity" (U.S. OMB, 1997). The standards were developed "to provide a common language for uniformity and comparability in the collection and use of data on race and ethnicity by Federal agencies." The standards set forth the following basic racial categories for organizing such data: American Indian or Alaska native, Asian, black or African American, Native Hawaiian or other Pacific Islander, and white. There are two categories for data on ethnicity: "Hispanic or Latino," and "Not Hispanic or Latino" (OMB, 1997). Sociologist Michael Omi observes, "Directive 15 has become the de facto standard for state and local agencies, the private and nonprofit sectors, and the research community" (Omi, 1997, p. 21). This dynamic reinforces what Omi has characterized as an "interesting dilemma" facing scientists in the United States: "On the one hand," Omi asserts, "scientists routinely use racial categories in their research and regularly make comparisons between races with respect to health. . . . On the other hand, many scientists feel that racial classifications are meaningless and unscientific" (Omi, 1997, p. 7).

Producing and Organizing Genetic Information: Federally Sponsored Genetic Databases

This dilemma is likely to become ever more problematic when biomedical researchers and clinicians are using the social categories of race mandated by Directive 15 alongside of purportedly genetic population groupings produced and organized by federally sponsored genomic initiatives. As genetic research has grown over the past three decades, the federal government has sponsored or co-sponsored an array of databanks that collect, store, and classify genetic information for use by biomedical researchers. Such data banks include the National Institute of General Medical Sciences (NIGMS)–Coriell Human Genetic Variation Collections, the National Human Genome

Institute's DNA Polymorphism Discovery Resource (PDR), the National Center for Biotechnology Information's dbSNP database, and the International Haplotype Map Project (also known as the HapMap). Each of these databases organizes genetic information into highly problematic population groupings that have been taken up and used by researchers as correlates for or equivalents of racial categories. Overarching these collections is a repository maintained by the federal government's National Center for Biotechnology Information (NCBI, 2005), known as GenBank, which contains a Web-based annotated collection of all publicly available DNA sequences. These databases are powerful not only because of the categories they use to organize genetic information internally, but also because, as examples of authoritative scientific knowledge, they provide working models of acceptable schemes of categorization by which any genetic data may be organized, wherever obtained or stored.

These federal databases mix and match crude categories that variously employ racial and ethnic constructs (e.g., U.S. Caucasians, African Americans, and Hispanics), geographic constructs (e.g., North Africa, East Africa), political nation-states (e.g., Russia and satellite republics), mixes of geography and nation-states (e.g., sub-Saharan nations bordering the Atlantic north of the Congo River), and mixes of all three (e.g., "All samples north of Tropic of Cancer. This would include defined samples of U.S. Caucasians, African Americans and Hispanics.") (NCBI, 2005). As biomedical researchers mine such data, they are accessing and organizing it in terms that juxtapose or directly classify genetic data in terms of race, ethnicity, nation, and/or geography. When used in studies or trials covered by federal mandates, the diverse and sometimes contradictory population classifications employed variously by the NIGMS, the PDR, and dbSNP databases cry out to be simplified and reclassified in terms of OMB's basic Directive 15 categories of race and ethnicity. Thus, for example, a locally specific genetic sample, originally designated as from an individual in Tokyo, Japan, is likely in subsequent practice to be conflated into the overarching category of "Asian."

All of these genetic databases are technologies of classification. Systems of classification, however, are artifacts that embody ethical choices. As Geoffrey Bowker and Susan Star note, "Each standard and category valorizes some point of view and silences another. This is not inherently a bad thing—indeed it is inescapable. But it is an ethical choice, and as such it is dangerous—not bad, but dangerous" (1999, pp. 5–6). In the realm of genetics, where such systems address the human body, new biologically based categories can profoundly affect people's identities, aspirations, and dignity. Genetic classification is powerful, but it is also dangerous because it involves biological categories that may be confused and conflated with race. Any resulting reification of social categories of race as biological constructs risks new forms of exclusion and stigma (Duster, 1990; Foster, 2002; Lock, 1999; Marks, 1995).

Bowker and Star argue that "politically and socially charged agendas are often first presented as purely technical and they are difficult even to see. As layers of classification system become enfolded into a working infrastructure, the original political intervention becomes more and more firmly entrenched. In many cases, this leads to a naturalization of the political category. . . . It becomes taken for granted" (1999, p. 196). Genetic databases and OMB Directive 15 are seemingly "technical" methods of categorization, but such apparent neutrality is precisely what drives and lends the aura of legitimacy to the casual and often reflexive conflation of race and genetics in a variety of biomedical contexts.

These federal mandates have a profound effect upon the use of racial categories in biomedical research, clinical practice, product development, and health policy. At the most basic level, they create incentives to introduce race into biomedical contexts, regardless of their relevance. Once introduced, racial categories can take on a life of their own and become exploited in new and unanticipated ways, with unforeseen and potentially harmful consequences.

BiDil: Portent of Things to Come

This has already begun to happen. In June 2005, a drug called BiDil became the first drug ever approved by the FDA with a race-specific indication: to treat heart failure in African Americans (FDA, 2005a). Underlying the new drug application (NDA) submitted for this drug to the FDA is a race-specific methods patent: to use the drug for treatment of heart failure *in an African American patient*. The patent is premised on underlying assumptions regarding race and the genetic basis of heart disease. By granting such a patent, the Patent and Trademark Office (PTO) is giving the imprimatur of the federal government to the use of race as a genetic category (Kahn, 2004). BiDil's history is complex and has been explored in detail (Kahn, 2003, 2004, 2005; Lee, this volume). It reveals a story of how race and ethnicity were exploited in conjunction with patent law and the drug approval process to bring a new drug to market. In short, BiDil became a racially marked drug more because of law and commerce than because of medical evidence (Kahn, 2004).

All indications seem to show that the drug is highly effective at treating heart failure. The FDA approval, however, was based on results from A-HeFT, the African American Heart Failure Trial, that were published the previous November in the *New England Journal of Medicine* (Taylor et al., 2004). The trial design, approved by the FDA, was itself path-breaking because it included *only* self-identified African Americans. The results, therefore, give the impression that BiDil works *only* in African Americans. This is clearly not the case. The trial investigators themselves concede that BiDil will work in people regardless of race. Without a comparison population, the investigators cannot even claim that the drug works differently in African Americans than in any other group. Nonetheless, NitroMed, the corporate sponsor of BiDil, applied for and received FDA approval for the drug with a race-specific indication to treat heart failure only in African Americans. Pervasive media coverage of the announcement of the results and the FDA approval has also focused on the racial specificity of the drug, often explicitly claiming that this shows race is genetic (Kahn, 2004, 2005). Thus, for example, in addition to casual references to a race-specific genetic basis for BiDil's efficacy in the popular media, articles published in such scientific and professional journals as *Genome Biology*, the *British Medical Journal*, and *Health Affairs* have also incorrectly asserted a genetic variation more prevalent in self-identified African Americans to underlie BiDil's efficacy (Petsko, 2004; Rahemtulla and Bhopal, 2005; Carlson, 2005).

BiDil is the prototypical example of patenting race in a genomic age. Underlying the trial design is a race-specific patent that is premised on a genetic conception of race. The PTO issued the patent on October 15, 2002.[2] It confers intellectual property protection for the method of using the drug to treat heart failure in African Americans until 2020. This is thirteen years longer than a previous patent issued in 1987 to the same inventor for the same method of using the same drug in the general population without regard to race. In this case, bringing race into the patent system allowed the inventor to gain a substantial extension of his intellectual property monopoly. With a projected annual revenue stream of one to three billion dollars, the additional thirteen years amounts to a tremendous windfall for NitroMed (Kahn, 2004; Sankar & Kahn, 2005). BiDil's race-specific patent provided the underlying support that drove NitroMed's subsequent development of a race-specific trial design, its campaign to raise capital (first through private venture funding and later through a public offering of stock in 2004), the approach to the FDA for race-specific approval, and its massive marketing campaign to third-party payers, individual doctors, and the public at large. But the broader implications of using race to obtain patent protection and drug approval have only begun to be explored.

Both the patent and the drug trial for BiDil explicitly relate their race-specific design to a search for genetic markers underlying the disease (Taylor et al., 2004). On the one hand, this reflects an approach, largely sanctioned by many in the field of pharmacogenomics, of using race instrumentally as a surrogate to get at underlying genetic variation that could be ultimately identified without reference to race. On the other hand, for the foreseeable future, it presents the immediate reality of race being used as a quasi-genetic category to obtain patents and drug approval.

Is BiDil an anomaly? Discussions of similar race-specific trials for the cancer drug Iressa and the statin Crestor, among others, would seem to indicate that BiDil is ushering in a new era of race-based medicine (Herper, 2005). As Tate and Goldstein observe in this volume, there are already numerous drugs on the market that claim to have shown differential efficacy among different races. While they note that most of these claims are not well supported by the evidence, this merely underlines the importance of examining more closely what other factors may be providing incentives to see race as relevant. Similar dynamics are at work in Europe. In June 2005, over the strenuous objections of the European Council of Human Genetics, the European Patent and Trademark Office upheld a patent owned by Myriad Genetics relating to testing for the BRCA2 genetic mutation "for diagnosing a predisposition to breast cancer in Ashkenazi Jewish women" (Kienzel, 2005). Opponents of the patent noted that the test is currently available from other sources for all women regardless of ethnic or religious background. As a practical matter, this new patent means that individuals identified as Ashkenazi Jews will either have to pay a premium for the test or deny their identity. As with BiDil, here Myriad apparently is marking an ethnic group as genetically distinct primarily in order to extend patent protection—with potentially profound consequences (Gessen, 2005).

A recent report from the Royal Society in the United Kingdom asserted that the promise of truly individualized pharmacogenomic therapies remains decades away (Royal Society, 2005). In the gap between present reality and future promises there may be various strategies for capitalizing on emerging genetic knowledge relating to drug response and efficacy. Targeting a racial audience presents a particularly attractive interim option because at this point the technology and resources do not exist to scan efficiently every individual's genetic profile. Instead, businesses may market the product to a particular social group that is hypothesized to have a higher prevalence of a relevant genetic variation. Patent protection provides an essential underpinning for such commercial ventures. As race is becoming more relevant to marketing drugs, it is becoming a salient component of underlying biotechnology patents.

Patent law provides a focused and dynamic site in which to identify and examine emerging examples of race and genetics being mobilized in tandem to serve both biomedical and commercial projects. As researchers derive new inventions based on mining existing genetic databases, patent law provides powerful commercial incentives to conflate race and genetics. In approving such uses, the U.S. Patent Office gives the imprimatur of the federal government to the reification of race and ethnicity as genetic categories.

It also puts the weight of federal authority behind such uses by placing the burden of disputing such reification upon those (if any) who have the time, money, expertise, and inclination to challenge the patents.

The Rise of Racial Patents

A modern patent is a "government issued grant which confers upon the patent owner the right to exclude others from 'making, using, offering for sale, or selling the invention throughout the United States or importing the invention into the United States' for a period of 20 years ending from the filing date of the application" (Chisum, Nard, Schwartz, Newman, & Kieff, 2001, p. 2, citing 35 U.S.C. § 154). This authority derives from the U.S. Constitution, Article I, section 8, which states: "The Congress shall have power to . . . promote the progress of science and useful arts, by securing for limited times to authors and inventors the exclusive rights to their respective writings and discoveries." All patent applications must meet several statutory requirements. The most prominent of these are known as "useful[ness]" (or utility) (35 U.S.C. § 101), "novelty" (35 U.S.C. §102), "non-obvious[ness]" (35 U.S.C. § 103), and "specification" (35 U.S.C. § 112). The usefulness, or utility, requirement can be met by a showing that the claimed invention has a specific, substantial, and credible utility. Specificity requires the use to be specific to the character of the claimed subject matter. The novelty requirement is met if the invention is not "anticipated" (described in its relevant particulars) in a single reference of "prior art" (e.g., another patent or a published scholarly paper). The non-obviousness requirement is met if "the differences between the subject matter sought to be patented and the prior are such that the subject matter as a whole" would not be perceived as obvious to a "Person Having Ordinary Skill in the Art" (known in patent lingo as a PHOSITA, 35 U.S.C. § 103[a]). Specificity requires a written description of the invention that is adequate to enable a PHOSITA to make and use the invention (Elliott, 2002).

Patent law is premised on legally constructing a divide between nature and society. In affirming the patentability of a genetically engineered bacterium, the U.S. Supreme Court asserted that patentable subject matter included "anything under the sun made by man" (*Diamond v. Chakrabarty*, 1980). As discussed above, the U.S. patent system recognizes that genes can be patented to the extent that the genetic material is legally understood as having been isolated and purified in a manner that effectively takes it out of the realm of nature and into the realm of society as an artifact of human creation (Kahn, 2003).

Ironically, and ominously, when race is used in a gene-related patent, a reverse of this transformation may occur. In such patents, race begins as a social category, often derived from categories specified by OMB directives. Biomedical professionals may link race to genetic categories with the goal of somehow facilitating their research or practice. But when a gene-related race-identified patent issues, it legally marks race as, at least in part, a genetic category—i.e., *the patent takes the social category of race and transforms it into a "natural" category grounded in genetics.* DNA, however, is not patented simply to claim title to a nucleotide sequence. It is patented in order, ultimately, to bring some DNA-related product to market. When that product is a drug, federal guidelines mandate that clinical trial data be collected with reference to social categories of race and ethnicity that are promulgated by the Office of Management and Budget (U.S. FDA, 2003, 2005b).

Patent law is supposed to promote the invention of new and useful products. In recent biotechnology patents race and ethnicity are being exploited in new ways that do not spur the invention of a new product, but rather spur the reinvention of an existing product as racial or ethnic. In so doing, patent law both racializes the space of intellectual property, transforming it into a terrain for the re-naturalization of race as some sort of "objective" biological category, and commodifies race and ethnicity as goods to be patented and subjected to the dictates of market forces.

A review of "claims" and "abstract" sections of gene-related patents and patent applications[3] filed since 1976 indicates a significant trend toward using race in gene-related patents with a marked increase in just the past few years. This rise is clearly coincident with an increase in genetic information being produced through the federally sponsored Human Genome and HapMap Projects and also with rising federal emphases on requiring the use of racial and ethnic categories in the collection of data relating to clinical trials and drug applications. A typical patent is divided into several sections. The claims section presents a primary focus for investigation because it is the legal heart of a patent. The claims specify the legally operative scope of the patent, defining the formal legal "metes and bounds" of the territory covered by an invention.[4] The abstract is the basic summary presentation of the central purpose of the patent. Other sections typically include a "background" or "description of invention," plus drawings or other technical support data. A review of the claims or abstract sections of patents that employ OMB Directive 15 categories of race and ethnicity[5] in a manner that implies or asserts a genetic component to or basis for race[6] reveals a remarkable trend toward the increasing use of racial and ethnic categories in relation to patenting gene-related biomedical innovations (see table 7.1).

Of the twelve granted patents identified, the earliest specifies the use of the term "Caucasian." It relates to diagnostic testing for the BRCA1 genetic mutation for breast cancer and was granted only in 1998. Two of the remaining patents concern the drug BiDil, discussed above, the first of which

TABLE 7.1
The Rise of Racial Patents

Category	Patents Issued, 1976–2004	Patent Applications Filed Since 2001
Race	2	15
Ethnic	0	2
African American/black	4	11
Alaska Native	0	0
Asian	0	13
Caucasian/white	6	18
Hispanic/Latino	0	3
Native American	0	2
Pacific Islander	0	1
Total	12	65

was granted in 2002. In the 4 years since 2001, there has been a fivefold increase in the use of racial and ethnic categories in gene-related patent applications over existing patents issued in the 29 years since 1976. This is not because race has not previously been used in biomedical research, but rather because it is taking on increasing significance in the commercial world of biotechnology patenting. While there are some overlapping references (i.e., patents that use more than one OMB category), the trend remains powerful and clearly parallels the availability of vast new amounts of genetic information being produced and classified in federally sponsored databases. For example, on November 20, 2003, Tony Frudakis of DNAPrint Genomics filed an application, "Compositions and methods for inferring a response to statin," which explicitly bases some of its race-specific claims on samples taken from the Polymorphism Discovery Resource.[7] The application looks at allele frequencies in a "Caucasian" population to infer a differential race-specific response to statin, a blockbuster class of cholesterol-lowering drugs.

How exactly is race being used in these patents? At the most pragmatic level, many patent applicants appear to be invoking race in a strategically defensive manner to provide added protection against possible patent challenges. The structure of a typical claims section of a patent begins with claim 1 being as broad as possible. Successive claims generally provide narrower and narrower focus to the territory covered by the patent. The idea here is that if the broadest claim is struck down by the patent examiner or a subsequent challenge, the narrower claims may still survive. Patent claims are thus structured something like a medieval castle, with an outer ring encompassing the most territory with successively smaller rings providing additional layers of protection back to the core area of the castle keep.

A patent application for "Detection of susceptibility to autoimmune diseases," filed on July 1, 2004, exemplifies the use of concentric rings of race to provide maximum protection for its claims.[8] Its first three claims are as follows:

1. A method for determining an *individual's* risk for type 1 diabetes comprising: detecting the presence of a type 1 diabetes-associated class I HLA-C allele in a nucleic acid sample of the individual, wherein the presence of said allele indicates the individual's risk for type 1 diabetes.
2. The method of claim 1, wherein the individual is of *Asian* descent.
3. The method of claim 1, wherein the individual is of *Filipino* descent [emphasis added].

Claim 1 is not race-specific, referring only to an "individual's" risk. Claim 2 takes a smaller subset of humanity which it marks as "Asian." Claim 3 takes yet a smaller subset of the group "Asian," which it marks as "Filipino." In each case the categories are clearly linked to genetic alleles, forcefully implying a genetic basis to the specific racial groups.

The logic of connecting race and genetics in this context, however, is not driven by science so much as by the commercial imperatives of patent law. The body of the patent, which generally describes the invention and its background, reviews the scientific literature underlying the claims. In this, the less legally potent portion of the patent, the terms "Asian" and "Filipino" are invoked in terms of variable distributions of HLA allele frequency across populations. The "description" section of the patent compares, in particular, the incidence of type 1 diabetes in Japan and China to that of populations in the United States and Europe. It goes on to discuss the frequency in the Philippines as well. In this context, the boundaries of the racial or national categories being employed are not hard and fast. It is not that "Asians" per se have different genes from "Europeans." Rather, it notes that there appear to be variable allele frequency and disease incidence across certain populations. Such uses of population categories may remain problematic in their broad generality, but they are not essentially genetically reductive because they deal with relative allele frequencies that are acknowledged to exist across populations. In contrast, in the "summary of the invention" section, the application states, "The individual can belong to any race or population. In one embodiment, the individual is an Asian, preferably a Filipino." An embodiment refers to the formal metes and bounds of the patent delineated by the claims. Like the claims themselves, the summary sets forth definitively bounded categories that mark specified races as (genetically) distinct. The legal and commercial imperatives of effective patenting have here promoted the transmutation of variable genetic frequencies across populations that nonetheless all share common alleles into bounded genetic categories that are marked as distinct and functionally different.

Also of note is that the category "Asian" is apparently derived from studies only of Japanese and Chinese subjects, thus conflating two national populations with an entire continent. Moreover, there are separate claims regarding Asians and Filipinos, implying some distinctive genetic basis to Filipinos that distinguishes them from other populations encompassed by the larger category "Asian." This separate claim is apparently based on a study of 90 Filipinos discussed in the body of the patent.

Many of the patents invoke race when the inventors construct a perceived departure from an unstated white norm (e.g., of disease or allele prevalence) in a non-white group. In such contexts, the term "individual" or "human" implicitly stands for "white" in the claims. As Rene Bowser notes, "In nearly all racialized research published in the United States, the comparison group has been the majority (White) population. Far from being a neutral category, this approach consolidates Whites as the group with which all 'others' should be compared; it also disregards research that demonstrates the value of studying variations in health among, say, Blacks, as opposed to always comparing them with White Americans. The norm in racialized research is and has always been an unspoken but taken-for-granted White norm" (Bowser, 2001, p. 111). This is particularly evident in one of the BiDil patents. Issued on October 15, 2002, patent #6,465,463 refers in claim 1 to "a method of reducing mortality associated with heart failure . . . in a *black patient*." Claim 2 goes on to specify "the method of claim 1, wherein the black patient has a less active rennin-angiotensin system relative to a white patient." Here white is the norm from which black deviates. As elaborated in the "Background of the Invention," the patent goes on to assert that "heart failure in black patients has been associated with a poorer prognosis than in white patients. In diseases such as hypertension, Blacks exhibit pathophysiologic differences and respond differently to some therapies than Whites." The body of the patent thus pathologizes blackness as both biologically distinct from and less healthy than the white norm.

Another typical example of the unstated white norm may be seen in a patent application for a "method of identifying a polymorphism in CYP2D6," filed November 11, 2003.[9] (CYP2D6 is of particular interest to pharmaceutical corporations because it is involved with drug metabolism.) The first claim specifies "a method of determining a cytochrome P-450 2D6 genotype of an individual." Claim 12 specifies "the method, as claimed in claim 1, wherein said individual is Asian." The focus on an "Asian" individual is explained in the "Background to the Invention," which notes that "differences between Caucasians and Asians are explained by an unequal distribution of CYP2D6 alleles." The application asserted different population-based allelic frequencies between Caucasians and Asians; but this is a two-sided difference, that is, each differs from the other. But it is only Asians that are specified in the claims as a subset of the broader term "individual." Caucasians logically could be but here are not similarly marked out. On the one hand, this appears to be a failure of legal imagination to take advantage of an additional defensive claim. On the other hand, it seems to indicate an uncritical assumption that the category "Caucasian" and "individual" in the first claim were co-extensive and that only non-white races counted as distinct subgroups to be marked out as the basis for defensive claims.

Perhaps most incongruous, yet illustrative of the strategic reification of race and ethnicity in the context of biotechnology patents, are the few applications, such as one for "manganese superoxide dismutase gene polymorphism for predicting cancer susceptibility," filed April 8, 2004, that invoke "Hispanic" as a genetic term.[10] This particular application is distinctive both for its foregrounding of ethnicity in its first claim and for its genetic reification of the ethnic category "Hispanic," which generally does not have the same pronounced history of reification as racial categories such as "African" or "Asian." Indeed, as an ethnic category, Hispanic is so diffuse and diverse that it does not even have the purported link to continental ancestry that sometimes undergirds justifications for using racial categories as surrogates for ancestral descent populations. Using "Hispanic" as a catch-all genetic category risks both reifying ethnicity and providing misleading and conceptually muddled scientific data.

As the table of race-specific patents indicates, there are many references to "Caucasians." Several of these involve disease conditions and so may be understood implicitly to pathologize whiteness as well. Upon closer examination, however, two qualifying characteristics mark "whiteness" in several of these patents as neither deviant nor pathological. First are situations where the overwhelming majority of test subjects in the studies underlying a patent were white and so whiteness becomes a sort of default category for an additional defensive claim. Second are the patents that mark whiteness out of concern for a heightened efficacy of a potential medical treatment.

In a patent application for "genetic diagnosis of depression," filed July 8, 2004, the first claim specifies "a method of identifying individuals predisposed to major depressive disorder."[11] Its race-specific claim 4 specifies "the method of claim 1, wherein said subject is Caucasian." At first blush the patent might seem to be pathologizing whiteness by associating it with depression. Upon closer examination, however, it turns out that the studies underlying the invention were conducted in an exclusively "Caucasian" population (which the patent defines rather broadly as members of "the white race consisting of individuals of European, north African, or southwest Asian ancestry"). The clinical studies alluded to in the body of the patent did not show anything distinctive about Caucasians that would identify them as having race-specific markers for depression that differentiated them from any other race. The invocation of race here does not logically follow from the clinical evidence. Rather, following the commercial logic of patent law, rather than science, the patent drafters have employed race defensively to protect against possible challenges to the patent. Such strategic reification of race has been facilitated as race has come more commonly to be understood and accepted as a legitimate and salient category both in genetic research and in patent strategy. Having learned to "see" race as relevant to patent protection, the inventors invoked the category "Caucasian" because it was the only available race they could extract from their data.

A patent for a "peptide-based vaccine for influenza," issued May 25, 2004, exemplifies the use of whiteness as a target for improved therapy.[12] Its first claim refers to "the NP380–393 epitope[13] according to SEQ ID NO: 5 that are the most prevalent HLA molecules in a *Caucasian* human population." As elaborated in the "Description of the Invention" section, the patent specifies that the vaccine will change "according to the population type" and asserts that "the CTL influenza epitopes are different in the Caucasian, the Asia- or Africa-originated population [*sic*]." It does not define these populations, but it refers to them elsewhere as "Caucasian and non-Caucasian," clearly privileging Caucasian as the norm. And, indeed, Caucasian is the only race specified in the legally enforceable claims section, making it a target of the invention.

Similarly, the patent application for a "mixture of peptides derived from e6 and/or e7 papillomavirus proteins and uses thereof," filed November 2, 2004, characterizes the invention of terms of an allele with a particular frequency in the Caucasian population.[14] Unlike the BiDil patent, this application does not use "Caucasian" as a term that deviates from a particular norm or has a pathological gene variation, but rather as a population with a gene variation that will enable it to take advantage of the proposed invention. Here Caucasian genes are positive and empowering.

One reason for this, of course, is that Caucasian genes are where the money is. This is made abundantly clear in a patent application for "methods for obtaining and using haplotype data," filed December 21, 2001, by scientists from Genaissance Pharmaceuticals, a biotech company that describes itself as "a world leader in the discovery and use of human gene variation for the development of a new generation of DNA-based diagnostic and therapeutic products" (Genaissance, n.d.).[15] The market model of Genaissance is built around capitalizing on human genetic variation. It has a stake in finding population-specific genetic differences. In developing new products, Genaissance mines existing federally maintained genetic databases as an exploitable resource—a resource that already employs a myriad of population categories that are ripe for being conflated with the OMB Directive 15 social categories of race.

Genaissance's application also explicitly capitalizes on the type of data being produced by the federally sponsored HapMap project. This application begins with a broad claim to "a method of generating a haplotype database for a population." It goes on in claim 8 to specify that the reference population may include an "ethnic population," thereby directly connecting ethnicity to genetics. In the body of the patent it becomes clear that Genaissance is using the terms "race" and "ethnicity" more or less interchangeably. Thus, for example, it notes that "the invention may also be used to link variations in DNA to personal identity and racial or ethnic background." In describing the "Field of the Invention" after the claims section, the patent marks pharmacogenomic uses as primary, noting that genetic haplotype information can be used "to predict an individual's susceptibility to a particular disease and/or their response to a particular drug." Here the patent invokes the pharmacogenomic promised land of personalized medicine, but the invention largely depends on using racial and ethnic categories as proxies for genetic variation precisely because the practical reality of widespread use of truly individualized therapy remains far in the future.

To this point, the patent uses race and ethnicity broadly, without singling out any particular group. Strikingly, however, in the "Detailed Description of the Invention," the patent elaborates on one particular embodiment of the invention, declaring, "Analysis of the candidate gene(s) (or other loci) requires an approximate knowledge of what haplotypes exist for the candidate gene(s) (or other loci) and of their frequencies in the general population. To do this, *a reference population is recruited, or cells from individuals of known ethnic origin* are obtained from a public or private source. The population *preferably* covers the major *ethnogeographic groups in the U.S., European, and Far Eastern pharmaceutical markets*" (emphasis added). This description weaves ethnicity into the concept of a "reference population." This is essential to a marketing strategy that exploits race in the gap between current realities and the promised future benefits of individualized pharmacogenomic therapies. Secondly, the description, whether intentionally or not, is a brazen declaration that ethnicity only matters where markets matter—the United States, Europe, and the Far East. Africa, South America, the Middle East, and South Asia apparently are irrelevant. The patent invokes ethnicity not solely as a shortcut to finding genetic correlations with particular population groups, but also, and inextricably, as a basis for developing drugs for major markets. Ethnicity here becomes a function not only of genes, but of genes plus markets.

Conclusion

In the cases of BiDil and the Myriad BRCA2 test, racialized patents have played a central role in the marketing strategy for the product. Both involve technologies that were already available and in use. Adding race to the patents did not change the technology so much as it provided an added incentive to market and extend monopoly control for the product. This moves beyond the use of race to defend a patent against potential challenges to an affirmative projection of race as a central component of product development and marketing. In the case of BiDil, race also unmistakably added publicity value to the product. A good deal of the publicity both produced and was produced by social understandings that reified race as genetic.

The striking rise of racialized biotechnology patents indicates that cases such as BiDil are paving the way for a new proliferation of patents and drug approvals that are producing new and highly problematic understandings of race as genetic. BiDil obtained its commodity value from the rebiologization of race in the regulatory process. Additional racial patents, for products not yet as prominent as BiDil, have secured the imprimatur of the state for using race as a genetic category. Like more traditional extractive industries, biotechnology corporations are mining the raw material of race as a social category and using the patent process to refine it into a natural construct that lends legal utility and novelty to their inventions. The patents are in place and proliferating, ready to be invoked to protect a product or extend a market. Genetic race literally is becoming a commodity as race-specific patents allow biotechnology corporations to raise venture capital and develop marketing strategies that present a reified conception of race as genetic to doctors, regulators, and the public at large.

NOTES

A longer version of this article can be found in J. Kahn (2007), "Race-ing patents/ patenting race: An emerging political geography of intellectual property in biotechnology," *Iowa Law Review, 92*, 353–416.

1. In the interests of economy and manageable syntax, in the remainder of the chapter I will often refer only to "race" when speaking generally of racial and ethnic categories. I am assuming both to be socially constructed categories that nonetheless have come to have biological implications as they play out in real-world biomedical contexts. I will use the terms "race" and/or "ethnic" when referring to specifically marked groups. Thus, for example, the U.S. Census codes "White" or "Asian" as racial categories and "Hispanic" or "Latino" as ethnic categories.

2. U.S. Patent No. 6,465,463.

3. Issued patents have been formally approved by the PTO. Patent applications are currently pending before the PTO for review. Under new policies applications are made available to the public 18 months after their initial filing while still pending review.

4. "Metes and bounds" is a legal description of a parcel of land that begins at a well-marked point and follows the boundaries, using directions and distances around the tract, back to the place of beginning.

5. The results are from searches of the U.S. PTO patent database conducted between August 25, 2005, and September 15, 2005, using the Web-based search engine available at www.uspto.gov. The search terms used included the following: race, racial, ethnic, ethnicity, Caucasian, Caucasoid, African, African-American, Negro, Negroid, Asian, Oriental, Mongoloid, Hispanic, Latino, Native American, Alaska Native, Pacific Islander. The terms "black" and "white" alone were too broad to be useful and so were qualified with the additional terms of "gene" or "genetic" or "nucleotide."

6. This is an admittedly subjective basis for sorting the patents. The categorization of patents that imply or assert a significant genetic component to race or ethnicity is meant to exclude those patents that use racial/ethnic categories as one or more of a longer list of general demographic characteristics, usually employed for information organization rather than for identifying or treating a particular physiological state. The categorization is meant to include those patents that use racial/ethnic categories as a basis for asserting a distinctive prevalence or etiology for a physiological condition, genetic variation, and/or drug response.

7. U.S. Patent Application No. 20030215819.

8. U.S. Patent Application No. 20040126794.

9. U.S. Patent Application No. 20030170651.

10. U.S. Patent Application No. 20040067519.

11. U.S. Patent Application No. 20040132062.

12. U.S. Patent No. 6,740,325.

13. An *epitope* is a single antigenic site on a protein against which an antibody reacts.

14. U.S. Patent Application No. 20040170644.

15. U.S. Patent Application No. 20040267458.

REFERENCES

Bowker, G., & Star, S. L. (1999). *Sorting things out: Classification and its consequences.* Cambridge: MIT.

Bowser, R. (2001). Racial profiling in health care: An institutional analysis of medical treatment disparities. *Michigan Journal of Race and Law, 7,* 79–133.

Carlson, R. (2005, October 11). *The case of BiDil: A policy commentary on race and genetics.* Retrieved November 5, 2005, from http://content.healthaffairs.org/cgi/content/abstract/hlthaff.w5.464v1.

Chisum, D., Nard, C., Schwartz, H., Newman, P., & Kieff, F. S. (2001). *Principles of patent law* (2nd ed.). New York: Foundation Press.

Diamond v. Chakrabarty. (1980). 477 U.S. 303.

Duster, T. (1990). *Backdoor to eugenics.* New York: Routledge.

Elliott, G. (2002). A brief guide to understanding patentability and the meaning of patents. *Academic Medicine, 77*, 1309–1314.

Federal Register. (2001). *Federal Register, 66*, 1093.

Food and Drug Modernization Act of 1997. (1997). 111 Stat. 2296.

Foster, M. (2002). *Ethical issues in developing a haplotype map with socially-defined populations.* Retrieved June 17, 2002, from http://www.nhgri.nih.gov/About_NHGRI/Der/haplotype/haplotypeELSI.html.

Genaissance. (n.d.) *Overview.* Retrieved November 7, 2005, from http://www.genaissance.com/aboutus/home.html.

Gessen, M. (2005, July 26). Jewish guinea pigs, *Slate.com.* Retrieved September 15, 2005, from http://slate.msn.com/id/2123397.

Herper, M. (2005, May 10). Race-based medicine arrives. *Forbes.* Retrieved November 15, 2005, from http://www.forbes.com/technology/2005/05/10/cx_mh_0509race medicine.html.

Jasanoff, S. (2002). The life sciences and the rule of law. *Journal of Molecular Biology, 319*, 891–895.

Kahn, J. (2003). What's the use? Law and authority in patenting human genetic material. *Stanford Law & Policy Review, 14*, 417–444.

Kahn, J. (2004). How a drug becomes "ethnic": Law, commerce, and the production of racial categories in medicine. *Yale Journal of Health Policy, Law & Ethics, 4*, 1–46.

Kahn, J. (2005, January/February). "Ethnic" drugs. *The Hastings Center Report.*

Kienzel, G. (2005, July 1). BRCA2 patent upheld. *The Scientist.* Retrieved September 15, 2005, from www.the-scientist.com/news/20050701/01.

Lee, S. S.-J. (2008 [this volume]). Racial realism and the discourse of responsibility for health disparities in a genomic age. In B. A. Koenig, S. S.-J. Lee, and S. S. Richardson (Eds.), *Revisiting race in a genomic age* (pp. 342–358). New Brunswick, NJ: Rutgers University Press.

Lock, M. (1999). Genetic diversity and the politics of difference. *Chicago-Kent Law Review, 75*, 83–112.

Marks, J. (1995). *Human biodiversity: Genes, race, and history.* New York: Aldine De Gruyter.

National Center for Biotechnology Information (NCBI). (2005). *GenBank overview.* Retrieved November 7, 2005, from http://www.ncbi.nlm.nih.gov/Genbank/index.html.

National Institutes of Health (NIH). (2001, October). *NIH policy and guidelines on the inclusion of women and minorities as subjects in clinical research.* Retrieved June 10, 2004, from http://grants2.nih.gov/grants/funding/women_min/guidelines_amended_10_2001.htm.

Omi, M. (1997). Racial identity and the state: The dilemmas of classification. *Law & Inequality Journal, 15*, 7–24.

Parke-Davis & Co. v. H. K. Mulford Co. (1911). 189 F. 95 (S.D.N.Y. 1911).

Petsko, G. (2004). Color blind. *Genome Biology, 5*, 119.

Rahemtulla, T., & Bhopal, R. (2005). Pharmacogenetics and ethnically targeted therapies. *British Medical Journal, 300*, 1036–1037.

Royal Society. (2005). *Personalised medicines: Hopes and realities.* London, UK: Royal Society.

Sankar P., & Kahn, J. (2005). BiDil: Race medicine or race marketing? Retrieved November 5, 2005, from http://content.healthaffairs.org/cgi/content/full/hlthaff.w5.455/DC1.

Tate, S. K., & Goldstein, D. B. (2008 [this volume]). Will tomorrow's medicines work for everyone? In B. A. Koenig, S. S.-J. Lee, and S. S. Richardson (Eds.), *Revisiting race in a genomic age* (pp. 102–128). New Brunswick, NJ: Rutgers University Press.

Taylor, A., Ziesche, S., Yancy, C., Carson, P., D'Agostino, R., Jr., Ferdinand, K., et al. (2004). Combination of isosorbide dinitrate and hydralazine in blacks with heart failure. *New England Journal of Medicine, 351*, 2049–2057.

U.S. Food and Drug Administration (FDA). (1999). *Population pharmacokinetics.* Retrieved November 1, 2005, from http://www.fda.gov/cder/guidance/1852fnl.pdf.

U.S. Food and Drug Administration (FDA). (2003). *Guidance for industry: Pharmaco-genomic data submissions.* Retrieved November 10, 2004, from http://www.fda.gov/cder/guidance/5900dft.doc.

U.S. Food and Drug Administration (FDA). (2005a, June 23). FDA approves BiDil heart failure drug for black patients. *FDA News.* Retrieved July 5, 2005, from http://www.fda.gov/bbs/topics/NEWS/2005/NEW01190.html.

U.S. Food and Drug Administration (FDA). (2005b). *Guidance for industry: Collection of race and ethnicity data in clinical trials.* Retrieved November 1, 2005, from http://www.fda.gov/cder/guidance/5656fnl.htm.

U.S. Office of Management and Budget (OMB). (1997). *Revisions to the standards for the classification of federal data on race and ethnicity.* Retrieved April 16, 2002, from http://www.whitehouse.gov/omb/fedreg/ombdir15.html.

U.S. Patent and Trademark Office. (2004). *Manual of Patent Examining Procedure, § 1504.01(e).* Retrieved November 10, 2004, from http://www.uspto.gov/web/offices/pac/mpep/documents/1500_1504_01_e.htm#sect1504.01e.

8

The Molecularization of Race

U.S. Health Institutions, Pharmacogenetics Practice, and Public Science after the Genome

DUANA FULLWILEY

In her 1986 book *How Institutions Think*, British social anthropologist Mary Douglas sought to outline how the natural world is often appropriated as a rhetorical resource for social categorization. Douglas wrote, "There needs to be an analogy by which the formal structure of a crucial set of social relations is found in the physical world, or in the supernatural world, or in eternity, or anywhere, so long as it is not seen as a socially contrived arrangement" (1986, p. 48). In other words, social structures are often maintained by tacit logics and assumptions, which, in language and practice, draw from *assumed* orders to institute them as their visible referents. Douglas continued, "When the analogy is applied back and forth from one set of social relations to another, and from these back to nature, its recurring formal structure becomes easily recognized and endowed with a self-validating truth" (1986, p. 48).

In what follows I argue that many basic scientists in the field of pharmacogenetics, and their funders at the National Institutes of Health (NIH), find themselves bound to U.S. institutionalized notions of race and ethnicity when classing human DNA through a dynamic quite similar to that described by Douglas. Despite the fact that the U.S. Office of Management and Budget (OMB) specifies that American racial descriptors "do not conform to any biological, anthropological, or genetic criteria,"[1] many scientists come to understand race as a "physical world" referent for sorting human genetic variation. How this happens is a complex process that begins with the very good intention to "include minorities and subpopulations" in government-funded research. In deploying culturally available monikers of human difference, and a clear policy definition of who "minorities and subpopulations" (NIH, 2001) in the United States are, the scientists in the case study presented herein molecularize race through a tautological "back and forth" whereby American racial taxa applied to genomic DNA becomes naturalized when the unequal distribution of genetic variants, no matter how small, are deemed to have fundamental biological importance. Such powerful attributions of human difference take place often without clear evidence that these patterns of variation are, in fact, biologically *meaningful*. Instead, the very appearance of genetic variation, when analyzed across the American racialized groups that entrained their very entry into the laboratory, corroborates some scientists' belief that the American census categories have a genetic basis to them. Such slippages take place when genetic nomenclature alone frames research questions and when human history and societal mechanisms that bear on population differences, as well as biological traits and their expressions, are not equally considered.

Race is a complex notion that relies on many spheres of human life. As such, when it is considered solely as a biological trait, or an assembly of genetic base pair changes, its cogency as a concept quickly dissolves. In what follows, there are moments when the researchers presented shortly find racialized DNA to have "inconsistent" DNA sequence characteristics when compared to other identically labeled samples (such as "Caucasians" who differ from *other* "Caucasians"). Such instances of genetic diversity that might de-racinate race thinking from science practice concerned with human DNA are detailed throughout. Despite the messy nature of biological race (in)consistency in the lab, many of the scientists in question, who are themselves often of mixed heritage and hail from international contexts where American racial taxa make little sense for them personally, tenaciously search for explanations of race purity in "nature" even as the cultural field presents them with various points of proof to the contrary.

From the Genome Project to Pharmacogenetics

When the draft of the human genome was completed in the summer of 2000, those leading the project made a concerted effort to speak to the commonality of the *human* race. Despite their many differences in vision, the heads of the public and private mapping effort left little room for doubt about our shared genetic makeup when they pronounced with exceeding precision that we are all 99.9% the same at the level of the genome. That same year the NIH launched the Pharmacogenetics Research Network (PGRN) with the goal of rationalizing human genomic variation data for pharmacy. In the latter research program, the 0.1% that we do not share, when 3 billion base pairs were at stake, was seen to be quite considerable. Thus, as the Human Genome Project (HGP) heads repudiated race as genetically significant, certain PGRN teams began to hypothesize its necessity for "rational medicine."[2]

The overwhelming presence of genomic points of sameness (compared to differences)—the 99.9% of our billions of nucleotides—might have razed previous taxonomies of race. Instead, the possibility of such a tabula rasa was precluded by habits and ways of reading socially understood racial difference, coupled with habits and ways of seeking points of variation as a basic practice in the life sciences (from zoology to molecular genetics). Hopeful that they were free of prejudice, with a goal to understand genetic variation in drug response to ameliorate human health, certain researchers of the PGRN instead structured their studies—from the collection of DNA to the organization, storage, and analyses of that DNA—by none other than a literal tabula *raza*. Through both logics and practice, DNA frequency differences and race often emerged as the two primary units of analysis. As such, each was routinely articulated through the other in tables, spreadsheets, and bioinformatic cells whereby racial delineation inspired a gaze of differentiation that conditioned scientific discovery.

In this chapter I examine the place of race in two related scientific efforts funded by the PGRN. I argue that for many working in this field race as an organizing principle emerges as a natural choice and referent for categorizing human individuals not because of any explicit commitment to a political agenda or racism, but because the hopeful curiosity that often spurs contemporary research into raced groups is set within institutional formal structures that give human variation its sense on multiple registers. These registers map onto what Nikolas Rose describes as "the framing of explanations at the molecular level," the use of artifacts fabricated there, and, most importantly, "a reorganization of the gaze of the life sciences, their institutions, procedures, instruments, spaces of operation and forms of capitalization" (2001, p. 13). Drawing upon Rose, I argue that the back-and-forth between DNA and its seemingly natural organization by societal descriptors of race works to *molecularize* race itself. This happens through practices of racialized recruitment of tissue donors, storage of DNA samples, and organization and reporting of research that rely on racial population differences defined by the U.S. Office of Management and Budget (OMB) and implemented through the National Institutes of Health (NIH) as a research funder.

My analysis is based on participant observation and interviews with leading pharmacogenetic researchers in the San Francisco Bay area. Drawing on ethnographic fieldwork between two collaborating labs, this chapter chronicles how DNA molecules are increasingly made to carry the self-reported U.S. racial descriptor of their donors as they leave his or her body and enter the laboratory. That racial label adheres to the DNA sample for the duration of its life in the lab and thereafter operates as a natural conveyor of racial distinction in other formal structures, such as scientific studies, medical practice, and referenced knowledge databases. Race not only becomes a substance discernable at the molecular level, it becomes naturalized there through databases structured by U.S. understandings of racial groups and the subsequent comparison of frequency differentials among DNA sequence base pairs in humans classed by social markings of race. Researchers who are part of or affiliated with the projects described here are quite successfully publishing, building databases, and inspiring prolonged debates in major journals on their assertions that racial differences exist at the level of the human genome biology. (See Risch, Buchard, Ziv, & Tang, 2002; Burchard et al., 2003; Burchard et al., 2005.) Their public presence and professional success marks a renewed form of social capitalization that has made this instance of molecularization possible.

Good Intentions of Health Institutions

For over a decade, the National Institutes of Health has been involved in making research more representative, to include minorities and women (Epstein, 2007). Today all grant seekers are told, "Members of minority groups and their subpopulations must be included in all NIH-supported biomedical and behavioral research projects involving human subjects. . . . NIH funding components will not award any grant, cooperative agreement, or contract or support any intramural project that does not comply with this policy."[3] As American sociologist Steve Epstein has shown, the characterization of the human body in terms of race and other points of difference, such as gender, sexuality, and age, was not a historical given in medical and scientific research (2004, 2007; cf. Kahn, this volume). In the late 20th century, women and minorities successfully contested the notion that white men, traditional research subjects, could stand in for all of humanity. Various forms of activism and explicit intervention on the part of key women scientists at the NIH and African American members of congress resulted in the passage of the NIH Revitalization Act of 1993, which mandated that publicly funded research include women and minorities as subjects. In implementing the mandate, the NIH, like other government agencies, adopted the OMB racial and ethnic categories initially defined for the census. Epstein writes, "The new emphasis on inclusion and the distrust of extrapolation across social categories were not without opponents. . . . Concerns were . . . raised about the problematic business of defining medically meaningful racial and ethnic categories" (2003, p. 183). One key African American scientist, Otis Brawley, who from 1995 to 2001 was assistant director of the National Cancer Institute for Special Populations Research and who presently serves on the NIH Advisory Committee on Women's Health, argued in 1995 that the act "may eventually do more harm than good for the minority populations that it hopes to benefit. The legislation's emphasis on potential racial differences fosters the racism that its creators want to abrogate by establishing government-sponsored research on the basis of the belief that there are significant biological differences among the races" (1995, p. 293). This is a view shared by a small number of mostly African American researchers (Epstein, 2003, p. 184).

The ethos of the Revitalization Act and the reality of NIH funding formed the backdrop for initial discussions of how the PGRN would collect "diverse" human subjects and categorize genetic difference. In an interview with Rochelle Long, director of the PGRN,[4] it became clear that she and others were concerned about using race/ethnicity as a biologically based sorting technology for pharmacogenetic discoveries on the part of PGRN scientists. Yet she indicated that in discussions with specialists in the fields of genetics and pharmacy invited to "advise" on the issue it was decided that it made sense to use these ascriptions despite their imperfections, given that NIH grants and reporting were already structured around them. As Long saw it, the PGRN was powerless to challenge the monopoly that race and ethnicity descriptors had on government reporting with regard to U.S. population differences, regardless of what investigators were describing. She understood that the OMB terms were meant to refer to "social-political constructs," but nonetheless thought that these realities hinted, however imperfectly, at ancestry. In essence, the continued use of race as a sorting technology was justified by its previous use.

The following is an excerpt from an interview with Rochelle Long, of NIH:

RL: It was agreed that when the information was collected for PharmGKB [the PGRN collective database], the official OMB terms would be used. And this also relates to government-funded work in general. The terms were changing at the time. They were actually separating descriptions of race, if you will, from ethnicity. So now you could be white-Hispanic, black-Hispanic; you could be either Asian or Pacific Islander. While I always agreed that these were not the greatest terms for describing everything about somebody's heritage, they were at least consistent with the ways that [scientists] already had to recruit and report research study enrollment to the NIH. And, they were valid for the purposes of the Office of Management and Budget. They had been running [discussion] groups and concluded that these are the standards that the government is going to use. So that's the way the data are reflected in the PharmGKB.

Race

As Geoffrey Bowker and Susan Leigh Star have shown in their exemplary work on classification, *Sorting Things Out*, race often operates via an Aristotelian binary system of mutually exclusivities (i.e., "you either are, or you are not") *and* a more abstract system of prototypical thinking (2000, pp. 61–63). Because of its existence on these multiple registers, "race" may be used without much reflection in many aspects of daily life. However, it can easily provoke active scrutiny (or confusion) when cases of ambiguity emerge—that is, where social and political circumstances, individual biography, or race itself resist the categories that emerge from the Aristotelian and prototypical logics of specific practices and societal contexts. Such disconnects can lead to experiences of cognitive dissonance where personal and lived experience are "torqued"—as Bowkler and Star would put it—by classification and vice versa (Bowkler & Star, 2000, p. 324). As several of my interviews with PGRN scientists reveal, an expected ease and recognition works to stabilize race as an Aristotelian classification whose effect is "to place any member of a population into one and only one class" (Bowker & Star, 2000, p. 62). Yet when these same researchers are asked to reflect on the use of racial categories for their science, they immediately point out inconsistencies in their own personal, local (U.S.) experiences of race. This is a moment where the certainty of such classification in the lab could begin to lose its force. The fact that it does not attests to the power of socially understood and seemingly objective cultural norms, despite the fact that individuals themselves do not always find themselves in such understandings.

The Labs

In early March 2003 I began a six-month fieldwork stay at the University of California, San Francisco (UCSF), Department of Biopharmaceutical Sciences, whose chair, Kathleen Giacomini, has a four-year PGRN grant to study the pharmacogenetics of cell membrane transporters, or PMT. Cell membrane transporters are vital to understanding the first and last phases of drug distribution since any drug—toxin or medication to which we are exposed—must interact with transporters before it can be metabolized. At the same time, I also began fieldwork in another UCSF lab headed by a physician-researcher of asthma genetics, Esteban Gonzáles Burchard. Burchard has built his career on researching what he terms "racial admixture" (Burchard et al., 2005) in Latino Americans after amassing a database of 2,000 Mexicans and Puerto Ricans from the United States, Mexico City, and San Juan. Through a subsidiary grant under the PMT project Dr. Burchard was also funded to collect and bank a repository of 500 individuals' DNA from "ethnically diverse populations." This DNA was collected locally in the

San Francisco Bay Area so that the researchers could call back subjects as needed after initial genotyping and analysis in order to conduct subsequent clinical studies on them.

A Natural Course of Evolution: A Brief History of PMT

To situate the discussion that follows, some discussion of the preceding science that took place in Giacomini's lab as it underpins both the PMT and subsidiary study is necessary. In 1997 a graduate student at the UCSF program in biopharmaceutical sciences cloned the first human organic cation transporter (OCT 1) from liver tissue (Zhang, Dresser, Gray, et al., 1997).[5] Although the human OCT 1 protein was found to be 78% identical to previously cloned rat OCT 1, the discovery of the human protein was important for understanding the involvement of OCT 1 in the absorption, distribution, and elimination of substances from a wide array of clinical classes, including antihistamines, skeletal muscle relaxants, antiarrythmics, and ß-adrenoceptor blocking agents (Zhang, Brett, & Giacomini, 1998, p. 431). OCT 1 has been theorized to interact with N-methyl-4-phenylpyridinium (MPP+ for short), which is believed to play a role in the development of Parkinson's disease. Later that same year, this star student went on to clone a novel variant of the rat OCT 1 protein characterized by a 104 base pair deletion (Zhang, Dresser, Chun, et al., 1997). Shortly thereafter the same protein was cloned in the pig, the mouse, and the rabbit.

As Giacomini and collaborators became increasingly interested in questions of comparative SNPs[6] and their pharmaceutical consequences across these species and within *Homo sapiens*, they would borrow this logical structure of comparing different forms of life in "nature" to look at different forms of life in society, i.e., in African Americans, Caucasians, Asians, and Mexicans. Simultaneously, the question of genetic function with regard to drug uptake, distribution, and elimination was gaining ground at the NIH, and in 1998 the PGRN developed its first request for applications.

Race as Organizing Principle

Before she submitted her PGRN proposal to the NIH, Giacomini and collaborators in the neurological sciences at UCSF purchased human genomic DNA from the Coriell Institutes for Medical Research Cell Repository, a New Jersey tissue bank, run by NIH's National Institute of General Medical Sciences (NIGMS), that is utilized by scientists worldwide. The UCSF researchers genotyped these Coriell samples for polymorphisms in two neurotransmitters: the serotonin transporter, SERT, and the vesicular monoamine transporter, VMAT2. In one of our interviews, Giacomini reported that the team found "interesting polymorphisms" that affect the function of these two proteins, but they were ultimately frustrated because, in their view, the DNA available from Coriell at the time lacked crucial information. The problem was that from their perspective as end users the samples were "ethnically unidentified." Although the donor groups were coded by "ethnicity" and each sample from the same group had a similar key, the researchers did not know for *which* ethnicities the codes stood. They noticed that certain polymorphisms happened only in some groups and that frequencies for other variants were not evenly distributed among all. Yet, because of the masked ethnicity, these prize polymorphic differences could not be "fully interpreted." As Giacomini put it, "It was a waste because we had no control for association studies. We wouldn't know who to compare the polymorphic samples to!"

Systems and Structures: The Tabula Raza

Giacomini was one of the first to be given PGRN funds, and in the first year of the four-year grant her team genotyped 24 membrane transporters in human DNA samples, using Coriell's (then) new racially labeled "Human Variation Panels," which consisted of "100 Caucasian Americans" (51 males, 49 females); "100 African Americans" (17 males, 83 females); and "30 Asians," of which 10 were Southeast Asians (5 males, 5 females), 10 were Japanese (4 males, 6 females), and 10 were Chinese (sex unavailable since this panel is now obsolete). They also obtained "10 Mexican" samples (3 males, 7 females) and "7 Pacific Islanders" (4 males, 3 females). These panel sizes were clearly uneven, but at the time the researchers were content to have clearly marked populations in order to genotype the membrane transporters by race/ ethnicity, which they theorized would prove the most telling organizing principle for variations.

All PMT genotyped data is formatted for a shared database wherein each of the now 52-plus transporters and their many variants are organized by gene and then by race. The hypertext of the online (restricted) database allows the viewer/researcher to click on the transporter's name, which then takes him or her to a list of options, two of which are salient for understanding the centrality of race for the project. The first option allows the researcher to click on a graphic of the coding sequence in which each exon is marked by shades that indicate what kind of changes genotyping has revealed it to contain (fig. 8-1).[7] Beneath each exon is a link with its position number that enables the viewer to get detailed information about what kind of change occurs in that coding sequence.

That position link in turn opens onto a table in which each row contains a crucial bit of information in the following order:

- Position of the exon number in the gene sequence,
- Order of the SNP,

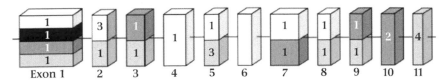

| Exon 1 | 2 | 3 | 4 | 5 | 6 | 7 | 8 | 9 | 10 | 11 |

FIGURE 8-I. A graphic used to represent the exons of a transporter gene, with various shades representing the type of genetic change found there in the sampled population.

From Fujita, Urban, Leabman, Fujita, and Giacomini (2006, p. 31).
Courtesy of Wiley Interscience, *Journal of Pharmaceutical Sciences.*

- Position where it occurred,
- Nucleotide change,
- Amino acid position in the protein,
- Any amino acid change brought on by the nucleotide change, and, finally,
- Racial distribution of the change.

The racial distribution consumes half of the page. The first row lists the total frequency of the genetic change in question in all groups combined. Five racially labeled rows follow: "AA" (African American), "CA" (Caucasian), "AS" (Asian), "ME" (Mexican), and "PA" (Pacific Islander) listing the variant's frequency in each racial group, each comprising its own row on the grid. Variants that occurred at a frequency of more than 1% in any one group are highlighted.

The second salient, and often consulted, option on the database was a topographical model of the two-dimensional protein structure of the transporter gene in question, constituted by its amino acids (fig. 8-2). Each amino acid that underwent a change was color-coded, whereby various colors depicted the type of change that took place (non-synonymous, synonymous, insertion, or deletion). If one clicked on the colored (changed) amino acids, they would then be taken to the link discussed above (table 8.1), which, again, gave the breakdown and frequency of the variant in question for each racial group.

SOPHIE: *Study of Pharmacogenetics in Ethnically Diverse Groups*

During the first year and a half of the study, the PMT team was dedicated to genotyping and learning how the set 1 genes (typed in the Coriell panel) actually functioned in cellular models, i.e, in frog eggs as well as in immortalized cancer cells used for such experiments, known as HeLa. They continued, however, to compile a second list of genes of interest to the scientific community. At the same time, PMT scientists began to imagine possibilities of doing human clinical trials regarding the transporter genetic variants that

TABLE 8.1
Tabula of Genetic Difference and Race

Exon	SNP #	Exon Position	Nucleotide Change	Amino Acid Position	Amino Acid Change	Total Frequency	AA Frequency	CA Frequency	AS Frequency	ME Frequency	PA Frequency
						n=494	n=200	n=200	n=60	n=20	n=14
1	1	(−38)	C ⇒ T	—	—	*0.023* n=472	0.005 n=190	*0.053* n=188	0.000 n=60	0.000 n=20	0.000 n=14
1	2	(−23)	C ⇒ T	—	—	*0.119* n=472	*0.053* n=190	*0.218* n=188	*0.050* n=60	*0.100* n=20	0.000 n=14
1	3	9	C ⇒ T	3	Synonymous change	*0.042* n=474	*0.105* n=190	0.000 n=190	0.000 n=60	0.000 n=20	0.000 n=14
1	4	38	C ⇒ G	13	Ser ⇒ Cys	0.004 n=474	*0.011* n=190	0.000 n=190	0.000 n=60	0.000 n=20	0.000 n=14

Note: The first exon of an unnamed transporter gene of interest to PMT as submitted, displayed, and consulted on their intranet database. All transporters and their exons were presented in this manner. Gene changes with a frequency of 0.010 or greater are indicated in boldface; n = the number of chromosomes in a sample set (each person has two chromosomes for each locus).

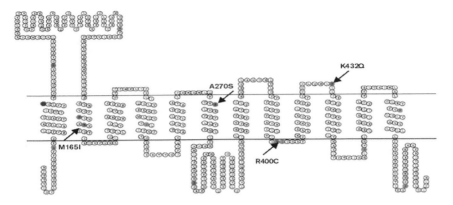

FIGURE 8-2. Topical representation of a transporter protein.
From Leabman et al. (2002, p. 399). Courtesy of Lippincott
Williams & Wilkins, *Pharmacogenetics and Genomics Journal.*

had since been characterized and tested in cellular phenotyping studies in
the lab.

Now the team began to place more importance on the fact that the
Coriell Human Variation panels were unevenly matched when one compared
the number of samples from each race. Furthermore, for the purposes of
clinical studies, the people who donated that DNA were almost as mysterious to the researchers as those in the PDR panel. With the exception of race,
which was actually "self-reported," as the researchers bemoaned on numerous occasions, the samples were anonymous and could never be called back
for clinical trials if found to have polymorphisms of interest. In an effort
to get beyond what was perceived as a potential research impasse, those at
the helm of PMT decided to call on a local physician-researcher, Esteban
Gonzáles Burchard, to help them build a resource of racialized research subjects. Together they wrote a supplemental grant to the PGRN, as part of PMT,
and were funded to build a genetic database specifically for PMT called the
Study of Pharmacogenetics in Ethnically Diverse Groups, or SOPHIE.

Once again, learning from the limitations of their previous DNA panels,
the researchers now emphasized that the point of SOPHIE was to enlist people from various U.S. racial groups who lived locally and who not only would
consent to give their DNA, but would also be willing to be called back for
future studies. The major difference, however, between the two data sets was
that Burchard was in many ways much more diligent than the NIGMS Coriell
tissue sample-takers. Coriell's "quality" of race identification is, as mentioned above, strictly based on "self-report." Burchard and his PMT collaborators were determined to collect, in his words, more "racially pure" DNA.
His method was to automatically exclude anyone who reported racial mixing
in their genealogies for the past three generations. In theory, subjects were

thoroughly screened to rule out such a possibility, while the donor, their parents, and their grandparents all had to identify with the ethnic/racial category in question. For the Chinese and the Mexicans it was stressed that a subject's grandparents had to also be from "the homeland."

In an interview with Dr. Giacomini, the place of race in PMT as the major organizing principle becomes clear. Other questions, such as age and gender, which surely impress upon how one distributes, transports, and metabolizes drugs, were included in initial discussions of the study design but ultimately abandoned. The following is an excerpt from an interview with Kathleen Giacomini:

DF: Was there any discussion when organizing SOPHIE to design the study in any way other than race as a phenotype? . . . Given some peoples' hesitancy in using race in science, was there any discussion about not using it?

KG: Well, we were thinking of a few things at the moment of SOPHIE. First, did we want normal, healthy volunteers? Or did we want a disease population? So that was one thing. Secondly, what about age? Did we want to include pediatric patients? Or, even, elderly patients? So we had to make a number of calls. So then, also race. Which race[s]? . . . We were pretty clear that we wanted European Americans, African Americans, Asian Americans—and defining that was a little bit difficult. Also [defining] Mexican Americans. [But] those were the four groups that we knew that we wanted.

Those recruited for SOPHIE were selected because the "socially contrived arrangement" of their perceived "physical world" purity was naturalized through a tautological assumption that their self-reported pedigree of "the same race" traduced biogenetic racial sameness. That this was still "self-report" and subject to an individual's perception of how they should self-classify, which has changed over time, especially for African Americans, seemed less important to Burchard, who considered it "pure enough."

Purity and DNA: Allelic Frequency Differences out of Place

When SOPHIE was finally compiled and the DNA extracted and stored, researchers from PMT began to take transporter genes in which they had found variants that altered protein function in the Coriell Human Variation Panels and genotype them in the SOPHIE cohort for eventual *in vivo* trials. One such gene was the human OCT 1. The student who initially cloned OCT 1 had long since graduated and in the intervening years a new student had become interested in OCT 1. His focus was specifically on the transporter's targeted activity in the elimination of the neurotoxin MPP+, as well as the role of OCT 1 in the development of Parkinson's disease. In the Coriell panels, PMT researchers found that human OCT 1 had 15 protein-altering variants. Five of the variants reduced or completely eliminated function. Three of those had been screened previously in a "European" population of 57 people with Parkinson's disease (Kerb et al., 2002, cited in Shu et al., 2003, p. 5903). Expanding on this work, PMT researchers obtained a panel of 470 Parkinson's patients (95% of whom were "Caucasian") from the Parkinson's Institute in Sunnyvale, California. The student then analyzed the Parkinson's donors' DNA against "a healthy control" to begin to theorize which variant could be associated with the disease. Using the Coriell "Caucasian" panel as the control, the student discovered that the Coriell panel differed from the Sunnyvale samples with respect to the frequency of one transporter variant, G465R, which codes for a nonfunctional OCT 1 protein. The student then genotyped the SOPHIE "Caucasian" panel with hopes that he would find the same variant and be able to conduct a further, informative study with SOPHIE tissue donors who had consented to be called back.

The student and the lab were surprised to find that SOPHIE "Caucasians" and Coriell "Caucasians," who were both healthy, differed *more* than the Coriell versus the Sunnyvale Parkinson's disease population for the variant in question. Moreover, the student added, this difference was "statistically significant."

Giacomini and the team wondered what had gone wrong. Her first reaction was to rule out error and re-genotype. Her next thought was that perhaps the difference was, in fact, "a population stratification issue," or that these were different populations with different mixes and genetic heritage. The team theorized that if the latter was the case, it must have been the Coriell panel that was problematic. The lab researchers faulted the Coriell samples for not being as "pure" as SOPHIE, since SOPHIE Caucasians "went back three generations." Yet for neither data set had compilers actually asked about European-specific ancestry, such as Welch, German, or Scot. In this regard, Coriell and SOPHIE were more alike than different. Giacomini posited that the Coriell data set might actually be "contaminated," by which she meant "mixed."

KG: The number one thing that I'm going to rule out is, Is there an artifact? The student used a different sequencing method, comparing his to that done at the genomics core facility here at UCSF. He was not comparing his to his. . . . The student is going to try to reconcile this [or at least rule out the possibility of it being an artifact]. Then, I did try and find out the ethnic stratification of the Caucasian DNA that we collected here in SOPHIE as well as in Coriell, but that's not easy to get. They're just "European Americans." So whether, in fact, the ones from Coriell came from Ireland or Finland, and ours are all from Italy and Spain, I don't know. . . . One of the main differences between Coriell and SOPHIE is the way that they were collected [individual self-report versus family history of identifying as that group for three generations]. I would say we should take a close look at this because people may not want to be using Coriell if it is contaminated.

When the student had the two data sets re-sequenced with a newer and less error-prone technology, the frequency discrepancy for G465R remained. This was only the first sign of an inconsistency between the two data sets. Other "Caucasian" as well as "African American" discrepancies were on the horizon.

A few short months later I witnessed an illustration of the extent to which notions of racial purity characterized SOPHIE, while the Coriell samples were increasingly seen as "questionably" pure. Not only were African Americans and Caucasians seen as different, they were tacitly understood to be two sides of a symmetrical arrangement in the physical world, of a kind that characterizes many patterns in nature but that runs counter to most accepted ideas of human genetic diversity. That is, each was perceived as the other's *opposite* race.

At issue was a second organic cation transporter, OCT 2, which is primarily found in the kidney rather than the liver. In this case, another of Giacomini's previous graduate students, now a key postdoctoral researcher on the PMT project, looked at variants in OCT 2 in both Coriell and SOPHIE. Drawing on 247 samples from the Coriell panel, she not only looked at single changes, but also constructed these into haplotypes (multiple genetic loci inherited together). Based on the genotypes that she found in the Coriell data set, she identified two common haplotypes that were seen in close to 60% of the Coriell samples. In the published paper these were called haplotype *1 and haplotype *2A (fig. 8-3).

Now the postdoctoral researcher proposed a study with "real SOPHIE people" to examine the possibility that OCT 2 could prove a valuable drug target for anti-diabetic agent metformin. As PMT researchers combed the SOPHIE genotyping data to identify possible study candidates who possessed OCT 2 variants, the researcher began looking at the SOPHIE DNA samples to compare the OCT 2 frequencies with those of Coriell and immediately noticed that the haplotype frequencies had not remained static, as she had assumed they would. This change in frequency distribution was perceived as alarming. On seeing the results, she confided: "Something's weird in OCT 2. Some of the haplotype frequencies have completely changed to the opposite race. Ones that were African American are now Caucasian, and ones

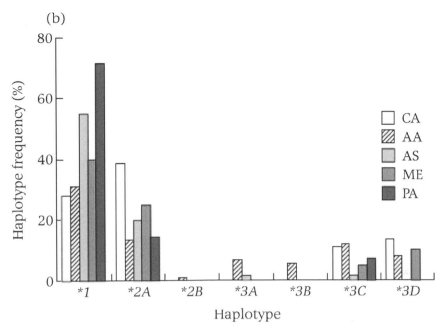

FIGURE 8-3. Frequencies for several OCT 2 haplotypes and their racial/ethnic coding. Haplotype *1, wherein "Caucasians" and "African Americans" both have frequencies between 25% and 30%, is the haplotype under discussion here.

From Leabman et al. (2002). Courtesy of Lippincott Williams & Wilkins, *Pharmacogenetics and Genomics Journal*, p. 400.

that were mostly Caucasian are now African American. We've already seen discrepancies between Coriell and SOPHIE for OCT 1. Maybe this is the same thing. Maybe some of those Coriell people were not really African Americans, or Caucasians. Maybe they were mixed, because Coriell was just based on self-report" (field interview).

When I inquired further, one of the haplotypes involved was one shared among all groups, haplotype *1. The researcher also admitted that the majority of haplotypes retained their earlier patterns. In going back to the published paper, where haplotype *1 was concerned, the frequencies differed (slightly) between African Americans and Caucasians, yet it was in no way a majority African American or Caucasian pattern, as both Asians and Pacific Islanders possessed the haplotype nearly twice as often as either blacks or whites. More to the point, however, African Americans and Caucasians shared haplotype *1 at *nearly the same frequency*, which hovered around 25% to 30%. What troubled the researcher was that the Caucasians in SOPHIE surpassed the African American's frequency for this haplotype, when the "referent" frequencies in Coriell were so close. Thus, her dismay was that the haplotype frequencies "switched to the opposite race."

The (In)Consistency of Genetic Variants
in Confirming Race Thinking

In both cases the research team perceived frequency changes to somehow be "matter out of place," which implies both an assumed order, or pattern, and "a contravention of that order" (Douglas, 2000/1966, p. 44). In the case of the OCT 1 G465R variant, the researchers made clear their assumptions about Caucasian stability, consistency, and integrality. Yet the present coherence of the term "Caucasian" only emerged in the 1920s when the idea of "so-called minor divisions of humanity" from Europe was "willingly erased" and an encompassing scientific idea of whiteness, enshrined in the term "Caucasian," took hold (Jacobson, 1998, p. 96). With the Johnson-Reed Act of 1924, which grossly restricted immigration by European individuals deemed "unfit," anxieties over inferior white races entering the United States began to be muted and the economic and ideological stakes that supported racial distinctions *among* whites (such as Irish, Slav, or Greek) began to dissipate. The catchall term "Caucasian," a political response to creating a shared identity among those desired by the U.S. government (and allowed in) at a certain historical moment, today may divert scientists from recognizing important mutations that affect members of this culturally defined group at the level of their genome biology. When asked, researchers in the Giacomini lab admitted that "whites," like all humans, surely possessed genetic variation when compared among themselves.

In the case of the postdoctoral researcher working on the OCT 2 haplotype *1, the same assumptions were operative since the Coriell Caucasians and the SOPHIE Caucasians again proved "inconsistent" with each other. Additionally, the African American story of discrepancy warrants its own analysis. Given that Africa contains the greatest human genetic diversity on the planet, the "coherence" of African Americanness at a biological level is, of course, problematic. It is not totally clear how this genetic complexity has shaped present-day African Americans' genomes, gene expressions, and biology, but one can safely say that assuming genetic homogeneity of black Americans as a people makes little sense.

The historical realities of Africans in the New World may shed some light on why an assumption of African American homogeneity is a problematic foundation on which to analyze genetic data sets that do not produce identical results. Historians and anthropologists of the transatlantic slave trade emphasize the distinct elements of African cultures, languages, and ethnicities that not only were extant in the New World, but, moreover, helped to facilitate the very success of the slave trade in Africa, as African governments, merchants, and others were involved in the capture and sale to Europeans of other Africans they considered "aliens" or "outsiders." (See

Eltis, 2000, p. 226, chap. 9; Greene, 2002, pp. 1018–1019; Lovejoy, 2000, chap. 8; Thornton, 1998, chap. 4; cf. Herskovits, 1936). Once in the New World, the first African Americans, as an "ethnically heterogeneous aggregate of individuals" (Mintz and Price, 1992, p. 14), in many (but not all) instances created unifying social structures and cultural practices (Gomez, 1998; Eltis, 2000, pp. 250–257). Furthermore, the very notion of African identity, rooted in reference to the African subcontinent, is a relatively recent invention (Eltis, 2000, p. 230; Mudimbe, 1988, p. 1), as is the idea of an encompassing black identity with roots in a place called Africa for those in the United States (Gomez, 1998, chap. 1).

When asked to reflect on a clear definition of the term "race" for her science, the researcher responsible for the OCT 2 study, who held the view that black and white races were "opposites," was less certain of a rigid racial structure, as becomes clear in an interview.

DF: How would you define race?

POSTDOC: That's a good question. I mean I—from my limited knowledge, and you know I'm still learning so much about the population genetics and how variation differs between groups—but I do feel as though there are ethnic-specific SNPs. Because it seems as though people—when people do these large-scale genomic screens, they see these frequencies in the groups. One SNP will only pop up in one group, and another SNP will pop up only in another group. So I do think that there's some genetic basis for ethnicity or race. In terms of defining a given race based on SNPs or genetic variation, I don't think we know enough yet to do that. But from what I've seen, I do think that there are genetic differences, or more global genetic differences between—and maybe it's just sub-populations. I don't know—I don't really *know* what defines a race or an ethnicity, and I get confused between "What is a race?" and "What is an ethnicity?" So . . .

DF: Before you got into SNP mapping, how would you have defined race?

POSTDOC: I think before I got into all of this, I would have thought, Oh, "Where are your parents from?" And you know, my dad came from India. I actually grew up in a very mixed background [*sic*]. My dad grew up in India until he was twenty-five, and my mom is part Czechoslovakian, so I'm of very mixed descent. So, if someone had two parents and they were second generation, and the parents came from China, I would say, oh, they are Chinese. I mean that was my simple kind of definition of some-body's race or descent. But I think of it much differently now.

In the Burchard lab, where the SOPHIE recruiters mainly resided, it was clear that researchers also mostly held static and immutable notions of

Caucasians and African Americans for research purposes. That said, however, specific tacit assumptions in the recruitment process were not lost on them. This is evident in one recruiter's musings on the category of Caucasian in a conversation about how it took nearly two years to recruit the African American, Chinese American, and Mexican American samples for SOPHIE, but only six months to recruit the Caucasians:

RI: They were the easiest group to recruit, and doesn't that all make sense? They are the ones studied the most often, the ones who are most involved and most aware, perhaps, you know? So that tells me a lot. It tells me a lot about conducting research and how difficult it is to get the other populations. [She then continued.] So, um, the other thing with Caucasians, though, is that it's [sic] anybody, basically. . . . There was really no set limit in terms of their background.

DF: Did you ask about the four grandparents and all of that?

RI: Yes, but there was no set protocol for them. They could've been Italian, German, and Jewish from anywhere.

DF: And were they?

RI: Yeah. I think parents and grandparents . . . a lot of these folks, I think their parents were born in the U.S., their grandparents may have been born in the U.S., maybe one grandparent was born somewhere else, we did get a variety of that. . . . But there was really no tough eligibility criteria for the Caucasians, whereas for the other groups they [grandparents] had to be *Chinese*, from *China*, and *Mexican*, from *Mexico*.

DF: Did you ever discuss that [with the team]?

RI: . . . I don't know if it was a big issue for [the head scientists in the Burchard lab] because ultimately what we want to see is more [on] the ethnic groups and how drug metabolism works for them, maybe in comparison to the Caucasians, but I think they've been studied so much that it wasn't a big deal. It wasn't a big issue for anybody. It was just like, get them and get them in.

Later in the interview, when I asked how she defined race in her work and other aspects of life, it became clear that the SOPHIE recruiter had insightful observations about how race, especially what she perceived to be her own "Latino race," was anything but immutable.

DF: How do you define race?

RI: Wow [*pause*]. That's a good one. . . . I think that there're many definitions out there. Race—people see it more as geographical, being from a certain geographic location, . . . like if you go to one area of our country,

you see diversity within our own race, within our own geographical location of Mexico, . . . within our own *race* of Latinos, if you want to call it
like that, there's such diversity, you can just see it in skin color and eye
color, and everyone's coming from different regions. . . . You go to Vera
Cruz, there are *black* people in Vera Cruz.

A second recruiter's discussion of the recruitment process also provides
insight into the social forces that, in part, defined the (unchallenged) contours of the tabula *raza* once SOPHIE was compiled. He refers to the method
of amassing samples for SOPHIE as one of "recruiting in crops." This was the
first point at which DNA was targeted through a racialized gaze, before it was
even extracted from donors' bodies:

R2: It would be ideal to have a . . . staff that was available year round of all
the ethnic groups so that we're not recruiting in crops. Because what
happened in the SOPHIE project is that we were essentially recruiting
in specific ethnic groups just because at the time we had people of that
ethnicity on staff [temporarily] to recruit [people of their same race/
ethnicity], and so we had to take advantage of that because those [stu-
dents] wouldn't be around. Some of them were just here for the summer;
others were here on a part-time basis.

Later, when specifically asked how he views race not only for this work but
also in everyday life, like the first recruiter his response encompassed more
than just DNA and biological purity:

R2: I think race is essentially very culturally specific. . . . I don't think that
there's a great definition for race other than culturally specific, culturally specific—oh, what's the word I'm looking for . . . ? Ideas and
thoughts that a person can identify with when in their racial group.
So, for me, specifically in a Latino racial group, as a South American
Argentine-Columbian, I would say that my background is very different
than a Mexican American who is also considered Latino. . . . And that
differs dramatically in . . . food, music, . . . everything across the board,
and I think that's not being addressed [in the United States]. We just
lump everyone together.

Despite the difficulties they saw in delimiting race by biology, and vice
versa, both recruiters nonetheless each posit "natural" rhetorical references,
or analogies, that seal the social processes they describe in a structure of
the physical world. The first recruiter links the complexity of Mexican racial
difference to geography, to land and diverse racial types and features in
specific locations of Mexico. The second recruiter refers to the targeted harvesting of DNA—a clear social process whereby recruiters of the perceived

and self-reported race of the donor would be responsible for collecting their DNA—as crops. Both are attempts to root the untidy topography of culture in a more pristine-seeming physical-world corollary. Other interviews with scientists reveal similar references, especially to geography, as if being from "the land" was a literal proof of nature cultivating genetic, thus racial, difference.

While they deployed porous definitions with regard to their personal views about race, culture, geography, and intra-Latino diversity (even as both referred to this as a "racial" group in part), both recruiters also deeply believed in the efficacy of the design strategy of SOPHIE. This is evident in the following assertion by the second recruiter: "I know there is debate . . . and there are pros and cons on the issue, but I'm just of the belief that [race] is very important. . . . I think [PMT] is a perfect opportunity to look at very specific racial groups and see how they differ with the same medication. It's something that I don't think has been done and needs to be done. And there will be differences. I'm sure of it." Indeed, with regard to the OCT 2 *1 haplotype "switching to the opposite race," this same recruiter commented: "It really might be because of the different recruitment processes. Coriell isn't as reliable as SOPHIE because we went back three generations for both African Americans and Caucasians. Both could actually be mixed in Coriell." Thus, even as he sees race as largely cultural, where his own experience as a Latino is concerned, this same recruiter avows racial biological purity when it comes to SOPHIE DNA.

Conclusion

In this chapter, I have detailed specific rationalities that link institutions invested in bettering and democratizing health care to cultural practices of race ascription in classing human DNA. In making analogies to nature in order to root social processes in the physical world, the societal tendency to comprehend Americans in reference to the racial and ethnic categories promulgated by OMB (and mandated by NIH in this case) provides an initial framework for the molecularization of race.

As Mary Douglas reminds us, the scientist who classifies "according to known and visible institutions saves himself the trouble of justifying the classification" (1986, p. 94)—whether or not those classifications are coherent and well formed in themselves. I have attempted to open up the space compressed in such instances through ethnographic inquiry and candid questioning of how scientists involved in PMT and SOPHIE understood their work, how and to what end they utilized race, and how they dealt with discrepancies in allele frequencies across racialized data sets. Each process reveals intricacies of social categories that draw on seemingly natural corollaries.

Research into pharmacological susceptibility by race and ethnicity continues to define a great deal of pharmacogenetics in the United States, and thus I find it useful to close with another insight from Douglas. She writes: "However much they try to insulate their work, scientists are never completely free of their own contemporary society's pressures, which are necessary for the creative effort. Scientific theory is the result of the struggle between classifications being developed for professional purposes by a group of scientists and the classifications [operating] in a wider social environment. Both are emotionally charged" (1986, p. 56). Clearly, this reflexive, back-and-forth dynamic between science and society is often generative, creative, and productive. Time will tell if the emotional charge of race, as it has been lodged in this dynamic in the contemporary United States, will be allowed to simply carry efforts seen as creative and productive, or if researchers can move beyond emotional attachments embodied in beliefs in race to query the correctness and, more important, the *meaningfulness* of the genetic results before them.

NOTES

This chapter is a revised, truncated version of a more extensive analysis taken from D. Fullwiley (2007), "The molecularization of race: Institutionalizing racial difference in pharmacogenetics practice," *Science as Culture, 16* (1), 1–29.

1. See http://www.census.gov/Press-Release/www/2001/raceqandas.html (accessed June 21, 2006).

2. The term "rational medicine" is used interchangeably with pharmacogenetics. For more on this idea and the "dictate" that race be used to guide rational medicine, see Evans and Relling (1999, p. 488).

3. See NIH, http://grants2.nih.gov/grants/policy/emprograms/overview/women-and-mi.htm.

4. Because they are high-profile researchers, or their research is extremely specific, I use lead scientists' real names. Signed consent and permissions to do so were obtained using a consent form and attribution statement.

5. Organic cations, by definition, are organic molecules that possess a positive charge with regard to physiological pH. Organic anions, also important in the world of transporter genetics, are negatively charged molecules.

6. Single nucleotide polymorphisms, or SNPs (pronounced "snips"), are simply single base pair changes in DNA sequence. Where most people might have a stretch of ATGCCTTA in a genetic sequence, a few might have ATGCCTTG, or a single polymorphic change. Such potential polymorphisms were observed to happen at about every 1,000 base pairs, or DNA letters. Analyzed alone, or taken together in blocks known as haplotypes, they are now the primary units of comparative genomics, and thus the focus of much pharmacogenetics.

7. An *exon* is a segment of a gene that is present in the final functional transcript (messenger RNA) from that gene. It is any non-intron section of the coding sequence of a gene; together, the exons constitute the mRNA and are translated into protein.

REFERENCES

Bowker, G. C., & Star, S. L. (2000). *Sorting things out: Classification and its consequences.* Cambridge: MIT Press.

Brawley, O. W. (1995). Response to "Inclusion of women and minorities in clinical trials and the NIH Revitalization Act of 1993—the perspective of the NIH clinical trialists." *Controlled Clinical Trials, 16*, 293–295.

Burchard, E. G., Borrell, L. N., Choudhry, S., Naqvi, M., Tsai, H. J., Rodriguez-Santana, J. R., et al. (2005). Latino populations: A unique opportunity for the study of race, genetics, and social environment in epidemiological research. *American Journal of Public Health, 95*, 2161–2168.

Burchard, E. G., Ziv, E., Coyle, N., Gomez, S. L., Tang, H., Karter, A. J., et al. (2003). The importance of race and ethnic background in biomedical research and clinical practice. *New England Medical Journal, 348*, 1170–1175.

Collins, F. S. (2004). What we do and don't know about "race," "ethnicity," genetics, and health at the dawn of the genome era. *Nature Genetics Supplement, 36*, S13–S15.

Douglas, M. (1986). *How institutions think.* Syracuse, NY: Syracuse University Press.

Douglas, M. (2000). *Purity and danger: An analysis of concepts of pollution and taboo.* London: Routledge. (Original work published 1966.)

Eltis, D. (2000). *The rise of African slavery in the Americas.* Cambridge: Cambridge University Press.

Epstein, S. (2004). Bodily differences and collective identities: The politics of gender and race in biomedical research in the United States. *Body & Society, 10*, 183–203.

Epstein, S. (2007). *Inclusion: The politics of difference in medical research.* Chicago: University of Chicago Press.

Evans, W. E., &. Relling, M. (1999). Pharmacogenetics: Translating functional genomics into rational therapeutics. *Science Magazine, 286*, 487–491.

Fujita, T., Urban, T. J., Leabman, M. K., Fujita, K., & Giacomini, K. (2006). Transport of drugs in the kidney by the human organic cation transporter, OCT2 and its genetic variants. *Journal of Pharmaceutical Sciences, 95*, 25–36.

Giacomini, K. M. (1998). Membrane transporters in drug disposition. *Journal of Pharmacokinetics and Biopharmaceutics, 25*, 731–741.

Gomez, M. A. (1998). *Exchanging our country marks: The transformation of African identities in the colonial and antebellum South.* Chapel Hill: University of North Carolina Press.

Greene, S. E. (2002). Notsie narratives: History, memory, and meaning in West Africa. *The South Atlantic Quarterly, 101*, 1015–1041.

Hedgecoe, A. (2004). *The politics of personalized medicine: Pharmacogenetics in the clinic.* Cambridge: Cambridge University Press.

Herskovits, M. (1936). The significance of West Africa for Negro research. *The Journal of Negro History, 21*, 15–30.

Jacobson, M. (1998). *Whiteness of a different color: European immigrants and the alchemy of race.* Cambridge: Harvard University Press.

Kahn, J. (2008 [this volume]). Patenting race in a genomic age. In B. A. Koenig, S. S.-J. Lee, & S. S. Richardson (Eds.), *Revisiting race in a genomic age* (pp. 129–148). New Brunswick, NJ: Rutgers University Press.

Leabman, M. K., Huang, C. C., Kawamoto, M., Johns, S. J., Stryke, D., Ferrin, T. E., et al. (2002). Polymorphisms in a human kidney xenobiotic transporter, OCT2, exhibit altered function. *Pharmacogenetics, 12*, 395–405.

Lee, S. S.-J. (2003). The meaning of race in the new genetics: The politics of identifying difference. Paper presented at invited session, A Critical Anthropology of Human Genetic Variation Research. 102nd annual meeting of the American Anthropological Association, Chicago, IL.

Lovejoy, P. E. (2000). *Transformations in slavery: A history of slavery in Africa* (2nd ed.). Cambridge: Cambridge University Press.

Mintz, S., & Price, R. (1992). *The birth of African-American culture: An anthropological perspective.* New York: Beacon Press.

Mudimbe, V. Y. (1988). *The invention of Africa: Gnosis, philosophy, and the order of knowledge.* Bloomington: Indiana University Press.

National Institutes of Health (NIH). (2001). *Office of extramural research, NIH policy and guidelines on the inclusion of women and minorities as subjects in clinical research, amended, October, 2001.* See http://grants.nih.gov/grants/funding/women_min/guidelines_amended_10_2001.htm.

Oudshoorn, N., & Pinch, T. (2003). Inclusion, diversity, and biomedical knowledge making: The multiple politics of representation. In N. Oudshoorn & T. Pinch (Eds.), *How users matter: The co-construction of users and technologies* (pp. 173–190). Cambridge, MA: MIT Press.

Risch, N., Buchard, E. G., Ziv, E., & Tang, H. (2002). Categorization of humans in biomedical research: Genes, race, and disease. *Genome Biology, 3,* 1–12.

Rose, N. (2001). The politics of life itself. *Theory, Culture & Society, 18,* 1–30.

Shreeve, J. (2004). *The genome war: How Craig Venter tried to capture the code of life and save the world.* New York: Knopf.

Shu, Y., Leabman, M. K., Feng, B., Mangravite, L. M., Huang, C. C., Stryke, D., et al. (2003). Evolutionary conservation predicts function of variants of the human organic cation transporter, OCT 1. *Proceedings of the National Academy of Science, 100,* 5902–5907.

Thornton, J. (1998). *Africa and Africans in the making of the Atlantic world, 1400–1800* (2nd ed.). Cambridge: Cambridge University Press.

Zhang, L., Brett, C. M., & Giacomini, K. M. (1998). Role of organic cation transporters in drug absorption and elimination. *Annual Review of Pharmacology and Toxicology, 38,* 431–460.

Zhang, L., Dresser, M. J., Chun, J. K., Babbitt, P. C., & Giacomini, K. M. (1997). Cloning and functional characterization of a rat renal organic cation transporter isoform (rOCT1A). *Journal of Biological Chemistry, 272,* 16548–16554.

Zhang, L., Dresser, M. J., Gray, A. T., Yost, S. C., Terashita, S., & Giacomini, K. M. (1997). Cloning and functional expression of a human liver organic cation transporter. *Molecular Pharmacology, 51,* 913–921.

9

Tracking Race in Addiction Research

MOLLY J. DINGEL AND BARBARA A. KOENIG

Addiction—a biobehavioral trait—provides a useful case to think with when considering the policy implications of behavioral genetics research targeted by race and linked with differential health outcomes. Stereotypes linking race and addiction are ubiquitous. The two concepts track together in U.S. policy debates. Whether real or convenient social fiction, the notion that certain populations are at increased risk for substance abuse shapes research agendas. When genomics data are added to the mix—as a means of specifying human population boundaries or clarifying a complex behavioral phenotype—the potential for negative social consequences escalates.

Using racial categories in genomic studies of addiction creates a paradox, and that paradox is the central concern of *Revisiting Race in a Genomic Age*: How should health research account for the possibility of biological differences among human populations from diverse continental origins while at the same time avoiding social harms to individuals or groups? Genomic studies of substance use have revealed, for example, the molecular basis for the "flush" reaction to alcohol, which is found more frequently in populations of Asian origin (Luczak et al., 2006). However, individual variability associated with genes is plugged into a research paradigm that assumes the importance of genomic comparison across human racial and ethnic populations, often in a simplistic fashion. Regardless of how the individual human body metabolizes substances like alcohol, nicotine, or heroin, there is inevitably a social and political component to drug use. The use of both legal and illicit substances plays a significant role in health disparities in the United States; there is much at stake in how addiction researchers conceptualize race.

Molecular genetic studies of addiction often stem from a desire to explain and alleviate health disparities, a major public health goal. This chapter begins with an exploration of the historical links between racialized populations and addictive substances in the United States. We next explore how molecular genetic research combines two uncertain and problematic concepts: the phenotype of addiction and that of race. Both must be specified in research; each is problematic. We explore how each phenotype is employed and present findings from interviews with scientists engaged in that research, findings that demonstrate the conceptual limitations of genomic approaches to addiction.

We conclude with an analysis of the policy consequences flowing from these endeavors: Research seeks to link the *behavior* of substance use with specific genetic signatures, providing an aura of molecular certainty. That certainty then adheres to human groups thought to be particularly vulnerable, bringing consequences such as discrimination or enhanced stigma along with the hope of treatment or prevention. But the addiction phenotypes invoked in genetic studies are inherently problematic; linking biological classifications with the social and political dimensions of drug use is challenging. Reducing addiction—a phenomenologically rich social concept—to a set of variables that can be assessed with a genome scan is fraught with difficulty. As argued throughout this volume, actual and meaningful human biological difference does not neatly fit the social categories of race dictated by research regulations. Combining these two complex phenotypes, both necessarily social constructs, engenders policy concern.

Racial Stereotypes and Addiction

Native American alcoholism, long a subject of public policy concern and scientific and medical scrutiny, presents an excellent case study. Alcohol use has caused much suffering and destruction in Native American communities since its introduction by Europeans. The costs of alcohol addiction among Native Americans remain evident today: though rates vary among tribes, and Native Americans have slightly lower average drinking rates than whites, blacks, or Hispanics, they consume more alcohol when they drink. Native Americans are more likely than whites or blacks to die from chronic liver disease and cirrhosis, alcohol-related motor vehicle accidents, and alcohol-related suicide, and they are more likely to commit alcohol-related homicide (Beauvais, 1998; Galvan & Caetano, 2003).

The historical association of Native Americans with alcohol overuse and intoxication gave rise to the "drunken Indian" stereotype, sometimes called the "firewater myth," a term used to describe the widespread belief that Native Americans were constitutionally (or biologically) prone to developing problems with alcohol (Lamarine, 1988; Mail, 2002). Folk theories about Native American drinking were translated into scientific hypotheses.

Research on alcoholism arose within a particular historical context, one espousing scientific theories claiming that racial minorities were inferior, both biologically and psychologically, to white Europeans, and within a medical tradition that has increasingly understood addiction as a biological process. Genetic studies of alcohol abuse and alcoholism, including comparisons by race, were a logical outcome of this historical trajectory.

The story is similar for cigarette and illegal drug use among other minority groups. Native Americans have a national average smoking rate of 35.6% (compared to 21.2% and 21.8% for whites and African Americans, respectively). Laotian American and Cambodian American men have an astonishingly high smoking rate of 72% and 71%, respectively, which may be responsible for a higher prevalence of lung cancer among Asian American and Pacific Islanders (Vilschick, 2001). Though the smoking rates of African Americans are in line with those of whites, evidence indicates that they suffer disproportionately from tobacco-related health problems, including higher rates of lung cancer (American Cancer Society, 2005). Minorities also suffer disproportionately from the consequences of illegal drug use, which is associated with the higher homicide and incarceration rates among African Americans and the disproportionate number of Hispanics and African Americans with AIDS (National Institute on Drug Abuse, 2003).

Drug use and the health issues surrounding it must also be placed within the context of the troublesome health disparities that exist in the United States, with minorities less likely to have health insurance and more likely to have poor health, especially when it comes to chronic and infectious illnesses (Institute of Medicine of the National Academies, 2003; Sharma, Malarcher, Giles, & Myers, 2004).

With these statistics as a backdrop, in recent years there has also been an explosion of new research in the genetics of addiction and pharmacogenetics, supported by the National Institutes of Health (NIH) and pharmaceutical companies. As demonstrated in this volume, research has developed in tandem with studies focusing on genetic differences among races—framed scientifically as uncovering the full range of human genetic variation and answering fundamental questions about human population history and migration (Cavalli-Sforza, 2000).

The speed with which individuals metabolize nicotine (or other drugs, legal or illegal) is an open and valuable scientific question. Metabolic rate might determine the degree of tissue or end-organ damage from using a substance, or explain why some individuals are more likely than others to find a particular drug pleasurable or distasteful. However, we argue that the social categories of race are not the best dividing lines to distinguish between individuals. Racial categories are presumed to be common sense, but as others in this volume argue, they are often forged from bureaucratic and political guidelines (Kahn, this volume). Once the categories are carved out, they in turn fuel further inquiry using the same guidelines.

Genetic Studies of Addiction

The paradigm of scientific inquiry in addiction has become increasingly reductionist, a move stimulated by the ascendance of genomics and neuroscience technologies like functional brain imaging. A further stimulus is undoubtedly the extreme recalcitrance of drug and alcohol problems; easy solutions are clearly lacking. This field of research is united by an understanding of addiction as a "disease of the brain," a specific pathology caused by a range of drugs showing similar effects on the brain, especially the mesolimbic system (Koob & Moal, 2005; Nestler, 2005). Following this logic, researchers seek variations in genes that correlate with the use of specific substances, like the molecular basis for the widely observed "flush" reaction to alcohol, mentioned above. There have also been inquiries to identify genes in Native Americans that predispose individuals to alcoholism, with many studies concluding that there are either protective or predisposing effects from various genes (Ehlers & Wilhelmsen, 2005; Wall, Carr, & Ehlers, 2003). Research does indicate that some individuals with Native American ancestry are especially susceptible to suffering ill effects from ingesting alcohol, a phenomenon attributed to variability in genes that control metabolic clearance. But the occurrence of this susceptibility is not unique to Native Americans (Mail, 2002).

Genetic research on alcohol, tobacco, and methadone use has also examined variations in cytochrome P450 (CYP450), a major drug-metabolizing enzyme that is primarily expressed in the liver. Rate of metabolism appears to correlate with smoking behavior, with slow metabolizers being both less likely to be regular smokers and more likely to successfully quit (Schoedel, Sellers, Tyndale, Hoffmann, & Rao, 2004). CYP450 also seems to affect alcohol (Zimatkin & Deitrich, 1997) and methadone metabolism (Gerber, Rhodes, & Gal, 2004). Moving immediately to comparative studies, researchers have concluded that CYP450 genetic variations (or polymorphisms) differ by racial category (Schoedel et al., 2004). Tate and Goldstein (this volume) describe this phenomenon for a range of licit drugs.

Scientific studies that legitimate the classification of people into distinct and purportedly biologically valid categories may reinforce stereotypes like the firewater myth. Researchers consistently reinforce these historically problematic categories by casually grouping people by self-reported race or by studying one subset of a racial category (i.e., Yoruba, Han Chinese) and uncritically applying results to the larger category (Fausto-Sterling, 2004). Samples from a small number of Nigerians, Chinese, or Native Americans come to stand for all Africans, Chinese, or Native Americans. Social and political groupings are transformed into biological categories.

The Importance of Good Research Design

However promising genetic studies may be, they represent only one dimension of why people begin drug use, become addicted, or have difficulty quitting. Two main concerns must be addressed when doing research on the genetics of drug use: first, methods must be sophisticated enough to account for social, cultural, and environmental influences; second, great care must be exercised when characterizing the addiction phenotype in research design.

The Social Environment and Drug Use

Focusing on genetics turns our gaze away from the social world, yet drug use can only be fully understood in context. For example, though Japanese men are more likely than white American men to be "slow metabolizers" of nicotine, and therefore "genetically" at lower risk of becoming addicted to nicotine, they have a much higher smoking rate (Mwenifumbo et al., 2005; Schoedel et al., 2004). Similarly, though Native Americans and Asians share similar physiological reactions to alcohol, Native Americans have some of the highest and Asians some of the lowest rates of alcoholism (i.e., Wall, Carr, & Ehlers, 2003). These examples demonstrate that molecular biology alone is inadequate to determine why people smoke or drink, yet media and public funding support and publicize genetic studies to the exclusion of myriad social, cultural, and individual pressures that contribute to addictions (Conrad & Weinberg, 1996).

Returning to our example of Native Americans and alcohol, some have posited that the use of alcohol in Native American communities is not the result of a biological predisposition to alcoholism, but instead is the result of these groups "learning" how to consume and misuse alcohol from early North American European settlers (Beauvais, 1998; Frank, Moore, & Ames, 2000; French & Hornbuckle, 1980). Because Native Americans had no experience with alcohol, unlike Europeans, they did not have established mores or taboos on excessive alcohol use; in fact, before contact with Europeans, mind-altered states were positively associated with a quest for enlightenment, powers of healing, and making war (Frank et al., 2000).

Moreover, federal policy undermined Native Americans' ability to create meaningful taboos. When introduced to alcohol, not only did the "lawless edges of the new society" (Frank et al., 2000, p. 348) provide the model for alcohol use adopted by Native Americans, but they also encouraged abuse through the continued use of alcohol in treaty negotiations (Beauvais, 1998;

French & Hornbuckle, 1980). Alcohol abuse was further exacerbated by 350 years of genocide, the introduction of deadly diseases, and federal policies that undermined and sought to destroy Native American societies. These policies ranged from expropriation of territory to the General Indian Intercourse Act, a law that prohibited alcohol use by Native Americans from 1832 to 1953 and thus promoted a forced abstinence that Native Americans felt compelled to challenge (French & Hornbuckle, 1980).

Similarly, in the case of smoking, a variety of studies indicate a wide range of environmental factors involved in smoking initiation, including cultural norms (Nichter, 2003), socioeconomic status (Fernandez et al., 2006; Graham, Francis, Inskip, & Harman, 2006), smoking status of peers and parents (Biglan, Duncan, Ary, & Smolkowski, 1995), racism (King, 2004), and tobacco advertising techniques (Pierce, Choi, Gilpin, Farkas, & Berry, 1998; Pierce, Gilpin, & Choi, 1999), including those advertisements targeting minorities and working-class communities (Barbeau, Wolin, Naumova, & Balbach, 2005; Jain, 2003; White, White, Freeman, Gilpin, & Pierce, 2006).

Focusing exclusively on biology directs our attention away not only from the social factors that shape patterns of drug use, but also from the contentious politics and history of racial inequality. Indeed, using Native American biology to explain alcohol use in their communities is easier than remediating the consequences of centuries of genocide, racism, and oppression that government policies have forced upon tribes. Likewise, forces of commerce also intrude upon these processes—a pill or medical cure is an easier sell than trying to undo or provide adequate reparations for these social wrongs.

It is clear that biology alone cannot predict the differences among racially and ethnically defined groups. Given the social dynamics of drug use, good research design must explore alternative ways of pursuing studies of group differences in addiction. What other ways might we have of discovering population differences in metabolism? Are there additional, yet undiscovered, reasons behind differing rates of drug use among communities or nations? Eliminating an exclusive focus on race allows for creativity and exploration of other pertinent categories, comparisons, and explanations. Privileging racial explanations focuses too much attention on biology.

Defining the Phenotype

Technologies to characterize human genetic variation have dramatically decreased in price and increased in specificity. No matter how precise the technology, however, an accurate description of phenotype remains an essential—and necessarily human-made—element of good genetics research. Well-defined phenotypes are the foundation of genetics research; without them there is no way to assess whether molecular findings represent true phenomena or are simply artifacts of the ways we categorize the world at a particular historical moment.

Defining a phenotype for a complex behavior like addiction is particularly challenging (Press, 2006). Positive associations between particular genetic signatures and the use of nicotine have proven difficult to replicate because associations are contingent on how the smoking phenotype is characterized (Wilhelmsen, 2006). For example, the same genetic loci may be associated in several studies with measures such as "time to first cigarette in the morning," but not with other ways of defining smoking behavior, like continued use despite substance-related problems.[1] As another example, it is difficult to pinpoint when casual substance use becomes "problematic use," particularly for legal drugs such as alcohol and nicotine. Who counts as an addict? Similarly, it is difficult to characterize a "non-user" of a particular substance. Must an individual have had some exposure to a drug in order to prove they are unlikely addicts?

Social scientists working on alcohol use have long struggled with categorization of use, abuse, and addiction. While there are clearly times when heavy drinking becomes a medical problem (alcohol poisoning or liver disease, for example), these do not necessarily indicate a long-term problem or arise in time to intervene meaningfully (Conrad & Schneider, [1980] 1992). Population surveys have measured alcohol use in a variety of ways, from early surveys that only distinguished those who drank from those who did not to later surveys that tried to assess volume or timing of drinking (Midanik & Room, 2005). But common measures of *amount* of alcohol consumed do not always correlate with negative effects, indicating that patterns of drinking are also important. T. Norstrom found that an extra liter of alcohol drunk per year by adults in northern Europe causes three times greater mortality than the same amount drunk by adults in southern Europe (cited in Midanik & Room, 2005), a difference that has caused researchers to focus greater attention on customary patterns of drinking in different countries. This example also points to the difficulty in differentiating "problematic" drug use from "nonproblematic" drug use.

It is also extremely difficult to characterize the phenotype of a "non-addicted" individual. Yet such comparisons are the foundation of genetics studies, based on classic genetic approaches that contrast "affected" and "wild type" individuals.[2] Consider the case of smoking. How much exposure to cigarettes is required to define the non-user in genetic studies of addiction? Clearly, someone who has never tried tobacco is not an acceptable control, since their non-use may simply be lack of experience. Is a single cigarette or drink evidence of exposure? The actual definitions used in nicotine research (i.e., a "life-time exposure" of under 100 cigarettes to define someone as a non-smoker) are inevitably arbitrary.

Likewise, the boundary between harmful and non-harmful drug use is blurry, adding further complexity to phenotype measurement and assessment. Several historical examples reveal that degree of harm is contingent; adverse affects are not dictated solely by a substance's chemical or pharmacological qualities. For example, in many cultural contexts drugs that are today considered extremely hazardous and highly addictive were used for medicinal or spiritual purposes with few negative side effects. Social controls dictated moderate use of powerful substances (Courtwright, 2001; Stinson, 2006). Opium was used to treat digestive problems in the East, coca was used to deal with hunger and fatigue in the Andes, and tobacco was used in religious and medicinal ceremonies by Native American tribes (Courtwright, 2001; Stinson, 2006). Courtwright (2001) has documented how folk remedies and substances used in ritual practice were transformed into modern addictions. Global trade encouraged the spread of many drugs (tobacco was introduced to Europe, alcohol to the New World), and with increased demand, drugs were processed to increase their potency (liquor was distilled into spirits and the industrial cigarette was created). The result was a manufactured product: in the case of cigarettes, consumption increased dramatically with mass production and aggressive marketing (Brandt, 2007). Increased potency and availability yielded abuse, and with abuse came a myriad of social problems that prompted legislation to restrict or ban substance use (Courtwright, 2001). One could argue that addiction is "caused" by the changed social context of use rather than by a drug's essential characteristics or the genetic vulnerabilities of certain human populations. However, as with many contemporary social problems, there are diverse forces promoting addiction's "biomedicalization," a form of medicalization that is linked to biotechnological innovation, a focus on health management and surveillance of individuals, and commercial forces (Clarke, Fishman, Fosket, Mamo, & Shim, 2003).

Neurogenetic and molecular biology inquiries into drug use seek to create certainty by transforming the social classification of addiction into a biological classification, something that can be observed and measured with the precision of molecular technology. Similar transformations have occurred in other areas. What was once defined as "senility," a common consequence of aging, is now classified as Alzheimer's disease, including a focus on the neuropathology of plaques and tangles (Wachbroit, 2006). Research on these biological markers has had some success because a reliable phenotype was identified. But biology alone cannot define the phenotype of addiction. The above discussion demonstrates that genetic studies *must* account for complex political and economic factors to define adequately an addiction phenotype. Nonetheless, much genetics literature fails to include a discussion of social dynamics, leaving readers with the impression—even

if unintended—that addiction can be adequately addressed through biological means (e.g., Koob & Moal, 2005; Nestler, 2005). We argue that social
dimensions must be included; the problem then becomes how to elucidate
a phenomenon where social worlds are intricately intertwined with biology
(Carey & Gottseman, 2006).

Problems with Racial Categories

The second problematic phenotype is race; scientific research on human
diversity in drug response and addiction relies on categorization of human
groups. Though human variation seems readily ascertainable, in practice
biological race categories are not obvious or easy to apply (Duster, 2001).
Race is not a natural category, but is instead created by politics, law, art,
history, religion, and science (Omi and Winant, 1994).

There is little precision in how race is "measured" or assessed in addiction studies. Racial categories are often taken for granted; terms like "Caucasian" or "African American" are generally not defined (Shields et al., 2005).
In addition, genetic work on addiction and race oversimplifies the extent
and complexity of human variation. Common categories also assume a pris-
tine genetic purity, but ignore the role of gene flow resulting from migration
and intermarriage, particularly in immigrant societies like the United States.
Studies indicate that self-identified African Americans have significant European ancestry (Parra et al., 1998).[3] Given the history of genocide and years
of government policies that sought to incorporate Native Americans into
Western culture, it is likely Native peoples are similarly genetically diverse.
Though the existence of "pure" races would make the science easier, these
do not exist. Instead, human genetic variation is best described as a continuum, not as discrete categories with clear boundaries (Barbujani, 2005;
Manica, Prugnolle, & Balloux, 2005).

Racial Categories Used in Addiction Research: Interviews with Scientists

To understand how scientists themselves perceive these problems, we interviewed 19 scientists involved in genetic research on nicotine addiction.[4] We
asked these scientists to discuss current research, including studies making
comparisons across racial groups. Responses indicate that racial categories
are applied uncritically in addiction research and that few recognize potential negative social consequences. Almost half of the scientists who spoke
of human population differences saw it as a straightforward issue; the use
of racial and ethnic categories was seen as a legitimate, unproblematic way
to categorize data. One lead researcher from a large, nonprofit organization
noted: "[One of the] principles of pharmacogenetics is the presence of ethnic differences, and I don't think we fully understand or appreciate just yet

just how much of a role that plays across ethnicities. But we do know that there are differences, ethnic differences in drug metabolism in general."[5] While it is clear there is individual variability in rates of metabolism, this scientist thinks in terms of a population model in which these differences fall along ethnic lines, and takes it for granted that ethnicities are appropriate categories to use when examining human difference. A scientist who is studying the pharmacogenetics of nicotine cessation treatments echoed this sentiment: "We think our genetics of smoking [work] is quite impressive because it does have such a large effect, and there are these huge differences among races so that, for example, about 20% of the Japanese lack this enzyme, or have very low activity. . . . And then, depending on what we find in some ongoing work, it may be that African Americans have a rather different genetic basis for their enzyme activity, and then they may be at more risk."

Although the use of racial categorization in research was unquestioned, some scientists were concerned with how research results would be translated into clinical practice and, especially, how racialized findings would be communicated. A researcher who is exploring possible new drugs for tobacco cessation at a large Midwestern university stated: "The challenge is going to be, How do you communicate it to the consumers as well as the policy makers, or the people that are providing the programs, you know? To these individuals. How do you communicate something that could be potentially very complex in a way that's understandable, and a way that it can be interpreted accurately?" Another informant, the head of a genetics lab in a large southern university, noted, "We have a huge gap in our capacity to explain this sort of information to primary care docs as well as to patients."

It is clear that the intersection of racial categories and addiction studies has become routinized for these scientists in a way that prevents them from seeing potential harms or inaccuracies intrinsic to this research. Instead, these scientists hold to the view that the intent of the research is good; it is only when "pure" science is applied in the "real" world that problems arise and harms may develop. Our informants did not consider that scientific practices may reflect unconscious social biases or that scientific categories may reify questionable categories describing human difference. Though they seem aware of problems in translating research into practice, they see this as a communication problem, not as an inevitable consequence of performing research within a society riddled with racial inequality. An examination of the race, genetics, and addiction literature suggests that these sentiments are widespread in the scientific community: many researchers embrace (and publish) genetic research on race and addiction, and their methods of racial classification are often implemented with little reflection or explanation of how such classifications were made or why they may be pertinent to the topic at hand (Fullwiley, this volume; Lee, Mountain, & Koenig, 2001; Rebbeck & Sankar, 2005; Shields et al., 2005).

Racial Categories Used in Addiction Research: An Analysis of the Literature

As an indicator of how human racial phenotypes are constructed in addiction research, we reviewed research published between 1996 and 2006 on the genetics of race and addiction. We searched Medline using the broad search terms "substance related disorders," "ethnology," "genetics," "humans," "continental population groups," "ethnic groups," and "genetic predisposition to disease." This yielded a total sample of 72 articles. We then excluded documents that were not original research, did not mention racial categories in the title or abstract, dealt only with drug-related disease, or did not interrogate specific genes as determinants of drug use. The resulting sample was 41 research articles (see table 9.1).

Our analysis reveals considerable ambiguity in how human population groups were categorized. Twenty-seven of these studies either did not mention how they classified race or used self-report, a concept that was generally not further defined. An additional five studies used political or geographic categories, based on residence, to define race. Only eight of the studies tried to assess continental ancestry of individuals by gathering information about parents' or grandparents' race.[6] One study used ancestry-informative markers (AIMs) to classify subjects' race.[7] Some of the imprecision in racial categories is the result of U.S. Office of Management and Budget (OMB) guidelines that mandate the use of one specific set of racial categories— American Indian or Alaska Native, Asian, black or African American, Native Hawaiian or other Pacific Islander, and white—in all government-sponsored health research. These requirements lead to inappropriate use of political categories as proxies for biological variation and create conceptual confusion that leads to poorly designed science. The use of self-reported race without a full explanation of how research participants were questioned about background, or what methods were used to assess ancestry, leads to further imprecision. Those who identify as being of a specific race may actually have ancestry from several continents. To avoid research findings that create tautological links between genes and identifiable populations, it is vital not to pre-categorize human participants using social and political groupings.[8]

Tracking Race in Addiction: Policy Implications

The concerns we raise have implications not only for the practice and validity of science and possible treatments resulting from that science, but also for social policy. Linkages between race and addiction might decrease or

TABLE 9.1

Racial Classification of Humans in Studies on Addiction Genetics

Method of Classification	Twenty Studies Did Not Explain How They Measured Race	Seven Studies Used Self-Report	Eight Studies Used Grandparents' or Parents' Ethnicity	Five Studies Used Country/Region of Residence
Categories used (number of studies utilizing term, if more than 1)	African American (11), African descent, Ashkenazi Jews, Asian, Bedouins, Caucasian, Ethiopian Jews, European American (10), European ancestry, Han Chinese (3), Hispanic, Hispanic American, Japanese, Jewish, non-Hispanic Caucasian (2), Southwestern American Indian	African American (6), European American (6), mixed ancestry	African American, American Indian, Chinese, Chinese American, European American, European decent, European white, Hispanic American, Korean, Korean American, Mexican American, Mission Indian (2)	Ami, Atayal, Chinese, Finnish, German, Japanese, Korean, Southwestern Native American (2), Taiwanese Chinese
Studies	Beuten 2005; Comings 1999; Comings 2001; Dahl 2005[a]; Dahl 2005[b]; Drakenberg 2006; Gelernter 1999[a]; Gelernter 1999[b]; Gelernter 2006; Hoehe 2000; Lappalainen 2002; Li 2005; Luo 2006[a]; Luo 2003; Ma 2005; Peng 1999; Shi 2002; Spence 2003; Tsai 2005; Xuei 2006	Drgon 2006; Hinrichs 2006; Lou 2006; Liu 2005; Uhl 2001; Zhang 2006; Zhang 2004	Anney 2004; Ehlers 2004; Konishi 2004; Long 1998; Luczak 2001; Luczak 2006; Proudnikov 2006; Wall 2003	Kim 2006; Osier 1999; Radel 2005; Xu 2006; Yokoyama 2003

Note: One additional study used ancestry informative markers (AIMs) to racially classify subjects as either European American or African American (Luo 2006[b]).

increase the stigma traditionally associated with drug use, affect individuals' access to health care, and have implications for current public health programs.

Stigmatization, Racial Groups, and Group Harms

Stigma can be defined as an "undesired differentness" that negatively influences those an individual comes into contact with (Goffman, 1963). With regard to the intersection of race and drug abuse, stigma stems from a belief that certain racial groups are more likely to become addicted or to overuse drugs or alcohol. Native Americans have long suffered stigmatization due to the "firewater myth": the stereotype that Native Americans are biologically prone to excessive alcohol consumption. This myth was at the heart of the historical prohibition of Native American alcohol use by the U.S. government and conforms to a long history of research that ties disease to race (Lee et al., 2001). A variety of diseases, including sickle cell anemia and pneumonia, have been used to "prove" the biological inferiority of blacks to whites (King, 2004; Tapper, 1999). More recently, some groups of Ashkenazi Jews have been reluctant to engage in clinical trials investigating genetic links to breast cancer because they worry about being targeted as susceptible to breast cancer (Lee et al., 2001), a response common to other groups facing similar labeling (Rothstein & Epps, 2001).

Racialized genetic knowledge may have unintended consequences. Frequency of disease-associated alleles does vary across the globe. But findings that link specific genetic variations to certain racial groups may be incorrectly extended to individuals. It is difficult to establish legal or social protections against discrimination that results from group membership, especially when the stigma accrues to a "visible" minority group. The current U.S. regime of human subjects' protection seeks to shield individuals from harm through a focus on informed consent; assuring safeguards against group harms is challenging when individuals are asked to participate in research linking race with a trait like addiction. Current procedures—asking subjects' permission one by one—are unlikely to be helpful in protecting *communities* of people. "Community consultation," suggested by human subject's protection advocates, is largely unproven as a solution and raises questions of who can act as a legitimate community representative and what constitutes effective engagement (Quinn, 2004; Weijer & Miller, 2004). As Reardon (this volume) points out, calls for community consultation assume that populations are neatly organized into coherent groups; instead, the reality is that in order to solicit community input on potentially harmful research, agencies often must *constitute* those groups. In addition, this kind of community consultation fails to protect individuals from the broadest social harms of this research—the biologization not just of addiction, but of human difference itself. As Nikolas Rose points out, what is at stake in this scientific research is the very way we come to understand and govern "individual and collective human identities" (2006, p. 385).

Genetic studies of addiction that report research results by (racialized) population groups may lead to public health programs targeting those same groups. The social effects of such programs have not been systematically examined and, in fact, may even be valued as demonstrating "cultural competence," but it is likely that targeted programs will increase stigmatization or reinforce existing cultural stereotypes that link certain groups—often those with less power in society—with substance abuse. Because of stereotypes linking blacks with illicit drug use, African Americans are less likely to be prescribed pain medication than whites with the same ailments (Anderson et al., 2004; Tamayo-Sarver, Hinze, Cydulka, & Baker, 2003), and recent research indicates that regulatory surveillance of prescription drug use reduces legitimate use of legal narcotic drugs by African Americans (Pearson et al., 2006). If historically racialized communities are labeled as having a genetic predisposition to alcoholism, smoking, or illicit drug use, an unintended outcome may be less access to legitimate and helpful drugs for individuals from those communities.

Access to Health Care

There are well-known, if often poorly understood, disparities in access to high-quality health care in the United States. Native Americans, for example, are less likely to have health insurance than any other U.S. population group except those self-identifying as Hispanic (U.S. Census Bureau, 2004). Genetic studies may bring with them new categories of people who are labeled as "healthy," "less expensive," "hard to treat," "less profitable," or "more expensive" (Rothstein & Epps, 2001). Linking specific genetic variations to different races risks providing health care companies and employers an (inaccurate) "shorthand" to use to sort out medical risks. Genetic testing of actual vulnerabilities may be expensive or legally proscribed. But if authoritative scientific research labels certain groups as prone to addiction—a condition notoriously difficult to treat that leads to additional expensive health problems—individuals in those communities may have even greater difficulty obtaining health insurance (or employment that provides entrée to insurance) because of perceived risk. This result may be unrelated to well-documented scientific findings of *actual* genetic variation among human populations, variations that are associated with disease or disease risk.

Setting Priorities in Research Funding and Public Health Programs

When considering funding priorities for research seeking to decrease the health effects of addiction, a key question emerges: Should research target genes or the social environments inhabited by populations suffering the consequences of addiction? Or more realistically, how should the two approaches be balanced? Some have argued that genetic studies may divert attention from existing approaches known to be effective in reducing health disparities (Merikangas & Risch, 2003; Carlsten & Burke, 2006; Sankar et al., 2004). Diverting public research funding to projects focused on identifying genes may reduce our national interest in funding wide-scale public health programs, such as raising tobacco taxes. Focusing only on biological correlates of addiction may shift attention away from known environmental and social causes of drug use, particularly those concentrated in resource-poor minority communities. As noted previously, we do not advocate the cessation of genetic research on addiction or an abandonment of studies that incorporate human genetic variation in scientifically sound ways; we recognize that this research could provide new addiction therapies and a better understanding of the neurogenetics of addiction. But currently the most successful and cost-effective solutions to common addictions are based on traditional public health strategies. Smoke-free ordinances and steep taxes on cigarettes have had well-known positive effects on smoking quit rates (Chaloupka, Wakefield, & Czart, 2001; U.S. Department of Health and Human Services, 2006). However, powerful economic forces are at play: public health measures do not support a growing biotechnology-based economy; they are not as immediately profitable as promised pharmaceutical "cures." A quick fix for addiction has undeniable appeal.

Next Steps

A genomic age brings with it new challenges and new questions about the nature of human difference. Will race disappear and be replaced by new genetic categories, or will it be further reified by a plethora of genetic research utilizing racial categories carelessly? Will addiction follow a path of medicalization and biomedicalization, as has happened with many psychiatric conditions and even aging? Will we develop new methodologies to incorporate pertinent social variables into genetic research? How can we balance the benefits of genetic research—potential new therapies and better understanding of human bodies—with the harms of racial reification and extreme biological reductionism? Biological differences across populations with varied biogeographical ancestry do exist; gene frequencies governing metabolism of addictive substances vary across human populations, and understanding the mechanisms of how the body responds to drugs is valuable. The policy challenge is mining the information provided by genetics without reifying racial categories.

To begin to answer these questions, we must first understand that the reasons people become addicted, stay addicted, or quit using a substance are complex and include not only genetic/biological variables, but also psychosocial stress, targeted advertising, price of the drugs, socioeconomic status, and cultural factors. Clearly, reducing all these "variables" to a single phenotype—one that can be plugged into a whole genome analysis—is an enormous challenge, even if these variables did not intersect with the problematic category of race. Social factors shape our biology; racism interacts with our bodies to determine disease. High blood pressure and heart disease are two examples of illnesses disproportionately affecting African Americans, but research also indicates environmental aspects, including racism and social support networks, have a significant impact on these illnesses (Krieger & Sidney, 1996). The stress resulting from racism and discrimination is implicated in health disparities (Williams, Neighbors, & Jackson, 2003) and substance abuse (Gee, Delva, & Takeuchi, 2007). It is important that our inquiries integrate complex factors rather than denying or simplifying them. If racism increases stress, and stress levels affect smoking and drinking behavior, then we would expect that minorities and others in our society who experience discrimination will be more likely to abuse drugs regardless of any genetic polymorphisms associated with addiction. This logic implies that the same individuals would be at different risk for addiction depending on whether or not they lived in a society in which they occupied the social status of oppressed minority. This vision of addiction is absent in genetic studies but serves as an example of the complexity of combining an addiction phenotype with a racial phenotype.

While it is impossible to prevent the misuse of science, it is essential for researchers to give heightened attention to how their research will be used in the world beyond the laboratory (see Shields et al., 2005). Researchers, social scientists, and governmental bodies must work to prevent inappropriate use of racial categories, to anticipate responses from the communities under study, and to identify the potential harms flowing from the use of racial categories. Not only should scientists frame studies with an eye to possible social factors, but we should engage in public discussion that seeks to keep our attention focused on *all* causes of drug use, not just genetic ones.

Second, we must reconsider the use of race in all health research, particularly when genetic data are collected. Some have called for purging the term *race* from all public health research (Fullilove, 1998). However, eliminating the term entirely would deny the importance of racism in producing health disparities and would do a disservice to vulnerable populations. Indeed, studies documenting racism must continue to use race categories that are based in political—not genetic—realities. Researchers should reframe the use of race in two ways: by thinking about race, as Stevens (this volume) suggests, as "something to be studied in its immediate context" instead of as hereditary, and by abandoning the use of the term *race* when discussing genetic variation. Categories such as continental origin or biogeographical ancestry are preferable, and journals should require precision in their application. In research, we must always question whether racial comparisons are appropriate, and, if they are, question how they are done. Race *is* an appropriate term to use in studies about the impact of racism on individual or population health (Benkert & Peters, 2005; Krieger & Sidney, 1996; Piette, Bibbins-Domingo, & Schillinger, 2006). In studies of the molecular biology of addiction, race must be invoked as a polysemic category, reflecting both social and biological aspects, racism as well as genes.

What is the best way to design and carry out genetic studies of addiction that do not thoughtlessly reify racial categories and that avoid potential harms? NIH and governmental guidelines, though created to support diversity and inclusiveness in clinical research, have instead encouraged over-simplified categories of race (Roberts, 2006). Strict guidelines that lack conceptual clarity (like those promulgated by OMB), even if well-motivated, have the potential for harm. Kahn (this volume) outlines how the law and the patent process further normalize these social categories. To counterbalance current regulations, Stevens (this volume) suggests a federal oversight committee to deal with harmful or incorrect use of racial labels in research. However, regulatory guidelines may impinge upon academic freedom and be bureaucratically cumbersome (American Association of University Professors, 2000). Instead, we agree with Shields and colleagues that NIH inclusion guidelines must be reviewed and refined, that funders should support research that incorporates social dimensions, and that we should create policies encouraging stronger methodologies in genetic studies (Shields et al., 2005).

Will genetic research enhance our knowledge of alcohol use in Native American communities? Possibly. But studying addiction through an exclusively biological lens allows us to ignore the long history of oppression that undoubtedly plays into alcohol use. Given the inherent challenges of defining "the phenotype" in behavioral genetics research, we must ask whether combining two problematic phenotypes—race and addiction—yields harm or benefit.

NOTES

This research was supported by National Institute on Drug Abuse (NIDA) grant R01 DA 14577, "Genetics of Nicotine Addiction: Examining Ethics and Policy."

1. The *Diagnostic and Statistical Manual* (DSM) defines dependence as in part based on continued use despite significant substance-related problems, but this definition

is used infrequently to assess tobacco dependence because of its poor validity and weakness in predicting inability to quit. Other measures, like the Fagerstrom Test for Nicotine Dependence (FTND), are used much more frequently in genetic and other research on tobacco dependence (Beseler et al., 2006; Hughes, 2006). A central question of the FTND is "time to first cigarette in the morning," with shorter times consistently predicting a higher level of dependence.

2. In genetics, the term *wild type* refers to the predominant, "normal," or most common variant of a gene in a population.

3. As Templeton (1999) points out, the term *admixture* used by Parra et al. (1998) presupposes that there are (or were) distinct African and European races that are "pure" and can be "mixed." Here we simply cite these data to illustrate the disconnect between "commonsense" racial categories and the diverse genetic heritage of people in these categories.

4. These data come from the NIDA study noted above. We conducted interviews with scientists, clinicians, prevention workers, pharmaceutical company representatives, and health insurers (N=85). The data presented here are from open-ended, audio-recorded interviews with scientists conducted between February 2004 and June 2006.

5. Many informants used the term *ethnicity*, not *race*, when discussing human population differences. We believe that replacing the term *race* with *ethnicity* is frequently an effort to be politically correct. In this case, the term *ethnic* is applied to the concept of human biological difference even though the word is generally defined as reflecting cultural characteristics of a group.

6. These assessments were made slightly differently for each study. One based race on country of birth across three generations (Anney, Olsson, Lotfi-Miri, Patton, & Williamson, 2004), and other studies included participants if "three of four biological parents [were] of the same ethnicity" (Konishi, Calvillo, Leng, Lin, & Wan, 2004) or if all biological grandparents were of the same race (Luczak, Wall, et al., 2001; Luczak, Shea, et al., 2006). Other studies included participants if they were 1/16 Native American (Ehlers & Wilhelmsen, 2004), or at least 50% Native American (Wall, Carr, & Ehlers, 2003), but no indication was made as to how these percentages were assessed; another study relied on elder tribal members for information on family structures (Long et al., 1998). One study indicated that "individuals from mixed ethnicities (reporting mother and father from different ethnic groups) were not included in the study" (Proudnikov et al., 2006). Little or no additional information about the process is included in these articles.

7. There is considerable debate about the utility of using such markers to approximate social race categories. See Bolnick (this volume) for a discussion of the assumptions underlying computer algorithms such as those used by the *structure* program.

8. Some scholars, for example, Neil Risch and colleagues (2002), argue that self-reported race is a good proxy for genetic variation. We do not share this view.

REFERENCES

American Association of University Professors. (2000). *Institutional review boards and social science research*. Retrieved June 5, 2007, from http://www.aaup.org/AAUP/About/committees/committee+repts/CommA/protecting.htm.

American Cancer Society. (2005). *Cancer facts & figures.* Retrieved March 2006, from http://www.cancer.org/downloads/STT/CAFF2005f4PWSecured.pdf.

Anderson, K. O., Mendoza, T. R., Payne, R., Valero, V., Palos, G. R., Nazario, A., et al. (2004). Pain education for underserved minority cancer patients: A randomized controlled trial. *Journal of Clinical Oncology: Official Journal of the American Society of Clinical Oncology, 22* (24), 4918–4925.

Anney, R.J.L., Olsson, C. A., Lotfi-Miri, M., Patton, G. C., & Williamson, R. (2004). Nicotine dependence in a prospective population-based study of adolescents: The protective role of a functional tyrosine hydroxylase polymorphism. *Pharmacogenetics, 14*, 73–81.

Barbeau, E. M., Wolin, K. Y., Naumova, E. N., & Balbach, E. (2005). Tobacco advertising in communities: Associations with race and class. *Preventive Medicine, 40* (1), 16–22.

Barbunjani, G. (2005). Human races: Classifying people vs understanding diversity. *Current Genomics, 6*, 215–226.

Beauvais, F. (1998). American Indians and alcohol. *Alcohol Health & Research World, 22* (4), 253–259.

Benkert, R., & Peters, R. M. (2005). African American women's coping with health care prejudice. *Western Journal of Nursing Research, 27* (7), 863–889.

Beseler, C., Jacobson, K. C., Kremen, W. S., Lyons, M. J., Glatt, S. J., Faraone, S. V., et al. (2006). Is there heterogeneity among syndromes of substance use disorder for illicit drugs? *Addictive Behavior, 31* (6), 929–947.

Beuten, J., Ma, J. Z., Payne, T. J., Dupont, R. T., Crews, K. M., Somes, G., et al. (2005). Single and multilocus allelic variants within the GABA receptor subunit 2 (GABAB2) gene are significantly associated with nicotine dependence. *American Journal of Human Genetics, 76*, 859–864.

Biglan, A., Duncan, T. E., Ary, D. V., & Smolkowski, K. (1995). Peer and parental influences on adolescent tobacco use. *Journal of Behavioral Medicine, 18* (4), 315–330.

Bolnick, D. A. (2008 [this volume]). Individual ancestry inference and the reification of race as a biological phenomenon. In B. A. Koenig, S. S.-J. Lee, & S. S. Richardson (Eds.), *Revisiting race in a genomic age* (pp. 70–85). New Brunswick, NJ: Rutgers University Press.

Brandt, A. M. (2007). *The cigarette century.* New York: Basic Books.

Carey, G., & Gottseman, I. I. (2006). Genes and antisocial behavior: Perceived versus real threats to jurisprudence. *Journal of Law, Medicine, and Ethics, 34* (2), 342–351.

Carlsten, C., & Burke, W. (2006). Potential for genetics to promote public health. *Journal of the American Medical Association, 296* (20), 2480–2482.

Cavalli-Sforza, L. L. (2000). *Genes, peoples, and languages.* New York: North Point Press.

Chaloupka, F. J., Wakefield, M., & Czart, C. (2001). Taxing tobacco: The impact of tobacco taxes on cigarette smoking and other tobacco use. In R. L. Rabin & S. D. Sugarman (Eds.), *Regulating tobacco* (pp. 39–71). New York: Oxford University Press.

Clarke, A. E., Fishman, J., Fosket, J., Mamo, L., & Shim, J. (2003). Biomedicalization: Technoscientific transformations of health, illness, and U.S. biomedicine. *American Sociological Review, 68*, 161–194.

Comings, D., Gonzalez, N., Wu, S., Saucier, G., Johnson, P., Verde, R., et al. (1999). Homozygosity at the dopamine *DRD3* receptor gene in cocaine dependence. *Molecular Psychiatry, 4*, 484–487.

Comings, D. E., Wu, S., Gonzalez, N., Iacono, W. G., McGue, M., Peters, W. W., et al. (2001). Cholecystokinin (*CCK*) gene as a possible risk factor for smoking: A replication in two independent samples. *Molecular Genetics and Metabolism, 73*, 349–353.

Conrad, P., & Schneider, J. W. ([1980] 1992). *Deviance and medicalization: From badness to sickness.* Philadelphia: Temple University Press.

Conrad, P., & Weinberg, D. (1996). Has the gene for alcoholism been discovered three times since 1980? A news media analysis. *Perspectives on Social Problems, 8,* 3–25.

Courtwright, D. T. (2001). *Forces of habit: Drugs and the making of the modern world.* Cambridge, MA: Harvard University Press.

Dahl, J. P., Doyle, G. A., Oslin, D. W., Buono, R. J., Ferraro, T. N., Lohoff, F. W., et al. (2005[a]). Lack of association between single nucleotide polymorphisms in the corticotropin releasing hormone receptor 1 (CRHR1) gene and alcohol dependence. *Journal of Psychiatric Research, 39,* 475–479.

Dahl, J. P., Weller, A. E., Kampman, K. M., Oslin, D. W., Lohoff, F. W., Ferraro, T. N., et al. (2005[b]). Confirmation of the association between a polymorphism in the promoter region of the prodynorphin gene and cocaine dependence. *American Journal of Medical Genetics Part B Neuropsychiatric Genetics, 139B* (1), 106–108.

Drakenberg, K., Nikoshkov, A., Horvath, M. C., Fagergren, P., Gharibyan, A., Saarelainen, K., et al. (2006). Opiod receptor A118G polymorphism in association with striatal opiod neuropeptide gene expression in heroin abusers. *Proceedings of the National Academy of the Sciences USA, 103* (20), 7883–7888.

Drgon, T., D'Addario, C., & Uhl, G. R. (2006). Linkage disequilibrium, haplotype and association studies of a chromosome 4 GABA receptor gene cluster: Candidate gene variants for addictions. *American Journal of Medical Genetics Part B Neuropsychiatric Genetics, 141B,* 854–860.

Duster, T. (2001). The sociology of science and the revolution in molecular biology. In J. R. Blau (Ed.), *The Blackwell companion to sociology* (pp. 213–226). Malden, MA: Blackwell Publishers.

Ehlers, C. L., & Wilhelmsen, K. C. (2005). Genomic scan for alcohol craving in Mission Indians. *Psychiatric Genetics, 15* (1), 71–75.

Fausto-Sterling, A. (2004). Refashioning race: DNA and the politics of health care. *Differences, 15* (3), 1–37.

Fernandez, E., Schiaffino, A., Borrell, C., Benach, J., Ariza, C., Ramon, J. M., et al. (2006). Social class, education, and smoking cessation: Long-term follow-up of patients treated at a smoking cessation unit. *Nicotine and Tobacco Research, 8* (1), 29–36.

Frank, J. W., Moore, R. S., & Ames, G. M. (2000). Historical and cultural roots of drinking problems among American Indians. *American Journal of Public Health, 90* (3), 344–351.

French, L. A., & Hornbuckle, J. (1980, July). Alcoholism among Native Americans: An analysis. *Social Work,* 275–280.

Fullilove, M. T. (1998). Comment: Abandoning "race" as a variable in public health research—an idea whose time has come. *American Journal of Public Health, 88* (9), 1297–1298.

Fullwiley, D. (2008 [this volume]). The molecularization of race: U.S. health institutions, pharmacogenetics practice, and public science after the genome. In B. A. Koenig, S. S.-J. Lee, & S. S. Richardson (Eds.), *Revisiting race in a genomic age* (pp. 149–171). New Brunswick, NJ: Rutgers University Press.

Galvan, F. H., & Caetano, R. (2003). Alcohol use and related problems among ethnic minorities in the United States. *Alcohol Research and Health, 27* (1), 87–94.

Gee, G. C., Delva, J., & Takeuchi, D. T. (2007). Relationships between self-reported unfair treatment and prescription medication use, illicit drug use, and alcohol dependence among Filipino Americans. *American Journal of Public Health, 97* (5), 933–940.

Gelernter, J., Kranzler, H., & Cubells, J. (1999[a]). Genetics of two mu opioid receptor gene (OPRM1) exon I polymorphisms: Population studies and allele frequencies in alcohol- and drug-dependent subjects. *Molecular Psychiatry, 4* (5), 476–483.

Gelernter, J., Kranzler, H., & Satel, S. L. (1999[b]). No association between D-2 dopamine receptor (DRD2) alleles or haplotypes and cocaine dependence or severity of cocaine dependence in European- and African-Americans. *Biological Psychiatry, 45* (3), 340–345.

Gelernter, J., Yu, Y., Weiss, R., Brady, K., Panhuysen, C., Yang, B.-Z., et al. (2006). Haplotype spanning *TTC12* and ANKK1, flanked by the *DRD2* and *NCAM1* loci, is strongly associated to nicotine dependence in two distinct American populations. *Human Molecular Genetics, 15* (24), 3498–3507.

Gerber, J. G., Rhodes, R. J., & Gal, J. (2004). Stereoselective metabolism of methadone N-demethylation by cytochrome P4502B6 and 2C19. *Chirality, 16* (1), 36–44.

Goffman, E. (1963). *Stigma: Notes on the management of spoiled identity.* New York: Simon & Schuster, Inc.

Graham, H., Francis, B., Inskip, H. M., & Harman, J. (2006). Socioeconomic lifecourse influences on women's smoking status in early adulthood. *Journal of Epidemiology and Community Health, 60* (3), 228–233.

Hinrichs, A. L., Wang. J. C., Bufe, B., Kwon, J. M., Budde, J., Allen, R., et al. (2006). Functional variant in a bitter-taste receptor (*hTAS2R16*) influences risk of alcohol dependence. *The American Journal of Human Genetics, 78*, 103–111.

Hoehe, M. R., Kopke, K., Wendel, B., Rohde, K., Flachmeier, C., Kidd, K. K., et al. (2000). Sequence variability and candidate gene analysis in complex disease: Association of mu opioid receptor gene variation with substance dependence. *Human Molecular Genetics, 9* (19), 2895–2908.

Hughes, J. R. (2006). Should criteria for drug dependence differ across drugs? *Addiction, 101*, 134–141.

Institute of Medicine of the National Academies. (2003). *Unequal treatment: Confronting racial and ethnic disparities in health care.* Washington, DC: The National Academies Press.

Jain, S.S.L. (2003). "Come up to the Kool taste": African American upward mobility and the semiotics of smoking menthols. *Public Culture, 15* (2), 295–322.

Kahn, J. (2008 [this volume]). Patenting race in a genomic age. In B. A. Koenig, S. S.-J. Lee, & S. S. Richardson (Eds.), *Revisiting race in a genomic age* (pp. 129–148). New Brunswick, NJ: Rutgers University Press.

Kim, J.-W., Park, C.-S., Hwang, J.-W., Shin, M.-S., Hong, K.-E., Cho, S.-C., et al. (2006). Clinical and genetic characteristics of Korean male alcoholics with and without Attention Deficit Hyperactivity Disorder. *Alcohol & Alcoholism, 41* (4), 407–411.

King, P. A. (2004). Reflections on race and bioethics in the United States. *Health Matrix, 14* (1), 149–153.

Konishi, T., Calvillo, M., Leng, A.-S., Lin, K.-M., & Wan, Y.-J. Y. (2004). Polymorphisms of the dopamine D2 receptor, serotonin transporter, and $GABA_A$ receptor Beta$_3$ subunit genes and alcoholism in Mexican-Americans. *Alcohol, 32*, 45–52.

Koob, G. F., & Moal, M. L. (2005). Plasticity of reward neurocircuitry and the "dark side" of drug addiction. *Nature Neuroscience, 8* (11), 1442–1444.

Krieger, N., & Sidney, S. (1996). Racial discrimination and blood pressure: The CARDIA study of young black and white adults. *American Journal of Public Health, 86* (10), 1370–1378.

Lamarine, R. J. (1988). Alcohol abuse among Native Americans. *Journal of Community Health, 13* (3), 143–155.

Lappalainen, J., Kranzler, H. R., Malison, R., Price, L. H., Van Dyck, C., Rosenheck, R. A., et al. (2002). A functional neuropeptide Y *Leu7* pro polymorphism associated with alcohol dependence in a large population sample from the United States. *Archives of General Psychiatry, 59*, 825–831.

Lee, S. S.-J., Mountain, J., & Koenig, B. A. (2001). The meanings of "race" in the new genomics: Implications for health disparities research. *Yale Journal of Health Policy, Law and Ethics, 1*, 33–76.

Li, M. D., Beuten, J., Ma, J. Z., Payne, T. J., Lou, X. Y., Garcia, V., et al. (2005). Ethnic- and gender-specific association of the nicotinic acetylcholine receptor alpha 4 subunit gene (CHRNA4) with nicotine dependence. *Human Molecular Genetics, 14* (9), 1211–1219.

Liu, Q.-R., Drgon, T., Walther, D., Johnson, C., Poleskya, O., Hess, J., et al. (2005). Pooled association genome scanning: Validation and use to identify addiction vulnerability loci in two samples. *Proceedings of the National Academy of Sciences USA, 102* (33), 11864–11869.

Long, J. C., Knowler, W. C., Hanson, R. L., Robin, R. W., Urbanek, M., Moore, E., et al. (1998). Evidence for genetic linkage to alcohol dependence on chromosomes 4 and 11 from an autosome-wide scan in an American Indian population. *American Journal of Human Genetics, 81*, 216–221.

Lou, X.-Y., Ma, J. Z., Payne, T. J., Beuten, J., Crew, K. M., & Li, M. D. (2006). Gene-based analysis suggests association of the nicotinic acetylcholine receptor Beta1 subunit (*CHRNB1*) and M1 muscarinic acetylcholine receptor (*CHRM1*) with vulnerability for nicotine dependence. *Human Genetics, 120*, 381–389.

Luczak, S. E., Shea, S. H., Hsueh, A. C., Chang, J., Carr, L. G., & Wall, T. L. (2006). ALDH2*2 is associated with a decreased likelihood of alcohol-induced blackouts in Asian American college students. *Journal of Studies on Alcohol, 67* (3): 349–353.

Luczak, S. E., Wall, T. L., Shea, S. H., Byun, S. M., & Carr, L. G. (2001). Binge drinking in Chinese, Korean, and white college students: Genetic and ethnic group differences. *Psychology of Addictive Behaviors, 15* (4), 306–309.

Luo, X., Kranzler, H. R., Zhao, H., & Gelernter, J. (2003). Haplotypes at the OPRM1 locus are associated with susceptibility to substance dependence in European-Americans. *American Journal of Human Genetics, 120B*, 97–108.

Luo, X. G., Kranzler, H. R., Zuo, L. J., Lappalainen, J., Yang, B. Z., & Gelernter, J. (2006[a]). ADH4 gene variation is associated with alcohol dependence and drug dependence in European Americans: Results from HWD tests and case-control association studies. *Neuropsychopharmacology, 31* (5), 1085–1095.

Luo, X., Kranzler, H. R., Zuo, L., Wang, S., Schork, N. J., & Gelernter, J. (2006[b]). Diplotype trend regression analysis of the *ADH* gene cluster and the *ALDH2* gene: Multiple significant associations with alcohol dependence. *The American Journal of Human Genetics, 78*, 973–987.

Ma, J. Z., Beuten, J., Payne, T. J., Dupont, R. T., Elston, R. C., & Li, M. D. (2005). Haplotype analysis indicates an association between the DOPA decarboxylase (DDC) gene and nicotine dependence. *Human Molecular Genetics, 14* (12), 1691–1698.

Mail, P. D. (2002). Multiple perspectives on alcohol and the American Indian. In P. D. Mail, S. Heurtin-Roberts, S. E. Martin, & J. Howard (Eds.), *Alcohol use among American Indians and Alaska natives: Multiple perspectives on a complex problem. NIAAA*

Research Monograph No. 37 (pp. 3–23). Bethesda, MD: U.S. Department of Health and Human Services.

Manica, A., Prugnolle, F., & Balloux, F. (2005). Geography is a better determinant of human genetic differentiation than ethnicity. *Human Genetics, 188,* 366–371.

Merikangas, K. R., & Risch, N. (2003). Genomic priorities and public health. *Science, 302,* 599–601.

Midanik, L., & Room, R. (2005). Contributions of social science to the alcohol field in an era of biomedicalization. *Social Science & Medicine, 60,* 1107–1116.

Mwenifumbo, J. C., Myers, M. G., Wall, T. L., Lin, S. K., Sellers, E. M., & Tyndale, R. F. (2005). Ethnic variation in CYP2A6*7, CYP2A6*8, and CYP2A6*10 as assessed with a novel haplotyping method. *Pharmacogenetics and Genomics, 15* (3), 189–192.

National Institute on Drug Abuse. (2003). *Drug use among racial/ethnic minorities.* Bethesda, MD: U.S. Department of Health and Human Services/National Institutes of Health.

Nestler, E. J. (2005). Is there a common molecular pathway for addiction? *Nature Neuroscience, 8* (11), 1445–1449.

Nichter, M. (2003). Smoking: What does culture have to do with it? *Addiction, 98* (Suppl. 1), 139–145.

Omi, M., & Winant, H. (1994). *Racial formation in the United States: From the 1960s to the 1990s* (2nd ed.). New York: Routledge.

Osier, M., Pakstis, A. J., Kidd, J. R., Lee, J.-F., Yin, S.-J., Ko, H.-C., et al. (1999). Linkage disequilibrium at the ADH2 and ADH3 loci and risk of alcoholism. *American Journal of Human Genetics, 64,* 1147–1157.

Parra, E. J., Marcini, A., Akey, J., Martinson, J., Batzer, M. A., Cooper, R., et al. (1998). Estimating African American admixture proportions by use of population specific alleles. *American Journal of Human Genetics, 63,* 1839–1851.

Pearson, S.-A., Soumerai, S., Mah, C., Zhang, F., Simoni-Wastila, L., Salzman, C., et al. (2006). Racial disparities in access after regulatory surveillance of benzodiazepines. *Archives of Internal Medicine, 166,* 572–579.

Peng, G.-S., Wang, M.-F., Chen, C.-Y., Luu, S.-U., Chou, H.-C., Li, T.-K., et al. (1999). Involvement of acetaldehyde for full protection against alcoholism by homozygosity of the variant allele of mitochondrial aldehyde dehydrogenase gene in Asians. *Pharmacogenetics, 9,* 463–476.

Pierce, J. P., Choi, W. S., Gilpin, E. A., Farkas, A. J., & Berry, C. C. (1998). Tobacco industry promotion of cigarettes and adolescent smoking. *Journal of the American Medical Association, 279* (7), 511–515.

Pierce, J. P., Gilpin, E. A., & Choi, W. S. (1999). Sharing the blame: Smoking experimentation and future smoking-attributable mortality due to Joe Camel and Marlboro advertising and promotions. *Tobacco Control, 8* (1), 37–44.

Piette, J. D., Bibbins-Domingo, K., & Schillinger, D. (2006). Health care discrimination, processes of care, and diabetes patients' health status. *Patient Education and Counseling, 60* (1), 41–48.

Press, N. (2006). Social construction and medicalization: Behavioral genetics in context. In E. Parens, A. R. Chapman, & N. Press (Eds.), *Wrestling with behavioral genetics: Science, ethics, and public conversation* (pp. 131–149). Baltimore, MD: Johns Hopkins University Press.

Proudnikov, D., LaForge, K. S., Hofflich, H., Levenstien, M., Gordon, D., Barral, S., et al. (2006). Association analysis of polymorphisms in serotonin 1B receptor (HTR1B) gene with heroin addiction: A comparison of molecular and statistically estimated haplotypes. *Pharmacogenetics and Genomics, 16* (1), 25–36.

Quinn, S. C. (2004). Protecting human subjects: The role of community advisory boards. *American Journal of Public Health, 94* (6), 918–922.

Radel, M., Vallejo, R. L., Iwata, N., Aragon, R., Long, J. C., Virkkunen, M., et al. (2005). Haplotype-based localization of an alcohol dependence gene to the 5q34 gamma-aminobutyric acid type A gene cluster. *Archives of General Psychiatry, 62*, 47–55.

Reardon, J. (2008 [this volume]). Race without salvation: Beyond the science/society divide in genomic studies of human diversity. In B. A. Koenig, S. S.-J. Lee, & S. S. Richardson (Eds.), *Revisiting race in a genomic age* (pp. 304–319). New Brunswick, NJ: Rutgers University Press.

Rebbeck, T. R., & Sankar, P. (2005). Ethnicity, ancestry, and race in molecular epidemiologic research. *Cancer Epidemiology Biomarkers and Prevention, 14*, 2467–2471.

Risch, N., Burchard, E., Ziv, E., & Tang, H. (2002). Categorization of humans in biomedical research: Genes, race, and disease. *Genome Biology, 3* (7), 2007.1–2007.12.

Roberts, D. (2006). Legal constraints on the use of race in biomedical research: Toward a social justice framework. *Journal of Law, Medicine, and Ethics, 34* (3), 526–534.

Rose, N. (2006). *The politics of life itself.* Princeton, NJ: Princeton University Press.

Rothstein, M. A., & Epps, P. G. (2001). Ethical and legal implications of pharmacogenomics. *Nature, 2*, 228–231.

Sankar, P., Cho, M. K., Condit, C. M., Hunt, L. M., Koenig, B., Marshall, P., et al. (2004). Genetic research and health disparities. *Journal of the American Medical Association, 291* (24), 2985–2989.

Schoedel, K. A., Sellers, E. M., Tyndale, R. F., Hoffmann, E. B., & Rao, Y. (2004). Ethnic variation in CYP2A6 and association of genetically slow nicotine metabolism and smoking in adult Caucasians. *Pharmacogenetics, 14* (9), 615–626.

Sharma, S., Malarcher, A. M., Giles, W. H., & Myers, G. (2004). Racial, ethnic, and socioeconomic disparities in the clustering of cardiovascular disease risk factors. *Ethnicity and Disease, 14* (1), 43–48.

Shi, J., Hui, L., Xu, Y., Wang, F., Huang, W., & Hu, G. (2002). Sequence variations in the mu-opoid receptor gene (OPRM1) associated with human addiction to heroin. *Human Mutation, 19* (4), 459–460.

Shields, A. E., Fortun, M., Hammonds, E. M., King, P. A., Lerman, C., Rapp, R., et al. (2005). The use of race variables in genetic studies of complex traits and the goal of reducing health disparities: A transdisciplinary perspective. *American Psychologist, 60* (1), 77–103.

Spence, J. P., Liang, T., Eriksson, C.J.P., Taylor, R. E., Wall, T. L., Ethers, C. L., et al. (2003). Evaluation of aldehyde dehydrogenase 1 promoter polymorhisms identified in human populations. *Alcoholism: Clinical and Experimental Research, 27* (9), 1389–1390.

Stevens, J. (2008 [this volume]). The feasibility of government oversight for NIH-funded population genetics research. In B. A. Koenig, S. S.-J. Lee, & S. S. Richardson (Eds.), *Revisiting race in a genomic age* (pp. 320–341). New Brunswick, NJ: Rutgers University Press.

Stinson, D. (2006). Tobacco in the Native community. *The Circle: Native American News and Arts, 1*, 10–12.

Tamayo-Sarver, J. H., Hinze, S. W., Cydulka, R. K., & Baker, D. W. (2003). Racial and ethnic disparities in emergency department analgesic prescription. *American Journal of Public Health, 93* (12), 2067–2073.

Tapper, M. (1999). *In the blood: Sickle cell anemia and the politics of race.* Philadelphia: University of Pennsylvania Press.

Tate, S. K., & Goldstein, D. B. (2008 [this volume]). Will tomorrow's medicines work for everyone? In B. A. Koenig, S. S.-J. Lee, & S. S. Richardson (Eds.), *Revisiting race in a genomic age* (pp. 102–128). New Brunswick, NJ: Rutgers University Press.

Templeton, A. R. (1999). Human races: A genetic and evolutionary perspective. *American Anthropologist, 100* (3), 632–650.

Tsai, S.-J., Liao, D.-L., Yu, Y. W.-Y., Chen, T.-J., Wu, H.-C., Lin, C.-H., et al. (2005). A study of the association of (Val66Met) polymorphism in the *Brain-derived Neurotrophic Factor* gene with alcohol dependence and extreme violence in Chinese males. *Neuroscience Letters, 381,* 340–343.

U.S. Census Bureau. (2004). Income stable, poverty up, numbers of Americans with and without health insurance rise, Census Bureau reports. Retrieved April 2006, from http://www.census.gov/Press-Release/www/releases/archives/income_wealth/002484.html.

U.S. Department of Health and Human Services. (2006). The health consequences of involuntary exposure to tobacco smoke: A report of the Surgeon General. Retrieved July 2006, from http://www.surgeongeneral.gov/library/secondhandsmoke/report/.

Uhl, G. R., Liu, Q.-R., Walther, D., Hess, J., & Naiman, D. (2001). Polysubstance abuse-vulnerability genes: Genome scans for association, using 1,004 subjects and 1, 494 single-nucleotide polymorphisms. *American Journal of Human Genetics, 69,* 1290–1300.

Vilschick, J. (2001). Tobacco use: Health threat to Asian and Pacific Islander communities. U.S. Department of Health and Human Services. Retrieved March 2006, from http://www.omhrc.gov/assets/pdf/checked/Tobacco%20Use—Health%20Threat%20to%20Asian%20and%20Pacific%20Islander%20Communities.pdf.

Wachbroit, R. (2006). Normality and the significance of difference. In E. Parens, A. R. Chapman, & N. Press (Eds.), *Wrestling with behavioral genetics: Science, ethics, and public conversation* (pp. 235–253). Baltimore, MD: Johns Hopkins University Press.

Wall, T. L., Carr, L. G., & Ehlers, C. L. (2003). Protective association of genetic variation in alcohol dehydrogenase with alcohol dependence in Native American Mission Indians. *American Journal of Psychiatry, 160* (1), 41–46.

Weijer, C., & Miller, P. B. (2004). Protecting communities in pharmacogenetic and pharmacogenomic research. *Pharmacogenomics Journal, 4* (1), 9–16.

White, V. M., White, M. M., Freeman, K., Gilpin, E. A., & Pierce, J. P. (2006). Cigarette promotional offers: Who takes advantage? *American Journal of Preventive Medicine, 30* (3), 225–231.

Wilhelmsen, K. (2006). Genetics, neurobiology, and nicotine addiction: Scientific opportunities and public policy challenges. Paper presented at American Association for the Advancement of Science, February 2006, St. Louis, MO.

Williams, D. R., Neighbors, H. W., & Jackson, J. S. (2003). Racial/ethnic discrimination and health: Findings from community studies. *American Journal of Public Health, 93* (2), 200–208.

Xu, K., Lichtermann, D., Lipsky, R. H., Franke, P., Liu, X., Hu, Y., et al. (2006). Association of specific haplotypes of D2 dopamine receptor gene with vunerability to heroin dependence in 2 distinct populations. *Archives of General Psychiatry, 61,* 597–598.

Xuei, X., Dick, D., Flury-Wetherill, L., Tian, H.-J., Agrawal, A., Bierut, L., et al. (2006). Association of the κ-opoid system with alcohol dependence. *Molecular Psychiatry, 11,* 1016–1024.

Yokoyama, M., Yokoyama, A., Yokoyama, T., Hamana, G., Funazu, K., Kondo, S., et al. (2003). Mean corpuscular volume and the aldehyde dehydrogenase-2 genotype in male Japanese workers. *Alcoholism: Clinical and Experimental Research, 27* (9), 1395–1397.

Zhang, H., Ye, Y., Wang, X., Gelernter, J., Ma, J. Z., & Li, M. D. (2006). DOPA decarboxylase gene is associated with nicotine dependence. *Pharmacogenetics, 7* (8), 1159–1166.

Zhang, P. W., Ishiguro, H., Ohtsuki, T., Hess, J., Carillo, F., Walther, D., et al. (2004). Human cannabinoid receptor 1: 5 exons, candidate regulatory regions, polymorphisms, haplotypes and association with polysubstance abuse. *Molecular Psychiatry, 9,* 916–931.

Zimatkin, S. M., & Deitrich, R. A. (1997). Ethanol metabolism in the brain. *Addiction Biology, 2* (4), 387–399.

PART THREE

Genetic Ancestry, Identity, and Group Membership

10

Genetic Ancestry and the Search for Personalized Genetic Histories

MARK D. SHRIVER AND RICK A. KITTLES

Public demand and the development of large public and private databases of genetic information across human populations have encouraged the development of the new and rapidly growing field of genetic ancestry testing. Both the promise of the science that underlies this field and a lack of a full understanding of its limitations have fuelled the increased public interest in genetic ancestry testing.

The popularity of ancestry and genealogical research has grown rapidly over the past 10 years. Recently called "America's latest obsession" (Hornblower, 1999, p. 153), genealogical research has become the fastest growing hobby in many communities in the United States. Increasingly, DNA technology is being used to supplement historical documents for researching genealogy. Genetic data in the form of DNA polymorphisms sampled from different human populations are powerful tools for inferring human population history, exploring genealogy, and estimating individual ancestry.

The estimation of personalized genetic histories (PGHs) is not just a pastime. The diverse genetic origins of human populations have created challenges for the medical and social sciences that are also driving interest in understanding and defining PGHs. For example, estimates of PGHs can be used to adjust for population stratification and admixture stratification[1] (Hoggart et al., 2003; Shriver et al., 2003) to deconvolute environmental and genetic effects on complex diseases. Moreover, PGH can have a role in medical risk calculation, admixture mapping[2] (Smith et al., 2004; Patterson et al., 2004; McKeigue, 1998, 2000), forensic investigations (see below), personal interest genomics, and the assessment of ancestry for sociopolitical purposes (for example, adoption record access, affirmative action qualifications, Native American tribal affiliation). For whatever purpose, it is clear that an increasing number of members of New World populations are seeking more information on their Old World ancestries. Perhaps the most prominent example is the desire of many African Americans to identify their ancestral communities (Lee, Mountain, & Koenig, 2001; Baylis, 2003; Rotimi, 2003; see also Nelson, this volume; TallBear, this volume). However, many others are also eager to learn more about their Jewish, European, Asian, African, and/or Native American ancestors (Brown, 2002).

Genetic ancestry testing is seen by some to be controversial (Elliott & Brodwin, 2002; Johnston & Thomas, 2003; Zoloth, 2003; Johnston, 2003). For the most part, these controversies have more to do with the complex history of race, discrimination, and injustice than the science behind PGH. Regardless of these perceived controversies, the wide public interest in PGH and the uses of genetic ancestry tracing highlight the need for an overview of this area. Moreover, the burgeoning number of PGH companies that now offer fee-for-service tests for genetic ancestry (see Greely, this volume) indicates the timeliness of an overview of the issues. Here, we aim to highlight why there is such strong public interest in PGH and discuss some of the misconceptions about this rapidly expanding field as well as its limitations. First, we summarize the two principal analytical approaches being applied for the estimation of PGH. We then discuss the constraints and limitations placed on PGH estimation and the issues that surround its integration into society before concluding with an assessment of the future of the field.

Lineage-Based Analyses

PGH tests can be divided into two types: lineage-based tests, which target mitochondrial DNA (mtDNA) and the non-recombining Y chromosome (NRY), and autosomal marker-based tests, which use ancestry informative markers (AIMs)[3] to estimate biogeographical ancestry (BGA).[4] At present, most PGH companies use lineage-based approaches.

Maternally (mtDNA) and paternally (NRY) inherited DNA have been useful for studying human evolution and genealogical inference (Stumpf & Goldstein, 2001; Cann, Stoneking, & Wilson, 1987; Vigilant, Stoneking, Harpending, Hawkes, & Wilson, 1991). These sources of markers, which are not subject to recombination, define haplotypes[5] that can easily be used to reconstruct maternal and paternal lineages. They also have a lower effective population size[6] than autosomal loci and so are more sensitive to the effects of genetic drift, which makes them better markers of population structure. And mtDNA has the added advantage that it evolves more rapidly than nuclear DNA because it has a higher nucleotide substitution rate.

At present, more than 400 well-characterized polymorphic loci on the NRY can be analyzed using PCR (polymerase chain reaction) (Jobling & Tyler-

Smith, 2003). These include SNPs, Alu insertion/deletion polymorphisms (Hammer, 1994), and microsatellites (Jobling, Heyer, Dieltjes, & de Knijff, 1999; Kayser et al., 2000). Although the informative stretches of mtDNA and NRY represent less than 1% of the entire human genome, they have proved to be powerful tools for identifying and defining maternal and paternal lineages that have improved our understanding of human migrations and solidified our appreciation that, ultimately, all humans share a small group of recent common ancestors.

Maternal and Paternal Lineages

Lineage-based ancestry tests are popular because mtDNA and NRY haplotypes can provide information that is regionally specific (Jobling & Tyler-Smith, 2003; Budowle, Allard, Wilson, & Chakraborty, 2003; Hammer et al., 2001; Hammer et al., 1997; Kayser & Sajantila, 2001; Underhill et al., 2000; Wallace, 1994). For this reason, companies can now offer tests to determine whether an individual has paternal or maternal lineages that originate from Native American, European, African, or Asian populations. The process is straightforward: DNA is extracted from buccal cells that are collected on swabs, and informative genetic markers are sequenced or genotyped. The genetic markers are then compared with a reference database of haplotypes that have been identified in specific populations to search for a match. A detailed discussion of the statistical details that underlie personalized genetic history analyses is beyond the scope of this article. Nonetheless, PGH is basically an application of population genetics, a highly statistical field of science. As such, we believe that most PGH companies have yet to properly address the uncertainties that surround PGH estimates and how these should be presented to their current and potential customers. For example, even a perfect match between a customer and a sample in the reference database is not a straightforward result. A tempting conclusion to reach from such a match would be that the customer shares an ancestor with the reference sample donor in the past few generations. However, such a conclusion is much too strong: many more markers than are typically used (or even available) are needed to estimate the maximum number of past generations in which two matching mtDNA or non-recombining Y chromosome haplotypes could share an ancestor (Walsh, 2001).

The accuracy of lineage matching depends on the size and sampling of the database that is used to match mtDNA or Y-chromosome lineages to particular populations or geographical regions. The level of geographical resolution depends on both the sampled haplotype and the populations included in the database (Stoneking & Soodyall, 1996). Many databases that are derived from published research are too small and lack samples in certain geographical regions. Nonetheless, the high regional specificity of mtDNA and Y-chromosome haplotypes means that some less common lineages can be traced to particular ethnic groups or locales (Stoneking & Soodyall, 1996; Soodyall, Nebel, Morar, & Jenkins, 2003). However, tracing the more common haplotypes to a particular location is problematic.

Surname Matching

NRY lineages are correlated with surnames (Sykes & Irven, 2000; Jobling, 2001), so these lineages can be compared among individuals with the same or similar surnames to identify previously unconnected relatives. The potential to do such surname matching has led to an increased interest in establishing PGH databases for this purpose. For many of these databases, identifiers linked to each surname are used to allow the person who submitted the surname and genetic information to be contacted anonymously. For many individuals, these databases can provide information about specific ancestors to help them get past "brick walls" in their genealogies.

The standard surname-matching procedure for PGH companies is to compare markers on the Y chromosome of two or more men to determine relatedness and, if possible, an estimate of when their most recent common ancestor lived (Stumpf & Goldstein, 2001; Walsh, 2001). However, there is no consensus among these companies about whether people in surname databases who share the same lineage can be contacted. The few publicly accessible databases maintain strict confidentiality of participants' information when linking individuals together in genetic genealogies, whereas some private surname-matching databases are not so well monitored.

Biogeographical Ancestry Analyses

One major disadvantage of PGH estimates that use lineage-based analyses is that they focus on a single maternal or paternal lineage and therefore neglect the contribution of the vast majority of an individual's ancestors to their genome (see below). One solution to this problem is to use many autosomal genetic markers that distinguish among ancestral groups. Ancestry informative markers (AIMs; also known as population-specific alleles) (Parra et al., 1998; Shriver et al., 1997), ethnic difference markers (Collins-Schramm et al., 2002), and mapping by admixture linkage disequilibrium markers (Smith et al., 2001; Dean et al., 1994)) are autosomal genetic markers that show substantial differences in allele frequency across population groups. These groups can range from relatively local clusters (for example, Southern European/Northern European) to larger continental distinctions (for example, African/non-African). Allele-frequency databases that catalogue AIMs have lagged far behind other genomic databases. Nonetheless, there has been a tenacious set of researchers that has searched for and developed AIMs (Collins-Schramm et al., 2002; Smith et al., 2001; Dean et al., 1994; Pfaff et al., 2001). The main impetus for the development of these AIM panels has been the promise of admixture mapping (Li & Nei, 1974; Chakraborty & Weiss, 1988).

Biogeographical ancestry is an expression coined to describe the aspects of PGH that can be computed using the ancestry information contained in AIMs (Shriver et al., 2003). Factors such as isolation by distance, range expansions, land bridges, maritime technologies, ice ages, and cultural and linguistic barriers have all affected human migration and mating patterns in the past and have therefore shaped the present worldwide distribution of genetic variation (Templeton, 2002). BGA in the broad sense is the quantitative representation of the effects of these factors. Specifically, BGA aims to estimate a person's ancestry in terms of the proportional representation of AIMs from a selection of ancestral populations. PGH analyses are based on statistical calculations whereby confidence intervals can and should be estimated to represent their precision or lack thereof. Clear calculation and presentation of the statistics and confidences of PGH measures is an important challenge that we feel PGH companies need to address.

In this example (fig. 10-1) of biogeographical ancestry estimates for five people from one family (see pedigree in inset), a database of 176 ancestry informative markers (AIMs) is used to calculate four possible three-way ancestry models with respect to the four parental populations included: West African (WA), Western European (WE), East Asian (EA) and Native American (NA). These four models (WA/WE/EA, WA/WE/NA, WA/EA/NA, WE/EA/NA) are calculated separately and then recombined into an unfolded tetrahedron projection in which each of the four surfaces (inner triangles) is one of the models. The dot indicates the most probable position for the multilocus AIM genotype of the person in question. To present information on the precision of the result, concentric rings delineate probability spaces relative to the maximum-likelihood estimate (2X, 5X, and 10X). For example, all of the values within the first (smallest) confidence ring are less than two times less likely than the maximum. These two-dimensional representations can be cut out and folded into three-dimensional tetrahedrons (four-sided pyramids).

The analysis illustrated correctly shows that the father is primarily of European decent, whereas the mother has substantial Native American ancestry and smaller amounts of Western European and West African ancestry. As we would expect, the three children are intermediate between the two parents and similar to one another. Also evident is some level of variability among siblings that is expected in recently admixed people as a

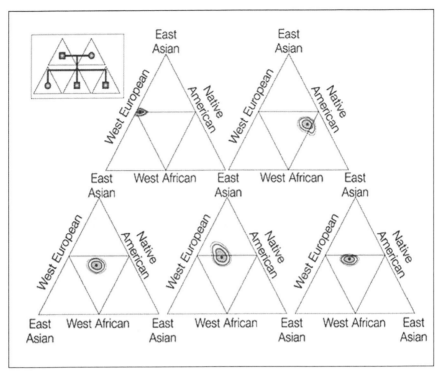

FIGURE 10-1. An example of biogeographical ancestry estimation.
Figure courtesy of T. Frudakis, DNA Print Genomics.

result of the independent assortment of large non-recombined chromosomal segments showing consistent ancestry.

One notable practical application of BGA methods is in forensic investigations in which ancestry estimates might provide police with vital clues and help to direct investigations. A recent example from the United States highlights the usefulness of these methods. By May 2003, the murders of five women in Louisiana had been linked through the analysis of forensic short tandem repeat marker panels; however, there were no hits when compared against the national Combined DNA Index System (CODIS) database of convicted felons. For months, on the basis of eyewitness accounts and psychological profiles, the police had restricted their investigations to white men, screening more than six hundred men in various DNA-based dragnets. Late in the investigation, the state police sought the assistance of a company, the services of which include commercially available PGH tests. The results of these tests revealed that the perpetrator was mainly of West African descent. This information caused the police to reconsider their focus: they broadened their investigation to include non-white suspects. Previously overlooked leads that indicated that an African American man might be involved were re-examined. This change in direction ultimately led to new evidence and the arrest of a suspect whose STR profiles matched those of the perpetrator.

Constraints and Limitations

Three main factors determine the accuracy and precision of PGH estimates: the number of markers used, the quality of the reference databases (that is, the geographical spread and number of samples included), and the levels of genetic differentiation among populations in the region(s) being considered.

Markers and References

In general, the more markers used, the more reliable are the PGH estimates. The decreasing expense of genotyping and sequencing technologies has allowed many PGH companies to increase the number of markers that they use. However, the addition of new markers also means that the database of reference samples should be typed for these new markers. Ideally, a standard set of genetic markers would be used to allow comparisons with data from the scientific literature and other public and private databases. To some extent, this is already happening: there is certainly significant overlap in markers that different companies use.

Lineage-based PGH companies use reference databases that range in size from a few thousand to tens of thousands of haplotypes. Most companies do not share database information. However, there have been several partnerships built around reference databases for specific populations, such as West Africans and Native Americans. Nonetheless, these partnerships are few and there is definitely room for increased sharing of reference-database information among companies and between the companies and their customers. In particular, most PGH companies do not provide details about the number and geographical spread of samples included. So, in general, it is difficult to assess exactly what limitations the quality of these databases put on PGH estimation.

Lineage-Based Analyses

Lineage-based products (mtDNA and NRY) provide a high level of ancestry information due to the lack of recombination, high mutation rates, and more restricted geographical distributions compared with autosomal markers. However, as previously mentioned, such tests provide information about only one line of descent (that is, one ancestor per generation) out of many that contributed to the present genetic makeup of an individual. For example, if you go back 14 generations (approximately 350 years), each person has a maximum of 16,384 direct ancestors (the actual number of ancestors might be much smaller as this calculation assumes infinite population size and non-consanguineous mating).

For many customers of lineage-based tests, there is a lack of understanding that their maternal and paternal lineages do not necessarily represent their entire genetic makeup. For example, an individual might have more than 85% western European "genomic" ancestry but still have a West African mtDNA or NRY lineage. In addition, lineage-based tests performed within a single family might reveal multiple lineages from diverse ethnic groups. Several lineage-based PGH companies now use professional genealogists to help to identify living relatives with different maternal/paternal lineages that can provide information about ancestors who have contributed autosomal DNA (but not mtDNA or NRY) to the present genetic makeup of a customer (see Woodtor, 1999; Winston & Kittles, 2004, for a discussion of traditional methods of tracing family history). However, we suggest that these companies need to put more effort into disclosing the types of results that a potential customer could expect to receive from such approaches and the statistical precision and power of any conclusions.

Biogeographical Ancestry Analyses

Similar to lineage-based analyses, BGA-based PGH estimates need to be considered in the context of the different potential sources of the genetic variants assayed. In particular, the less restricted geographical distributions of autosomal markers, relative to the sex-linked markers that are used for lineage-based analyses, mean that many potential ancestral patterns will be consistent with any given result. For example, a person might show 75% West African and 25% western European ancestry in a BGA estimate because three grandparents are from West Africa and one is from western Europe or because all four grandparents are of East African ancestry. Similarly, both recent and ancient admixture events can lead to intermediate BGA estimates in such analyses. For this reason, such results should be carefully and clearly presented and interpreted for customers, as should the parental population groups that provide the reference points for such analyses (see previous example). In particular, PGH providers should not misrepresent how informative such tests can be.

PGH and Society

PGH has the potential to affect ethnic, religious, racial, and family identities. Many of these types of identity overlap and might potentially conflict when individuals attempt to reconcile their social identities with their genetic histories. In particular, non-paternity, adoption, and multiple marriages are factors that can disengage PGH from social identity. However, these issues are not new, nor are they unique to PGH: traditional methods of estimating ancestry and identity also have to deal with the same problems.

One potential negative consequence of PGH testing is that the genetically defined ancestral categories that PGH companies use could be misinterpreted as indications of "real" racial divisions, even if they are explicitly acknowledged as being continuous and, to some extent, arbitrary groups (see Bamshad, Wooding, Salisbury, & Stephens, 2004). Underlying such misconceptions is a strong undercurrent of *genetic determinism* that is found in the public in general and is shared by some who work in the field and even some of the detractors of PGH. The challenge for PGH companies is to perform and present PGH tests in a way that educates the public and dispels such misconceptions.

The importance of, and interest in, PGH analyses varies among groups and individuals according to the values of the group and the individual. For example, the African component of the ancestry of African Americans has, in general, socially defined an individual in this group, regardless of the number of non-African ancestors that they might have (the one-drop rule). As a result, some African Americans want to use PGH analyses to reclaim their non-African ancestry. As this example illustrates, for some groups, PGH can be valued as another form of evidence that can be used to reclaim history, culture, and knowledge that is denied to individuals (Rotimi, 2003; Brown, 2002; Burroughs, 2001; Woodtor, 1999).

Although PGH is challenging, controversial, and has many limitations, it is important. PGH analyses will almost certainly cause some people to change the way that they think about themselves and their identities. However, it is probable that PGH tests will be used to supplement rather than supplant traditional genealogical tracing methods. The use of PGH to encourage a better understanding of family history has the potential to inform individuals about important issues such as hereditary diseases as well as to help to educate the public about the extent and meaning of genetic variation and human genetics in general.

Although in many ways the genetic analyses that PGH companies provide are just a supplement to traditional methods, these companies must be aware of the ethical and psychological implications of the services that they offer. Specifically, if individuals regard the meaning and significance of their genetic ancestry highly, the results of PGH tests could deeply affect them (Winston & Kittles, 2004). So, it might be important for PGH providers to provide useful psychological resources, such as access to a clinical psychologist who specializes in identity and psychological well-being counseling. In addition to the obligations that PGH companies might have to the psychological well-being of their customers, PGH testing raises several other ethical issues that have been addressed in detail elsewhere (Lee et al., 2001; Elliott & Brodwin, 2002; Zoloth, 2003; Winston & Kittles, 2004).

The Future of PGH

There are already many high-profile examples of studies of genetic history. These include studies that show that many southern African Lemba people share ancestry with Jewish priests (Thomas et al., 2000), that some European populations have multiple origins (Kittles et al., 1998; Semino et al., 2004; Semino et al., 2000), that Thomas Jefferson, or one of his male relatives, fathered offspring with Sally Hemmings, his enslaved servant (Foster et al., 1998), and that some African Americans are highly admixed with Europeans and differ in population substructure (Parra, et al., 1998; Parra et al., 2001). The appeal of such studies to many people who are interested in knowing more about their ancestry means that, in the future, PGH analyses might increasingly influence how individuals and groups define themselves. However, the inherent limitations in the informativeness of various databases might compromise future efforts to achieve optimal reliability and specificity in these analyses. Specifically, improvements to PGH tests will require larger databases of individuals for parental frequency estimates and lineage comparisons, larger panels of AIMs, and further development of analytical methods and models, such as estimating time since admixture (Patterson et al., 2004; Hoggart, Shriver, Kittles, Clayton, & McKeigue, 2004). In addition to these technical improvements, PGH companies must be prepared to meet the highest possible standards in communicating with their customers. The accuracy, meaning, and applications of PGH tests and their results must be clearly explained to these clients.

How do we ensure that PGH providers meet acceptable standards in practice? We suggest that organizations such as the American Association of Blood Banks for Parentage Testing and the National Forensic Science and Technology Center (International Organization for Standardization [ISO] accreditation) should jointly develop a code of ethical conduct for the PGH industry and should be responsible for accreditation of PGH laboratories. It is probable that this embryonic industry will continue to raise new ethical, social, and psychological issues in the future. However, the immediate challenge for PGH providers is to clearly articulate to the public the promise and the limitations of the science.

In summary, the strong public interest in genetic ancestry is grounded in historical and contemporary experiences. The identification of informative DNA markers in human populations indicates that PGH analyses will be able to address some of the many questions that individuals have about their ancestry. However, the most reliable method for inferring ancestry is to combine DNA evidence with other forms of genealogical and historical knowledge (that is, family and court documents). We recommend this mosaic approach to inferring ancestry as it clearly recognizes that identity is not simply that which is ascribed, based on phenotype (skin color and facial features) or quasi-genetics (the one-drop rule), but is multi-determined.

NOTES

This chapter was previously published as M. Shriver and R. Kittles (2004), "Genetic ancestry and the search for personalized genetic histories," *Nature Reviews Genetics, 5* (8), 611–618.

1. Population stratification refers to a population that contains subpopulations that differ in marker allele frequencies owing to nonrandom mating, finite population size, and/or geographical barriers. Admixture stratification is heterogeneity in a population that is composed of recent admixture of different ethnic groups that differ in marker allele frequencies.

2. Admixture mapping is the mapping of genes for traits or diseases that have different genetic risks in two or more populations that have admixed recently to form a third hybrid population: mapping by admixture linkage equilibrium.

3. Ancestry informative markers are genetic markers that show substantial differences in allele frequency across population groups.

4. Biogeographical ancestry refers to the aspects of personalized genetic histories that can be computed using the ancestry information that is contained in AIMs.

5. A haplotype is a specific segment of DNA sequence that is inherited as a unit.

6. The effective population size is the size of the ideal population in which the effects of random drift would be the same as observed in the actual population.

REFERENCES

Bamshad, M., Wooding, S., Salisbury, B. A., & Stephens, J. C. (2004). Deconstructing the relationship between genetics and race. *Nature Reviews Genetics, 5,* 598–609.

Baylis, F. (2003). Black as me: Narrative identity. *Developing World Bioethics, 3,* 142–150.

Brown, K. (2002). Tangled roots? Genetics meets genealogy. *Science, 295,* 1634–1635.

Budowle, B., Allard, M. W., Wilson, M. R., & Chakraborty, R. (2003). Forensics and mitochondrial DNA: Applications, debates, and foundations. *Annual Review of Genomics and Human Genetics, 4,* 119–141.

Burroughs, T. (2001). *Black roots: A beginner's guide to tracing the African American family tree.* New York: Fireside.

Cann, R. L., Stoneking, M., & Wilson, A. C. (1987). Mitochondrial DNA and human evolution. *Nature, 325,* 31–36.

Chakraborty, R., & Weiss, K. M. (1988). Admixture as a tool for finding linked genes and detecting that difference from allelic association between loci. *Proceedings of the National Academy of the Sciences USA, 85,* 9119–9123.

Collins-Schramm, H. E., Phillips, C. M., Operario, D. J., Lee, J. S., Weber, J. L., Hanson, R. L., et al. (2002). Ethnic-difference markers for use in mapping by admixture linkage disequilibrium. *American Journal of Human Genetics, 70,* 737–750.

Dean, M., Stephens, J. C., Winkler, C., Lomb, D. A., Ramsburg, M., Boaze, R., et al. (1994). Polymorphic admixture typing in human ethnic populations. *American Journal of Human Genetics, 55,* 788–808.

Elliott, C., & Brodwin, P. (2002). Identity and genetic ancestry tracing. *British Medical Journal, 325,* 1469–1471.

Foster, E. A., Jobling, M. A., Taylor, P. G., Donnelly, P., de Knijff, P., Mieremet, R., et al. (1998). Jefferson fathered slave's last child. *Nature, 396,* 27–28.

Greely, H. T. (2008 [this volume]). Genetic genealogy: Genetics meets the marketplace. In B. A. Koenig, S. S.-J. Lee, & S. S. Richardson (Eds.), *Revisiting race in a genomic age* (pp. 215–234). New Brunswick, NJ: Rutgers University Press.

Hammer, M. F. (1994). A recent insertion of an *alu* element on the Y chromosome is a useful marker for human population studies. *Molecular Biology and Evolution, 11,* 749–761.

Hammer, M. F., Karafet, T. M., Redd, A. J., Jarjanazi, H., Santachiara-Benerecetti, S., Soodyall, H., et al. (2001). Hierarchical patterns of global human Y-chromosome diversity. *Molecular Biology and Evolution, 18,* 1189–1203.

Hammer, M. F., Spurdle, A. B., Karafet, T., Bonner, M. R., Wood, E. T., Novalletto, A., et al. (1997). The geographic distribution of human Y chromosome variation. *Genetics, 145,* 787–805.

Hoggart, C. J., Parra, E. J., Shriver, M. D., Bonilla, C., Kittles, R. A., Clayton, D. G., et al. (2003). Control of confounding of genetic associations in stratified populations. *American Journal of Human Genetics, 72,* 1492–1504.

Hoggart, C. J., Shriver, M. D., Kittles, R. A., Clayton, D. G., & McKeigue, P. M. (2004). Design and analysis of admixture mapping studies. *American Journal of Human Genetics, 74,* 965–978.

Hornblower, M. (1999, April 19). Roots mania. *Time,* 153.

Jobling, M. A. (2001). In the name of the father: Surnames and genetics. *Trends in Genetics, 17,* 353–357.

Jobling, M. A., Heyer, E., Dieltjes, P. & de Knijff, P. (1999). Y-chromosome-specific microsatellite mutation rates re-examined using a minisatellite, MSY1. *Human Molecular Genetics, 8,* 2117–2120.

Jobling, M. A., & Tyler-Smith, C. (2003). The human Y-chromosome: An evolutionary marker comes of age. *Nature Reviews Genetics, 4,* 598–612.

Johnston, J. (2003). Resisting a genetic identity: The black Seminoles and genetic tests of ancestry. *Journal of Law, Medicine, and Ethics, 31,* 262–271.

Johnston, J., & Thomas, M. (2003). Summary: The science of genealogy by genetics. *Developing World Bioethics, 3,* 103–108.

Kayser, M., Roewer, L., Hedman, M., Henke, L., Henke, J., Brauer, S., et al. (2000). Characteristics and frequency of germline mutations at microsatellite loci from the human Y chromosome, as revealed by direct observation in father/son pairs. *American Journal of Human Genetics, 66,* 1580–1588.

Kayser, M., & Sajantila, A. (2001). Mutations at Y-STR loci: Implications for paternity testing and forensic analysis. *Forensic Science International, 118,* 116–121.

Kittles, R. A., Perola, M., Peltonen, L., Bergen, A. W., Aragon, R. A., Virkkunen, M., et al. (1998). Dual origins of Finns revealed by Y chromosome haplotype variation. *American Journal of Human Genetics, 62,* 1171–1179.

Lee, S. S.-J., Mountain, J., & Koenig, B. A. (2001). The meanings of "race" in the new genomics: Implications for health disparities research. *Yale Journal of Health Policy, Law, and Ethics, 1,* 33–75.

Li, W. H., & Nei, M. (1974). Stable linkage disequilibrium without epistasis in subdivided populations. *Theoretical Population Biology, 6*, 173–183.

McKeigue, P. M. (1998). Mapping genes that underlie ethnic differences in disease risk: Methods for detecting linkage in admixed populations, by conditioning on parental admixture. *American Journal of Human Genetics, 63*, 241–251.

McKeigue, P. M. (2000). Multipoint admixture mapping. *Genetic Epidemiology, 19*, 464–467.

Nelson, A. (2008 [this volume]). The factness of diaspora: The social sources of genetic genealogy. In B. A. Koenig, S. S.-J. Lee, & S. S. Richardson (Eds.), *Revisiting race in a genomic age* (pp. 253–268). New Brunswick, NJ: Rutgers University Press.

Parra, E. J., Kittles, R. A., Argyropoulos, G., Pfaff, C. L., Hiester, K., Bonilla, C., et al. (2001). Ancestral proportions and admixture dynamics in geographically defined African Americans living in South Carolina. *American Journal of Physical Anthropology, 114*, 18–29.

Parra, E. J., Marcini, A., Akey, J., Martinson, J., Batzer, M. A., Cooper, R., et al. (1998). Estimating African American admixture proportions by use of population-specific alleles. *American Journal of Human Genetics, 63*, 1839–1851.

Patterson, N., Hattangadi, N., Lane, B., Lohmueller, K. E., Hafler, D. A., Oksenberg, J. R., et al. (2004). Methods for high-density admixture mapping of disease genes. *American Journal of Human Genetics, 74*, 979–1000.

Pfaff, C. L., Parra, E. J., Bonilla, C., Hiester, K., Mckeigue, P. M., Kamboh, M. I., et al. (2001). Population structure in admixed populations: Effect of admixture dynamics on the pattern of linkage disequilibrium. *American Journal of Human Genetics, 68*, 198–207.

Rotimi, C. N. (2003). Genetic ancestry tracing and the African identity: A double-edged sword? *Developing World Bioethics, 3*, 151–158.

Semino, O., Magri, C., Benuzzi, G., Lin, A. A., Al-Zahery, N., Battaglia, V., et al. (2004). Origin, diffusion, and differentiation of Y-chromosome *Haplogroups* E and J: Inferences on the neolithization of Europe and later migratory events in the Mediterranean area. *American Journal of Human Genetics, 74*, 1023–1034.

Semino, O., Passarino, G., Oefner, P. J., Lin, A. A., Arbuzova, S., Beckman, L., et al. (2000). The genetic legacy of Paleolithic *Homo sapiens sapiens* in extant Europeans: A Y chromosome perspective. *Science, 290*, 1155–1159.

Shriver, M. D., Parra, E. J., Dios, S., Bonilla, C., Norton, H., Jovel, C., et al. (2003). Skin pigmentation, biogeographical ancestry and admixture mapping. *Human Genetics, 112*, 387–399.

Shriver, M. D., Smith, M. W., Jin, L., Marcini, A., Akey, J. M., Deka, R., et al. (1997). Ethnic-affiliation estimation by use of population-specific DNA markers. *American Journal of Human Genetics, 60*, 957–964.

Smith, M. W., Lautenberger, J. A., Shin, H. D., Chretien, J. P., Shrestha, S., Gilbert, D. A., et al. (2001). Markers for mapping by admixture linkage disequilibrium in African American and Hispanic populations. *American Journal of Human Genetics, 69*, 1080–1094.

Smith, M. W., Patterson, N., Lautenberger, J. A., Truelove, A. L., McDonald, G. J., Waliszewska, A., et al. (2004). A high-density admixture map for disease gene discovery in African Americans. *American Journal of Human Genetics, 74*, 1001–1013.

Soodyall, H., Nebel, A., Morar, B., & Jenkins, T. (2003). Genealogy and genes: Tracing the founding fathers of Tristan da Cunha. *European Journal of Human Genetics, 11*, 705–709.

Stoneking, M., & Soodyall, H. (1996). Human evolution and the mitochondrial genome. *Current Opinions in Genetics and Development, 6,* 731–736.

Stumpf, M. P., & Goldstein, D. B. (2001). Geneological and evolutionary inference with the human Y chromosome. *Science, 291,* 1738–1742.

Sykes, B., & Irven, C. (2000). Surnames and the Y chromosome. *American Journal of Human Genetics, 66,* 1417–1419.

TallBear, K. (2008 [this volume]). Native-American-DNA.com: In search of Native American race and tribe. In B. A. Koenig, S. S.-J. Lee, & S. S. Richardson (Eds.), *Revisiting race in a genomic age* (pp. 235–252). New Brunswick, NJ: Rutgers University Press.

Templeton, A. (2002). Out of Africa again and again. *Nature, 416,* 45–51.

Thomas, M. G., Parfitt, T., Weiss, D. A., Skorecki, K., Wilson, J. F., le Roux, M., et al. (2000). Y chromosomes traveling south: The cohen modal haplotype and the origins of the Lemba—the "Black Jews of Southern Africa." *American Journal of Human Genetics, 66,* 674–686.

Underhill, P. A., Shen, P., Lin, A. A., Jin, L., Passarino, G., Yang, W. H., et al. (2000). Y chromosome sequence variation and the history of human populations. *Nature Genetics, 26,* 358–361.

Vigilant, L., Stoneking, M., Harpending, H., Hawkes, K., & Wilson, A. C. (1991). African populations and the evolution of human mitochondrial DNA. *Science, 253,* 1503–1507.

Wallace, D. C. (1994). Mitochondrial DNA sequence variation in human evolution and disease. *Proceedings of the National Academy of the Sciences USA, 91,* 8739–8746.

Walsh, B. (2001). Estimating the time to the most recent common ancestor for the Y chromosome or mitochondrial DNA for a pair of individuals. *Genetics, 158,* 897–912.

Winston, C. E., & Kittles, R. A. (2005). Psychological and ethical issues related to identity and inferring ancestry of African Americans. In T. R. Turner (Ed.), *Biological anthropology and ethics: From repatriation to genetic identity* (pp. 209–230). Albany: State University of New York Press.

Woodtor, D. P. (1999). *Finding a place called home: A guide to African American genealogy and historical identity.* New York: Random House.

Zoloth, L. (2003). Yearning for the long lost home: L the Lemba and Jewish narrative of genetic return. *Developing World Bioethics, 3,* 128–132.

11

Genetic Genealogy

Genetics Meets the Marketplace

HENRY T. GREELY

The business of genetic genealogy is more than a possibility; it is a reality. At least sixteen different companies and one nonprofit organization are already offering genetic genealogical services to the public. One Web site boasts, "Genealogy by genetics is the greatest addition to Genealogy since the creation of the Family Tree!" (Family Tree DNA homepage, 2006). But commercial success involves selling, and selling often involves uneasy compromises for scientists. As those involved in the genetics revolution consider ways in which genomics may create consumer markets outside of medicine, genetic genealogy is a major potential market—and one that raises some serious concerns, which have received only limited attention (Shriver & Kittles, 2004; Brown, 2002). This chapter reports the results of a close examination of companies in this business, based on a survey of their Web sites. It first describes the services they offer and then discusses six concerns about what they provide—and how what they provide may differ from what customers think they are buying.

In the best published discussion thus far of genetic genealogy, two of the leading scientists in the area, Rick Kittles and Mark Shriver (each of whom is associated with a successful genetic genealogy company) discuss some limitations and ethical problems of this kind of work. (Shriver & Kittles, this volume). To their credit, they called for an ethical code for those in what they call the "personalized genetic history" industry. No code is currently followed, and a review of the existing company Web sites shows that most firms fall short even of their recommendations. And I believe their recommendations fall short of answering the problems raised by the firms' practices.

This chapter analyzes the current generation of genetic genealogy companies. It first describes the firms, the kinds of tests they offer, and the ways they suggest their customers can use them. It then discusses problems raised by this business. One set of problems involves race and is similar to—or identical with—some of those raised generally by the association of race with genetics. A second kind of problem concerns genealogy and how well the products these firms provide correspond to what interests, or perhaps should interest, genealogists. A third, and large, group of problems stems from the ways in which these firms offer "scientific" answers without the full disclosure of the work's limitations and risks that is the accepted norm in science and in medicine. Those problems may provide a preview of the broader tensions that may be expected from the anticipated field of *consumer genomics*—tensions caused by the disparity between the ethics of the marketplace and the ethics of science.

Genetic Genealogy Companies and Their Tests

As summarized in table 11.1, as of February 2006, at least fifteen firms offer consumers genetic genealogy services: African Ancestry, AncestryByDNA, DNA Consulting, DNA Heritage, Family Tree DNA, GeneBase, Gene Tree, Genelex, GeoGene, Niagen, Orchid Cellmark, Oxford Ancestors, Relative Genetics, Roots for Real, and Trace Genetics. (Some of these firms are jointly owned by other firms.) In addition, the "Genographic Project" of the National Geographic Society offers some DNA-based information about ancestry to donors to the project.

Each group analyzes a cheek-swab or saliva DNA sample sent by a paying consumer and to provide the consumer with some information of genealogical interest, usually in about six weeks. The prices vary, but most of the firms offer their mid-range products for about $100 to $300. Most of the firms are affiliated with genetics researchers, many of them professors at research universities. Fifteen of the sixteen groups analyze markers on either mitochondrial DNA (mtDNA) or the Y chromosome (or, for a customer willing to pay more, both). The sixteenth, AncestryByDNA, contracts with DNAPrint Genomics, Inc., which uses a different method, one that looks at single nucleotide polymorphisms (SNPs) throughout the nuclear DNA. At least four of the other firms also offer the DNAPrint service in addition to the mtDNA or Y chromosome tests. (The AncestryByDNA Web site says that it offers mtDNA and Y chromosome testing, but it contains no further information about them and provides no prices or way to order them.)

Mitochondrial and Y-Chromosome DNA Analysis
DNA segments from the mitochondria (organelles found inside each cell) and from most of the Y chromosome (found in the nuclei of cells of men only) have special characteristics that make them especially interesting for

genealogy. These stretches of DNA exist within one person in only one form, they do not undergo recombination, and they descend to the next generation from only one parent. The rest of a person's DNA generally comes in paired chromosomes, 22 pairs named 1 through 22 and, in women, the X chromosome. As sperm are created and eggs prepare for ovulation, these paired chromosomes swap stretches of DNA through what is called recombination. As a result, even though a person has one copy of, say, chromosome 12 from her mother and one from her father, the chromosome from each parent is a combination of the chromosome 12 from that parent's parents. (Men have only one X chromosome, but any child they pass it down to is a daughter, who has two X chromosomes, which will go through recombination in her eggs). Finally, the Y chromosome passes only from father to son (a person who has a Y chromosome normally *must* be male); the mitochondrial DNA in the zygote (and hence in the embryo, fetus, and adult) comes only from the egg and hence from the mother. Because of the process of recombination, DNA from the rest of our chromosomes is a mixture of some, but not all, sequences from all of our ancestors. A piece of distinctive sequence on almost any of a parent's chromosomes has a 50% chance of disappearing in any child, replaced by the sequence on the other copy of that chromosome. But a DNA sequence or pattern of sequences found at an unusual rate in people from a particular region or ethnic group will be preserved in the direct maternal line in mitochondrial DNA and in the direct paternal line in DNA from the non-recombining portion of the Y chromosome. This preservation of any special characteristic allows genetic genealogy firms to attempt to tie that DNA, and hence that ancestral line, to a particular geographical region or population. The Y chromosome also has the advantage that it is tightly associated in many cultures with family names, and thus it provides a link to the history of one ancestral line of particular interest to many genealogists— the last name they were given at birth.

All of the genetic genealogy firms except AncestryByDNA offer commercial analysis of markers found in mitochondrial or Y chromosome DNA. (Female customers who want to test their paternal ancestry need to submit a sample from their brothers, fathers, or others in their direct male line, as they have no Y chromosome themselves.) DNA Heritage (DNA Heritage Y, 2006) and GeneBase (GeneBase Y, 2006) offer only Y chromosome testing; Roots for Real offers only mtDNA testing (Roots for Real mtDNA, 2006). The others will test either or both mitochondrial and Y chromosome DNA. The firms generally test the one or both hypervariable regions of the mitochondrial DNA and a different number of Y chromosome markers—stretches of DNA that are known to differ among humans fairly often, also known as polymorphisms. The number of Y chromosome markers tested ranges from 9 to 46.

TABLE 11.1

Genetic Genealogy Companies

African Ancestry

    Y chromosome, 9 markers $349

    mtDNA, 400 base pairs $349 ($590 for both)

        Trace ancestry to find out where in Africa your family originated

AncestryByDNA (sometimes referred to as DNAPrint)

    full genome SNP analysis $219 (using DNAPrint)

    Euro-DNA SNP analysis $399 (using DNAPrint)

        Better understanding of heritage

DNA Consulting

    Y chromosome, 13 or 37 markers $200 or $310 (by Sorenson)

    mtDNA, one or two HPVs $300 or $400 (by Sorenson)

    whole genome SNP analysis $298 (by DNAPrint)

    Native American and other ethnic tests various prices

        Ancestral homeland and ethnicity

DNA Heritage

    Y chromosome, 23 to 43 markers $137–$199 (by Sorenson)

    Y SNP $99 (by Marligen Biosciences)

        Family searches with others, surname projects, database

Family Tree DNA

    Y chromosome, 12 or more markers $159–$289

    mtDNA $129–$189 (both for $229 up)

    Native American testing   (no price given)

        Family searches with others; surname projects, ancestral homeland, database search

GeneBase

    Y chromosome, 20 or 44 markers $119 or $199

        Broad genealogical company, surname searches

Gene Tree (owned by Sorenson Genomics)

    Y chromosome (by Relative Genetics) $245

    mtDNA (by Relative Genetics) $245

    whole genome SNP analysis $219 (through DNAPrint)

    Native American, mtDNA or Y $245 each (through Trace Genetics)

        Primarily paternity testing, but also offers genealogy

TABLE 11.1 *(Continued)*
Genetic Genealogy Companies

Genelex

    whole genome SNP analysis $395 (through DNAPrint)

    Native American, mtDNA or Y $395 each

Genographic Project

    Y chromosome $99.95 (through Family Tree DNA)

    mtDNA $99.95 (through Family Tree DNA)

        Be part of a research project

GeoGene

    Y chromosome $180

    mtDNA $180 ($306 for both)

        Discover your GeoFather and GeoMother

Oxford Ancestors

    Y chromosome   £180

    mtDNA   £180 (both for £340)

        7 Daughters of Eve, World Mothers, Y Clans, database searches

Relative Genetics (owned by Sorenson Genomics)

    Y chromosome, 18, 23, or 46 markers $95 to $195

    mtDNA, hypervariable 1 or 1 and 2 $125 or $195 (both $345 or $365)

    Native American, mtDNA or Y $245 each, $450 combined

    whole genome SNP analysis $219 (through DNAPrint)

        Ancestral origins, extended family relationships

Roots for Real

    mtDNA $275

        Ancestral origins

Trace Genetics (owned since 2005 by DNAPrint)

    Y chromosome $219

    mtDNA $219

    whole genome SNP analysis $219 (through DNAPrint)

    Native American test $219

        Specializes in Native American ancestry

Note: Each entry lists the name of the firm, the main services provided, their prices, and the reasons the Web site gives for using their services.

These firms give at least three different reasons why genealogists should buy their tests: determining geographical/ethnic origin, finding relatives through databases containing those already tested, and checking for genetic links between two or more people submitting DNA samples.

African Ancestry is the firm that most emphasizes the first reason. It caters to African American genealogists, who are almost always stymied by the Middle Passage, the terrible voyage of slaves from Africa to the Western Hemisphere. It offers to match a customer's results against its database of mitochondrial and Y chromosome sequences collected in Africa and to tell the customer in what country—and in some cases in what ethnic group—DNA is found that matches or comes closest to his or her sequences. Recently, African Ancestry received substantial press coverage (Henig, 2004; Poole, 2004) when civil rights hero and former Ambassador Andrew Young announced that his mother's family was from the Mende, a group whose population is now centered in Sierra Leone. A few other firms also hold out ancestral geographical information as one aspect of their services.

Oxford Ancestors uses a variant of geographical or ethnic variation, linking customers to various "clans" based on their mitochondrial DNA (36 clans around the world) (Oxford Ancestors Maternal, 2006) or Y chromosome DNA (Oxford Ancestors Y-Clan, 2006) (18 worldwide clans). These clans, popularized by Oxford Ancestors founder Bryan Sykes in his books *The Seven Daughters of Eve* and *Adam's Curse*, come complete with ancestral homelands, current common location, and names for the clan's last common ancestor, such as Ursula, Tara, and Jasmine. Thus, for example: "The clan of Tara (Gaelic for rocky hill) includes slightly fewer than 10% of modern Europeans. Its many branches are widely distributed throughout southern and western Europe with particularly high concentrations in Ireland and the west of Britain. Tara herself lived 17,000 years ago in the northwest of Italy among the hills of Tuscany and along the estuary of the river Arno" (Oxford Ancestors Maternal, 2006). These clan mothers and their histories are taken from Sykes's book; the details about them are fictional. GeoGene offers a similar service by providing you information about your "GeoMother" and "GeoFather," which appear to be quite similar to, if not the same as, the clans identified by Oxford Ancestors (GeoGene product, 2006).

At least four of the firms—DNA Heritage, Family Tree DNA, Oxford Ancestors, and Relative Genetics—also offer customers the chance both to check their DNA analyses against those of others in databases created by the firm and to add their own analyses to those databases. (The databases for Oxford Ancestors, Family Tree, and Relative Genetics are proprietary; DNA Heritage runs an open database of Y chromosome information.) Thus, customers would be told whether there are any existing perfect or near matches already in the company's files. They might be told only the location of the matching persons; for previous customers who had consented, the customers may also be told how to contact the matching parties. In addition, the customers can add their own analyses so that other customers can be notified of their existence if they match.

Five companies—DNA Heritage, Family Tree Genetics, GeneBase, Oxford Ancestors, and Relative Genetics—encourage customers to submit DNA from different people, usually analyzing the Y chromosome DNA to determine whether those who submitted the DNA are related in the paternal line. These can be submissions from two or three people who think they are related but are not sure or can be major "surname projects," looking for identical or highly similar sets of Y chromosome markers for large numbers of people.

In addition to the general testing, several firms also offer more specific testing of the mitochondrial or Y chromosome DNA. Several firms offer specific "Native American" testing of either or both types of DNA. At least one firm offers "the Cohanim chromosome" test (Family Tree DNA Jewish Ancestry, 2006), searching Y chromosome DNA for markers associated with last name "Cohen" or "Cohn," traditionally held among Jews to be the descendents of Moses's brother Aaron. DNA Consulting provides a specific "Melungeon" test (DNA Consulting Melungeon, 2006), which tests for connections to a group from the southeastern United States of unclear ancestry. It also provides a "Hindu" test (DNA Consulting Hindu, 2006), analyzing both male and female lines for membership in 1 of 49 *gotras*, or ancient lineages. And Oxford Ancestors has offered a test to tell men whether they are descended from Genghis Khan (Oxford Ancestors Genghis Khan, 2006).

The Genographic Project requires special mention. The National Geographic Society has embarked on this project with major funding from IBM and the Waitt Family Foundation (Genographic Project FAQs, 2006). The project hopes to collect DNA over five years from at least 100,000 indigenous people all over the world. The DNA will be analyzed and placed in a database with anthropological information; the database will be accessible to researchers for use in understanding human origins and migrations. Led by Spencer Wells, a geneticist who has previously been featured in National Geographic television programs, the Genographic Project can be seen as an attempt to realize the hopes of the failed Human Genome Diversity Project (Greely, 2001, 2005; Reardon, 2005 & this volume). The project also includes the sale of "Genographic Project Public Participation Kits" for $99.95. The net proceeds from the sale of those kits will fund both field research and a Genographic Legacy Project that will support cultural preservation projects among the indigenous peoples who participate in the project. A person who buys a participation kit and sends in a DNA sample will receive an analysis of the ancestral information in either his or her mtDNA or, if the participant is male, in his Y chromosome. (Participants can only choose one analysis.) The Web page says that participants will be "taking part" in the research project. But the project only collects the "personal information necessary to complete the credit card transaction and supply you with the Project's Public Participation Kit" (Genographic Project Participation, 2006), which makes it difficult to see how any useful research on human history can come from those samples—the Genographic Project will have no knowledge of the donor's ancestry.

Whole Genome SNP Analysis

AncestryByDNA takes a different approach. It does no mitochondrial DNA or Y chromosome analysis, but contracts with DNAPrint Genomics for whole genome analysis (AncestryByDNA full genome, 2006). DNAPrint types single nucleotide polymorphisms across the nuclear genome in order to produce an estimate of a customer's genetic admixture among four large population groups: the percentage of his or her ancestry that is Native American, European, East Asian, or African. The company charges $219 for each analysis. The Web site contains a discussion of the risks and benefits of the test that reads like an academic consent form (AncestryByDNA Interpretation, 2006). (DNA Heritage, Genelex, Gene Tree, Relative Genetics, and Trace Genetics, which was purchased by DNAPrint in 2005, also provide the genome-wide test from DNAPrint.) AncestryByDNA has recently begun to offer, also from DNAPrint, a "Euro-DNA" SNP analysis (AncestryByDNA Euro-DNA, 2006) for people who have already taken their basic test. The test is offered to customers who tested as at least 50% European ancestry and had ancestry from other regions below specified levels. The results state the percentage of their European ancestry that is Northern European, Mediterranean, Middle Eastern, or South Asian.

The AncestryByDNA Web site is the most detailed of the Web sites examined. It contains a link to the "manual" (AncestryByDNA Manual, 2006) for the SNP test, which includes extensive discussion of the method used, citations to relevant scientific literature, and explanations of apparently surprising results. It also includes a "Statement on Race" (AncestryByDNA Statement, 2006), which opposes racism and any attempts to assign racist significance to DNA test results. DNAPrint Genomics, Inc., and its scientists, notably Mark Shriver (this volume), claim to have identified (and have applied for patents on) specific SNPs that they call "Ancestry Informative Markers" (Frudakis & Shriver, 2004). The current version of the test analyzes 175 of these markers.

An earlier version of the Web site asked, "Who is interested in this test?" and gave the following answer:

We have sold the test to Genealogists with a desire to learn about ambiguous regions of their family trees. For example, say there is a gap in your tree, but the known branches are all Indo-European; if you learn that you are of 85% Indo-European and 15% Native American ancestry, you have learned something about the gap in your tree—some or all of those ancestors were Native American. The test is attractive for the adopted, for people that are simply curious and even for medical patients. One customer used the test to hone their search for an organ donor. Another suspected he was of significant Native American heritage, but had no way to prove it. The patent-pending AncestryByDNA test is the only scientifically rigorous method available for this purpose in existence today. (AncestryByDNA Why, 2006)

Whether and to what extent amateur genealogists will be interested in the kinds of information AncestryByDNA can provide remains to be seen.

Problems with Genetic Genealogy

Other chapters in this book discuss the poor fit between the culturally defined (and culturally variable) term "race" and the use of DNA to study race. Genetic genealogy presents both that problem and a slight variant of it.

Race, Ethnicity, and Genetic Genealogy

The whole genome test is, in effect, largely a test for race as traditionally understood in the United States, disguised as a test for *biogeographical ancestry*. It tests a person's percentage of European, East Asian, Sub-Saharan African, and Native American heritage. These may be regions of origin, not races, but the similarity to Caucasian, Mongolian, Negro, and Indian seems strong. A Boer's family may have lived in South Africa for 12 generations, as long as or longer than the families of his Bantu language–speaking neighbors, but he will not show up as "African" on that test, just as the vast majority of people born in the United States will not show up as "Native American." The test defines these four large regional ancestries, not coincidentally congruent with four large "races," by the presence or absence of particular genetic markers. The presence of the Duffy null allele helps determine African ancestry; the presence of the threonine polymorphism in the SLC25A5 gene helps determine European ancestry. If one takes at face value the disclaimers by AncestryByDNA that it is not a genetic test for race, one is left with a genetic test for something that seems indistinguishable from four traditionally understood races.

To the extent the mtDNA and Y chromosome tests are used to make statements of broad origin—European, Native American, African, etc.—they, too, serve as genetic tests of, and hence definitions of, race. But even when those variations are used not for broad ancestry but for more narrow ethnic ancestry, it promotes a similar view. Being Irish, or Yoruba, or Han becomes a matter of having particular genetic variations. A culturally defined group is transformed into a genetically defined one.

As others have argued, this confusion of culturally defined groups with genetically defined populations has at least two problems. The first is that it is not accurate. The cultural definition and the genetic definition will not always agree. Some people will be defined differently by the two tests, in both directions. Some people who are culturally accepted as part of the group will be excluded from it by the genetic tests; some people who are not members of the culturally defined group will be included in it.

A second problem is that reinforcing the idea that race is genetically defined has the potential to revitalize arguments that perceived behavioral differences between the traditionally defined races—or ethnicities—have a genetic basis. It is, of course, perfectly logical to argue that some traits are both correlated with race and genetically controlled, or at least influenced, but that behavioral traits are not among them. But by connecting race or ethnicity with genetics, one risks feeding racist views of genetic superiority and inferiority. Those conclusions need not operate only on a racial level, but individual ethnic groups may claim their own superiority; history certainly provides precedents for that.

Genetic genealogy is not likely to play a major role in associating genetics with race. More direct scientific studies, particularly those undertaken for medical purposes, such as pharmacogenomic studies, are likely to play a bigger role, particularly as they are extended to serve political and cultural ends. The influence of genetic genealogy will perhaps be stronger on genetics and ethnicity because of the ethnic-group specific uses of mtDNA and Y chromosome tests. Nonetheless, it will make a contribution, even if small, to the kinds of associations of race and genetics that may be socially dangerous.

Lack of Full Disclosure

The Web sites reviewed for this chapter generally do not seem to be inaccurate, but neither are they fully accurate. Instead, they fail to provide important information about both the results and the risks of their tests in at least four ways.

MATERNAL AND PATERNAL LINES. The biggest problem concerns the meaning of mitochondrial DNA or Y chromosome findings. These findings apply only to one line each of a genotyped individual's ancestry—the direct maternal line for mitochondrial DNA and the direct paternal line for the Y chromosome. Mitochondrial DNA is inherited only from one's mother, who inherited it from her mother, and so on back to the hypothesized mitochondrial Eve. Similarly, the Y chromosome is inherited only by a son, who got it from his father, who received it from his father, and so on. Thus, mtDNA and Y chromosome findings say absolutely nothing about one's paternal grandmother, maternal grandfather, or, in fact, the vast majority of one's ancestors. This places a very serious limitation on the value of such analyses for genealogy, one that increases exponentially as one goes back in time.

As a general matter, the number of one's ancestral lines increases by a power of two for each generation. Each of our 4 grandparents represents 1 ancestral line, as do each of our 8 great-grandparents. Going 200 years before the subject's birth reaches back about 8 generations, leading to 256 great-great-great-great-great-great-grandparents and, thus, 256 ancestral lines. Add another 50 to 60 years and the subject has 10 generations of ancestors—with 1,024 ancestral lines. (For most people, this number will, in fact, be lower as some individuals may appear as ancestors more than once.) Of these 1,024 lines, mtDNA provides information about exactly 1; the Y chromosome provides information about 1 more. It is possible for a person to have an eight-times-great-grandmother who was Japanese, an eight-times-great-grandfather who was Zulu, and 1,022 other ancestors in the 10th generation who were Irish. Mitochondrial DNA analysis would suggest that the subject *was* Japanese, or at least East Asian; Y chromosome analysis would suggest that the subject *was* Zulu, or at least sub-Saharan African. Neither would even hint at the Irish origin of 99.8% of the subject's ancestors.

None of the companies offering mitochondrial DNA or Y chromosome analysis offers a full explanation of this reality. (African Ancestry does approach this issue obliquely [African Ancestry Y, 2006] when it points out that 30% of its Y chromosome clients find that their Y chromosomes are European.) Many of the firms talk about the *direct* or *pure* maternal or paternal line but without explaining exactly what this means. None of these firms does the arithmetic to show possible customers what a small fraction of their ancestry may be explained by these findings. Only the AncestryByDNA site, which offers the different, and competing, whole genome SNP analysis, points this out. This would not be a problem if the potential customers were trained in genetics, but the average reader, even one interested in genealogy, may not recognize these facts.

ETHNIC OR GEOGRAPHICAL IDENTIFICATION. Firms that offer ethnic information based on mitochondrial DNA or Y chromosome DNA face another problem. There is no reason to believe that any mitochondrial DNA sequence or the set of Y chromosome DNA variations is unique to a single ethnic group or a single geographical region, however defined. At most, there may be allelic frequency differences between ethnic groups. One would expect these differences to be relatively small for groups that have lived— and loved—near each other. The number of ethnic groups sampled and the number of individuals sampled from those ethnic groups varies from company to company; in no case does it appear to be very large.

This is particularly a problem with African Ancestry. The current version of the African Ancestry Web site says that, in the case of a match, it provides information about where in Africa people can be found who share DNA markers with the customer. One of its frequently asked questions (African Ancestry FAQ, 2006) states: "Can a DNA analysis identify my racial or ethnic identity? There is no test for racial identification. Race is a social construct, not genetically determined. Similarly, ethnicity is more cultural than biological." A subsequent FAQ (African Ancestry FAQ, 2006) adds: "Can you tell what tribe I am from? We cannot necessarily tell you the particular tribe with which you share genetic ancestry. However, we can tell you the present-day country of Africa with which your lineage shares genetic ancestry." Its press clippings, however, are full of tribal affiliations, and its customer comments (African Ancestry Customer Comments, 2006) include comments such as "my DNA was a match to the Mende people of Sierra Leone."

When African Ancestry tells a customer that he or she is a "perfect match" for a Mende or Hausa genotype, it is actually saying that at least one person in its Mende or Hausa sample had the same genotype for the hypervariable region of the mitochondrial DNA or for the nine markers they type for the Y chromosome. Yet it is almost certainly the case that people in other ethnic groups have the same genotype, perhaps at a much higher frequency than the Mende or Hausa do. It might be, for example, that a particular genotype was found in 1.9% of the Mende, 0.7% of the Hausa, 2.1% of the Yoruba, 1.2% of the Fulani, and 0.1% of the rest of the world's population. To be accurately informative, the company should provide information about the number of matching samples in its database and the percentage those genotypes make up of each population. It should also provide information about exactly which populations are represented in its database and by how many subjects. The customer could then make an informed assessment of just how likely it was that his or her maternal or paternal ancestral line really was Hausa.

Instead, Rick Kittles, the firm's scientific advisor, has been quoted in newspaper stories saying the following:

> "My female line goes back to Northern Nigeria, the land of the Hausa
> tribe. I then went to Nigeria and talked to people and learned a lot

about the Hausa's culture and tradition. That gave me a sense about who I am. In a way, it grounded me."

"Two people there looked like cousins I have—they even behaved like them!" Kittles laughed. (Sailer, 2003)

Indeed, Kittles's connection to the Hausa is mentioned in at least 13 newspaper stories over the last three years. Of course, even if his maternal ancestor, 10 generations ago, was Hausa, and the people he saw were also descended from her, he would, as a result of this common ancestry, share about one one-millionth of his DNA polymorphisms with these 10th cousins.

NATIVE AMERICAN ANCESTRY AND OTHER SPECIFIC ETHNIC TESTS. The chapter by Kimberly TallBear (this volume) discusses at some length firms that offer "Native American" testing and some of the problems with those tests. I will touch on some other concerns they raise.

At least eight firms, DNA Consulting, Family Tree DNA, Genelex, Gene-Tree, Niagen, Orchid Cellmark, Relative Genetics, and Trace Genetics, offer tests specifically for Native American ancestry—"Native American Validation," as Relative Genetics calls it. (My information about Niagen and Orchid Cellmark comes from TallBear.) In addition, Oxford Ancestors also identifies some of its mitochondrial DNA and Y chromosome haplogroups as largely Native American. AncestryByDNA used "Native American" as one of its four geographical regions, so it also purports to be able to identify, as a percentage, a customer's Native American ancestry. DNA Consulting, Family Tree DNA, and Relative Genetics use mitochondrial DNA or Y chromosome analysis for their Native American testing, each of which can analyze only one of the customer's lines.

The sites for the firms that test mtDNA or the Y chromosome give no real information on the accuracy of their tests. DNA Consulting, Family Tree DNA, and Trace Genetics say nothing about it. Relative Genetics says only, "Nearly all full-blooded Native Americans belong to one of three Y chromosome lineages or haplogroups," without specific citation and without any discussion of less than "full-blooded" Native Americans, and provides no information about the mtDNA tests it offers (Relative Genetics Native American test, 2006). In addition, these tests have all the problems of any mitochondrial DNA or Y chromosome test. Again, using eight generations as an example, a customer might have 254 European ancestors and 2 Native American ancestors but be shown by both tests to be Native American. Similarly, a person who identifies as Native American and has 254 Native American ancestors in the eighth generation could test as purely European (or African or Asian) based on the two ancestors in that generation in his or her direct paternal or maternal line.

The tests for Melungeon ancestry, Hindu gotras, the Cohanim Y chromosome, or descent from (the presumably non-Cohanim) Genghis Khan share these problems. The Web sites involved provide little or no information about the validity and accuracy of the tests. The Genghis Khan test has its own special problem. No one has any genetic material from Genghis Khan's remains. He died in 1227 and was buried in an intentionally concealed grave. In 2003 a set of Y chromosome markers was found in an unusually high percentage of men in Central and East Asia, about 8% of the men in that region, which amounts to about 0.5% of all men in the world. An analysis (Zerjal et al., 2003) indicated that those Y chromosomes shared a common ancestor—a common grandfather—about a thousand years ago. As Genghis Khan supposedly had at least four acknowledged sons, some of whom had many sons (and many opportunities to beget children), the authors speculated that the modern Y chromosomes might have been derived from his. But, of course, it might also have belonged to a particularly handsome, or fertile, deputy—or musician, or stable boy—of the great Khan. Or, even if Genghis Khan carried an ancestral version of that chromosome, it might have become common in that area long before he was born. It could be that having many male relatives helped him to rise to leadership rather than that being a leader helped him have many male relatives. For all we know, Genghis Khan could have been sterile and his supposed sons adopted (or worse). But the scientific uncertainties detract from the romantic idea of carrying "Genghis Khan's Y chromosome" and thus are ignored.

An additional problem affects AncestryByDNA's testing for Native American ancestry. Unlike the other firms, it looks at the whole genome and not just the mitochondrial DNA or Y chromosome, but it finds that some people with no known or even plausible Native American background test as having some discernible Native American ancestry (AncestryByDNA Native America, 2006). Most such people have Southern European or Middle Eastern ancestry; the firm does not know why they test this way. The site does not talk about any false positives among its Native American customers.

The Trace Genetics site (Trace Genetics description, 2006) notes that its Y chromosome test cannot make any tribal match and that any tribal matches in its mtDNA test do not mean that you "belong to" that tribe. The Relative Genetics and Genelex sites contain no similar language. Not one of the sites discusses the extremely variable and complicated ways in which tribal membership and the right to various benefits of Native American status are determined.

FALSE PATERNITY. Historically, both clinical and research use of genetic testing within families quickly led to an unsettling discovery (unsettling at least to men): a significant percentage of children are not genetically related to the men who think they are their fathers. Much discussed in the profession, there are few if any published data on this result. Conversations with clinical geneticists and genetic researchers reveal a wide range of false paternity rates, varying by populations and perhaps by disease. Based on these discussions, a contemporary false paternity rate of 5% to 10% seems a decent guess.

At even 5% false paternity by generation, the chance that one's genetic ancestry matches one's *official* ancestry declines substantially with each generation. After 10 generations, the chance that the eight-times-great-grandfather supplied the subject's Y chromosome is about 60%. If false paternity rate is 10%, the same probability after 10 generations is just over 30%. (The chance that the *official* paternity is correct is 90% in each generation; over 10 generations, that probability is 0.9 to the 10th power, or about 30%.) This is a serious problem for surname searches, which use the Y chromosome markers to see if people with the same surname are related. People who may be legally, culturally, and historically related may turn out not to be genetically related. In family studies, which several of the companies encourage, this has the potential for causing serious family bitterness. What happens to the family ties of people who discover, to their surprise, that they did not really have the same grandfather? Note also that those responsible for the false paternity may still be alive and may not have given their consent to genetic testing that could have serious ramifications for them.

Genetic genealogy has the potential to harm living individuals and to break apart family bonds as well as to strengthen them. In scientific research with human subjects, federal regulations generally require that the research subjects be given clear and understandable information about the risks of participation; the common law of informed consent provides similar protections in medicine. Not one of the genetic genealogy Web sites, however, raises these thorny issues.

Conclusion: The Ethics of Science—the Morals of the Marketplace

Like the results of genetic genealogy tests, the story of genetic genealogy is interesting but arguably not very important. As a hobby, genealogy is an entertainment, perhaps a particularly mentally active one but one that is rarely of any real significance. Yet I believe the lack of complete disclosure in the genetic genealogy companies does have some broader significance. This is true even though this investigation has found little, if any, muck to rake. There seems to be no consumer fraud here. The science behind the companies is plausible and the Web sites are not dishonest. But, at the same time, neither are they completely honest (with the possible exception of the Web site of AncestryByDNA). I believe many consumers would be surprised to learn how little the results can really tell them about their ancestry, largely because the corporate Web sites usually do not tell them clearly and plainly the limits on what they can expect.

Of course, our commercial society thrives on not telling people clearly and plainly what products will do for them. Exaggeration, sex appeal, and hype are the basic stuff of marketing to consumers. Beer commercials sell sex to young men. The International Star Registry sells completely unofficial and meaningless names to stars invisible to the naked eye. For years, the Psychic Hotline flourished by selling fortune telling. The genetic genealogy companies are paragons of honesty by comparison. Even if one wanted to subject them to tougher standards of honesty in their advertising than other consumer firms face, the First Amendment protections for commercial speech might prevent it. (See, for example, Hoefges, 2003.)

Besides, apart from being a few hundred dollars poorer (and the possibility of an unexpected revelation of false paternity or adoption) what harm is done? If a customer feels good believing that he "is Mende" or that she "is from the clan of Tara," does it make any difference?

It may not make any difference to the customers, but it may make a difference to the scientists involved, and to science more broadly. Medicine, the first user of genetic information, has been, to some extent, less commercial than most of the economy, although advertisements for at least one medical genetic test, Myriad Genetics's tests for BRCA 1 and 2 mutations, have raised questions of hype. As we consider the possible ways in which genetics may move beyond the world of medicine, regulated by professional standards, legal rules of informed consent, and at least potentially by the Food and Drug Administration, we will need to think broadly about two implications of putting genetics in the world of consumer advertising.

The first implication concerns the geneticists involved. As scientists, their professional ethic insists, at least as an aspiration, on honesty, candor, and an acknowledgment of the limitations of their findings. Consumer advertising violates all those principles. Can a scientist be completely candid as a scientist while selling his or her name and talents to a firm that will not be completely candid about the science? Geneticists in industry thus far have generally not had to face this issue very directly. They work for pharmaceutical, biotechnology, or research tools firms, where the customers (or, in medicine, at least the effective customers—those who write the prescriptions) are physicians or other scientists. If it becomes common that other genetics products are sold directly to consumers, more geneticists will have to decide how to deal with this tension—and their less commercial colleagues will have to decide how to deal with them.

The second implication concerns the science involved. Genetic genealogy is selling the imprimatur of science. What makes it exciting to people is that it is *scientific* and therefore, they believe, true. But the real science of genetic genealogy is riddled with qualifications and limitations; it deals with varying degrees of probability and not with anything close to certainty. It looks at precise questions, precisely defined, like a direct paternal or maternal line. Genetic genealogy skips the caveats and in doing so promotes a false perception of science; it invokes science's power without accepting its limits. That seems wrong as a matter of principle, but it also may do some harm. That the false perception genetic genealogy promotes exaggerates the power of science should be no consolation to those who care about science. A public misunderstanding that science is about timeless certainty can easily become a source of public disillusionment with science when, as often happens, the science changes. Genetic genealogy is not likely, by itself, to undermine the public standing of science. But, at least as presently practiced, it will not help.

NOTE

The author would like to thank the participants in the January 2006 workshop, "Revisiting Race and Ethnicity in the Context of Emerging Genetic Research," and his research assistants, Jason Tarricone and Sean Rodriguez.

REFERENCES

African Ancestry Customer Comments. (2006). Retrieved March 1, 2006, from http://www.africanancestry.com/custcomment.htm.

African Ancestry FAQ. (2006). Retrieved March 1, 2006, from http://www.african ancestry.com/05faq.htm.

African Ancestry mtDNA description. (2006). Retrieved February 26, 2006, from http://www.africanancestry.com/mtdna.htm.

African Ancestry pricelist. (2006). Retrieved February 26, 2006, from http://www.africanancestry.com/shop.htm.

African Ancestry Y chromosome description. (2006). Retrieved February 26, 2006, from http://www.africanancestry.com/ychr.htm.

AncestryByDNA Euro-DNA SNP analysis. (2006). Retrieved February 26, 2006, from http://www.ancestrybydna.com/welcome/productsandservices/eurodna/ancestry kit/.

AncestryByDNA full genome SNP analysis. (2006). Retrieved February 26, 2006, from http://www.ancestrybydna.com/welcome/productsandservices/ancestrybydna/ ancestrykit/.

AncestryByDNA Interpretation of Results. (2006). Retrieved March 1, 2006, from http://www.ancestrybydna.com/welcome/productsandservices/ancestrybydna/ interpretationofresults/.

AncestryByDNA Manual. (2006). Retrieved March 1, 2006, from http://ancestrybydna.com/welcome/productsandservices/ancestrybydna/manual/.

AncestryByDNA Native American Affiliation in Mediterranean Europeans. (2006). Retrieved March 2, 2006, from http://www.ancestrybydna.com/welcome/products andservices/ancestrybydna/ethnicities/#medeuropeans.

AncestryByDNA pricelist. (2006). Retrieved February 26, 2006, from http://www.ancestrybydna.com/welcome/ordernow/.

AncestryByDNA Statement on Race. (2006). Retrieved March 1, 2006, from http://ancestrybydna.com/welcome/productsandservices/ancestrybydna/manual/#statement.

AncestryByDNA Why. (2006). Retrieved February 26, 2006, from cache at http://72.14.207.104/search?q=cache:XVwr2hatoIJ:www.ancestrybydna.com/profile.asp+%22say+there+is+a+gap+in+your+tree,+but+the+known+branches+are+all+Indo-European%22&hl=en&gl=us&ct=clnk&cd=1.

Brown, K. (2002, March 1). Tangled roots? Genetics meets genealogy. *Science, 295*, 1634–1635.

DNA Consulting African American test. (2006). Retrieved February 26, 2006, from http://dnaconsultants.com/Detailed/9.html.

DNA Consulting full genome SNP analysis. (2006). Retrieved February 26, 2006, from http://dnaconsultants.com/dna_tests/autosomal/index.html.

DNA Consulting Hindu. (2006). Retrieved March 1, 2006, from http://dnaconsultants.com/Detailed/26.html.

DNA Consulting Melungeon. (2006). Retrieved March 1, 2006, from http://dnaconsultants.com/Detailed/3.html and http://dnaconsultants.com/Detailed/5.html.

DNA Consulting mtDNA analysis. (2006). Retrieved February 26, 2006, from http://dnaconsultants.com/dna_tests/female/index.html.

DNA Consulting Native American test. (2006). Retrieved February 26, 2006, from http://dnaconsultants.com/Detailed/10.html.

DNA Consulting Y chromosome analysis. (2006). Retrieved February 26, 2006, from http://dnaconsultants.com/dna_tests/male/index.html.

DNA Heritage database. (2006). Retrieved March 1, 2006, from http://www.ybase.org/.

DNA Heritage full genome SNP analysis. (2006). Retrieved February 26, 2006, from http://www.dnaheritage.com/ysnps.asp.

DNA Heritage pricelist. (2006). Retrieved February 26, 2006, from http://www.dnaheritage.com/order.asp.

DNA Heritage Y chromosome analysis. (2006). Retrieved February 26, 2006, from http://www.dnaheritage.com/ydna.asp.

Family Tree DNA database. (2006). Retrieved March 1, 2006, from http://www.ysearch.org/.

Family Tree DNA homepage. (2006). Retrieved March 2, 2006, from http://www.familytreedna.com.

Family Tree DNA Jewish Ancestry. (2006). Retrieved March 1, 2006, from http://www.familytreedna.com/jgene.html.

Family Tree DNA Native American test. (2006). Retrieved February 26, 2006, from http://www.familytreedna.com/ngene.html.

Family Tree DNA products and pricelist. (2006). Retrieved February 26, 2006, from http://www.familytreedna.com/products.html.

Frudakis, T. N., & Shriver, M. D. (2004, November 18). Composition and methods for inferring ancestry. U.S. Patent Application No. 20040229231.

GeneBase Y chromosome analysis. (2006). Retrieved February 26, 2006, from http://www.genebase.com/ydna_howto.php.

Genelex full genome SNP analysis. (2006). Retrieved February 26, 2006, from http://www.healthanddna.com/ancestry.html.

Genelex Native American test. (2006). Retrieved February 26, 2006, from http://www.healthanddna.com/nativeamerican.html.

Genelex pricelist. (2006). Retrieved February 26, 2006, from http://www.healthanddna. com/orderform.html.

GeneTree Native American test. (2006). Retrieved February 26, 2006, from http://www. genetree.com/product/native-american-test.asp.

GeneTree pricelist and description of tests. (2006). Retrieved February 26, 2006, from http://www.genetree.com/product/population-assessment-dna-test.asp.

Genographic Project FAQs. (2006). Retrieved March 26, 2006, from https://www3. nationalgeographic.com/genographic/faqs_about.html#Q1.

Genographic Project Participation. (2006). Retrieved March 26, 2006, from https:// www3.nationalgeographic.com/genographic/faqs_participation.html.

Genographic Project Y chromosome and mtDNA analysis. (2006). Retrieved February 26, 2006, from https://www3.nationalgeographic.com/genographic/participate. html.

GeoGene pricelist. (2006). Retrieved February 26, 2006, from http://www.geogene.com/ highres/information/moneyback.html.

GeoGene product description. (2006). Retrieved March 1, 2006, from http://www. geogene.com/highres/product.html.

GeoGene Y chromosome and mtDNA analysis. (2006). Retrieved February 26, 2006, from http://www.geogene.com/highres/information/wallcharts.html.

Greely, H. T. (2001). Human genome diversity: What about the other human genome project? *Nature Reviews Genetics, 2*, 222–227.

Greely, H. T. (2005). Lessons from the HGDP? (book review). *Science, 308* (5728), 1554–1555.

Henig, R. M. (2004, October 10). The genome in black and white (and gray). *New York Times*, sec. 6.

Hoefges, M. (2003). Protecting tobacco advertising under the commercial speech doctrine: The constitutional impact of *Lorillard Tobacco Co. Communication Law and Policy, 8*, 267–311.

Oxford Ancestors database. (2006). Retrieved March 1, 2006, from http://www.oxford ancestors.com/members/.

Oxford Ancestors Genghis Khan. (2006). Retrieved March 1, 2006, from http://www. oxfordancestors.com/genghis_khan.html.

Oxford Ancestors Maternal. (2006). Retrieved March 1, 2006, from http://www.oxford ancestors.com/your-maternal.html.

Oxford Ancestors mtDNA analysis. (2006). Retrieved February 26, 2006, from http:// www.oxfordancestors.com/maternal-ancestry.html.

Oxford Ancestors Y chromosome analysis. (2006). Retrieved February 26, 2006, from http://www.oxfordancestors.com/paternal-ancestry.html.

Oxford Ancestors Y-Clan. (2006). Retrieved March 1, 2006, from http://www.oxford ancestors.com/service-yline.html.

Poole, S. M. (2004, August 22). African roots discovered with DNA tests: Former U.S. Ambassador Young traces his lineage with ancestry company. *Milwaukee Journal Sentinel*, sec. A.

Reardon, J. (2005). *Race to the finish: Identity and governance in an age of genomics*. Princeton, NJ: Princeton University Press.

Reardon, J. (2008 [this volume]). Race without salvation: Beyond the science/society divide in genomic studies of human diversity. In B. A. Koenig, S. S.-J. Lee, & S. S. Richardson (Eds.), *Revisiting race in a genomic age* (pp. 304–319). New Brunswick, NJ: Rutgers University Press.

Relative Genetics database. (2006). Retrieved March 1, 2006, from http://www.relative genetics.com/relativegenetics/surname_landing.jsp.

Relative Genetics full genome SNP analysis. (2006). Retrieved February 26, 2006, from http://www.relativegenetics.com/relativegenetics/product_categories/ancestral_origins.htm.

Relative Genetics mtDNA analysis. (2006). Retrieved February 26, 2006, from http://www.relativegenetics.com/relativegenetics/product_categories/maternal_analysis.htm.

Relative Genetics Native American test. (2006). Retrieved February 26, 2006, from http://www.relativegenetics.com/relativegenetics/product_categories/native_american.htm.

Relative Genetics pricelist. (2006). Retrieved February 26, 2006, from http://www.relativegenetics.com/relativegenetics/product_listing.jsp.

Relative Genetics Y chromosome analysis. (2006). Retrieved February 26, 2006, from http://www.relativegenetics.com/relativegenetics/product_categories/paternal_analysis.htm.

Roots for Real mtDNA analysis. (2006). Retrieved February 26, 2006, from http://www.rootsforreal.com/service_en.php.

Roots for Real pricelist. (2006). Retrieved February 26, 2006, from http://www.rootsforreal.com/ordering_en.php.

Sailer, S. (2003, April 28). African Ancestry Inc. traces DNA roots. *United Press International.* Retrieved March 1, 2006, from http://www.upi.com/inc/view.php?StoryID=20030428-074922-7714r.

Shriver, M. D., & Kittles, R. A. (2008 [this volume]). Genetic ancestry and the search for personalized genetic histories. In B. A. Koenig, S. S.-J. Lee, & S. S. Richardson (Eds.), *Revisiting race in a genomic age* (pp. 201–214). New Brunswick, NJ: Rutgers University Press.

Sykes, B. (2002). *The seven daughters of eve.* New York: W. W. Norton & Company.

Sykes, B. (2004). *Adam's curse: A future without men.* New York: W. W. Norton & Company.

TallBear, K. (2008 [this volume]). Native-American-DNA.com: In search of Native American race and tribe. In B. A. Koenig, S. S.-J. Lee, & S. S. Richardson (Eds.), *Revisiting race in a genomic age* (pp. 235–252). New Brunswick, NJ: Rutgers University Press.

Trace Genetics description and pricelist of all tests. (2006). Retrieved February 26, 2006, from http://www.tracegenetics.com/services_gene.html.

Zerjal, T., Xue, Y., Bertorelle, G., Wells, R. S., Bao, W., et al. (2003). The genetic legacy of the Mongols. *American Journal of Human Genetics, 72* (3), 717–721.

12

Native-American-DNA.com

In Search of Native American Race and Tribe

KIMBERLY TALLBEAR

As of 2005, roughly 15 companies in the United Kingdom, Canada, and the United States marketed DNA testing for so-called Native American DNA markers. At least 2 market the DNA fingerprint or "parentage test" directly to tribal and First Nation governments. I highlight 6 companies that are visible in the Native American DNA-testing market and/or in Native American–specific forums: GeneTree, Genelex, Niagen, DNAPrint Genomics, DNAToday, and Orchid Cellmark. Some companies market in venues that are broadly Native American in theme; they especially target the idea of tribal-specific identity (i.e., Genelex, Niagen, and GeneTree). Others market directly to U.S. tribes and Canadian First Nations (DNAToday and Orchid Cellmark). A few companies make direct claims about using genetics to access "Native American benefits" while others astutely steer clear of such claims. I pay attention to the sometimes marked differences between companies. Yet more savvy companies also use imagery that locates Native American identity in close relation to the DNA double helix. I provide overviews of varying depths for each company, focusing on the different technologies used and how each company foregrounds one of two crucial concepts, tribe or race, as they appeal to their particular customer constituencies.[1]

Before delving into the details of DNA testing, a short discussion of terminology is important. Historian Alexandra Harmon sums up the predicament of using always inadequate and hazardous words to talk about native peoples. Words contain their own histories, thus a writer unavoidably infuses her prose with contradictory and easily criticized meanings. Relying on the terms "Indian," "Native American," "tribe," or even a people-specific, self-ascribed name is to "[presuppose] the existence of the very groups whose creation, transformation, dissolution, or redefinition" one is documenting;

our lexical quandary is that we cannot "understand antiquated ideas about races and tribes without using antiquated language" (Harmon, 1998, p. 9). But I also use specific terms because communities describe themselves in such terms.

Second, a brief discussion of genetics terminology is in order. In technical language, "Native American DNA" is a set of genetic markers (nucleotides) that appear at different frequencies in different populations. The highest frequencies of so-called Native American markers have been observed by scientists in "nonadmixed" native populations in North and South America, therefore indicating probable Native American ancestry. But Native American DNA should not be understood simply as an objective molecular "thing." It is simultaneously a conceptual apparatus through which humans constitute and deploy life-organizing narratives: historical, national, and racial narratives; narratives about family and tribe and the origins of peoples—who individuals "really are."

Both seller and consumer stand to benefit culturally and financially from the reduction of Native American identity to DNA molecules. For tribes, the benefits are less clear. Molecular definitions of "Native American" need not reference the historical-regulatory paradigm of U.S. tribal governance or tribal community. Foregrounding molecules embeds assumptions about kinship, race, and individual identity that may undermine tribal group identities and governance.

Lineal Descendancy versus Tribal Blood

Let me now explain two distinct ways of reckoning kinship—lineal descendancy and tribal blood—as these inform my analysis of DNA-testing companies. Lineal descendancy is about measuring and tracing blood (or, more recently, molecules) between *individuals* in order to constitute family kinship or a distant lineal descendancy relationship. Certain "mainstream" kinship understandings that privilege lineal descendancy inform identity DNA testing. In *American Kinship* (1980), David Schneider argues that in (U.S.) American culture, kinship is biogenetic; "scientific discoveries—new facts about biogenetic relationship"—get incorporated into kinship reckonings as if new ways of seeing relatedness had always been so. Speakers who substitute "gene(tics)" for "blood," even when referring to kinship reckonings before "genetics" had entered our lexicon or worldview, demonstrate how new, scientific facts about kinship get incorporated as if they were always there. Ward Churchill (1999), for example, argues that in the late 19th century the federal government used the "white genetics" of mixed-bloods to determine which Indians were competent to individually hold land.

In Schneider's analysis, the blood relationship is considered "real" and "true": while "legal rights may be lost . . . , the blood relationship cannot be lost. It is culturally defined as being an objective fact of nature." It is believed that blood relatives share an identity, "a belief in common biological constitution," expressed as "being of the same flesh and blood" (1980, p. 25). Measuring and tracing blood establishes a multi-generational lineal descendancy relationship.

"Blood quantum" has been an important mechanism for determining membership in most "federally recognized tribes" since official rolls were first constructed in the late 19th and early 20th centuries. Scholars agree that the technique of blood quantum emerged in response to the General Allotment Act of 1887 and the division of tribal communal land into individually held parcels (Barker, 2000, 2003; Churchill, 1994, 1999; Harmon, 1998, 2001). In order to distribute land to every Indian head of household, federal agents had to have a list of "tribal members." Degree of Indian blood (i.e., the number of Indian parents or grandparents one was known to have) was a principal way in which rolls were constructed. Those with too few Indian ancestors and thus too little "Indian blood" were not enrolled. However, scholars disagree on whether blood quantum and its associated racial-cultural understandings were simply imposed on native peoples by a racist, colonial government (Churchill, 1999; Edmunds, 1995; Jaimes, 1992), or whether the idea of blood quantum also exhibits particular pre-European attachments to blood as symbolic of kin and identity (Harmon, 1998, 2001; Meyer, 1998, 2005). Today, the majority of federally recognized tribes view degree of blood as an important measure of one's closeness to the tribe; that degree, usually one-quarter (Thornton, 1997), should reflect a significant proportion of one's genealogically documented blood relations.

Lineal descendancy and tribal blood use symbolic blood to organize Native American identity in different ways. Blood-quantum critics such as Ward Churchill privilege lineal descendancy—that blood link to an individual known (or thought) to have had Native American blood. That is the blood link described by Schneider—an individual, linear link back through a genealogical line. On the other hand, the technique of blood quantum focuses on documented blood links between an individual and the *group* via blood links to a multiplicity of documented tribal individuals. Both positions locate identity firmly in blood, and both can be seen as shaped or partially shaped by American dialogues on race. Both are potentially racist. But to refer to reservation blood-quantum rules as simply racist or discriminatory misunderstands and de-legitimizes the meanings accorded to "blood" and "tribe" that support a land-based, group identification as the locus of Native American identity.

Beatrice Medicine, an anthropologist who worked among Dakota, Lakota, and Nakota people and other U.S. native peoples, and who was herself a tribal member, sheds light on the complex meanings ascribed to "blood" by tribal people. Like Schneider, Medicine sees the blood relationship as an important relationship of kinship and identity, but she is concerned with a different kind of blood kinship—one that incorporates the nuclear family and direct lineages, but that is preoccupied with a broader level of relatedness. Medicine focuses on the identity that emerges from one's relatedness to multiple, tribally affiliated individuals: "The identity of a person of Indian descent is tied to tribalness—that is, to a social grouping based upon biological relationships. Thus, the very nature of a Native female is biologically defined by virtue of enrollment in her tribe, her blood quantum, and recognition by her natal community" (2001, p. 138). Medicine's interpretation that emphasizes enrollment in and recognition by a "natal community" indicates that the blood-*infused* social relationship of an individual to a tribe, through multiple biological relationships reflected in blood quantum, is more fundamental to one's Indian-ness than is the *simply*-blood relationship to individuals through a genealogical line. However, in emphasizing group sociality, Medicine notes that tribal rootedness can override low degrees of Indian blood. The presence of any blood at all, *if* combined with a more than biological relationship, matters more than blood alone (Medicine, 2001, p. 298). In short, Medicine emphasizes a networked set of social and cultural relations based on biological relatedness.

Schneider's analysis of the shared heredity of individuals stands in contrast to Medicine's analysis of the blood relatedness shared by an individual with her tribe. Both kinship paradigms subdivide blood "with each reproductive step away from a given ancestor" (Schneider, 1980), but the tribal individual must name, add, and divide a multiplicity of tribal relations in order to ascertain his or her own blood quantum and thus the degree of tribal relatedness. Contemporary tribal membership accounts for today's nuclear family and for direct lineages, but the tribal blood relationship goes beyond the nuclear family and single lineages to include the tribe, a critical cultural-biological kinship identity that is not reflected in the dominant American kinship system.

Both classes of DNA testing, Native American DNA and the DNA fingerprint, reflect a lineal-descendancy understanding of kinship and race that is focused on relations between individuals. The molecular knowledge produced by DNA tests does not account well for group kinship that is central to tribes. Marketing such tests to tribes and individuals foregrounds the idea that being Native American is more about individual genetics (i.e., biological race) than it is about group life lived at multiple levels, both biological and social.

DNAPrint Genomics

DNAPrint Genomics' patented AncestrybyDNA™ test provides the most detailed and visible example of how the concept of race is employed by an ancestry DNA-testing company. Other companies also sell the test along with their own products. From a scientific perspective, AncestrybyDNA™ is unique because it is an autosomal ancestry DNA test that surveys for "an especially selected panel of Ancestry Informative Markers" (AIMs) across all 23 chromosome pairs. Shriver et al. define AIMs as "genetic loci showing alleles with large frequency differences between populations"; they propose that "AIMs can be used to estimate biogeographical ancestry (BGA) at the level of the population, subgroup (e.g. cases and controls) and individual" (2003, p. 387). Because AIMs are found at high frequencies within certain groups and at lower frequencies in others (sometimes very rarely and other times not so rarely), DNAPrint surveys the markers and uses a complicated algorithm to estimate an individual's BGA percentages. Other ancestry DNA tests look only at markers on the Y chromosome or on the mtDNA, markers that are genealogically very informative but limited in the ancestral lines they survey.

The most recent test version, AncestrybyDNA™ 2.5, looks at approximately 175 such AIMs. One such marker is the Duffy Null allele. It is found in virtually all sub-Saharan Africans, reaching 100% frequency in some of those populations, probably because it confers resistance to malaria caused by *Plasmodium vivax*, a malarial parasite (Livingston, 1984; Tournamille, Colin, Cartron, & Le Van Kim, 1995; Hadley & Peiper, 1997). Research shows the marker to be rare in populations outside of Africa (Hamblin & Di Rienzo, 2000; Hamblin, Thompson, & Di Renzo, 2002). The assumption is that if an individual has the Duffy Null allele, he or she must have relatively recent ancestry, i.e., on the order of several thousand years, in sub-Saharan Africa, an area of Africa where *Plasmodium vivax* is common.

But the Duffy Null allele is the only substantive example of an AIM provided on DNAPrint's Web site. The full range of AIMs used to calculate "ancestral proportions" is proprietary information, and the Duffy Null allele is not typical of AIMs. It is genetically selected for because it confers an important trait, resistance to a potentially life-threatening parasite. But Deborah Bolnick explains that the "vast majority of human genetic variation does not follow this pattern" (2003, p. 5) of radical differences in marker frequency between populations. AncestrybyDNA™ "emphasizes the very few markers that may do so." DNAPrint surveys 175 AIMs in calculating BGA percentages, and the company regularly looks to increase the markers surveyed. But the total number of base pairs in the human genome is between 3 and 4 billion, so 175 markers is a minute sample. It should be subject to methodological criticism.

Consider DNAPrint marker 1141, otherwise known as SGC30055*1. It occurs in 75.3% of Native Americans sampled, just 5.4% of African Americans sampled, and 51.1% of European Americans sampled. It is good for distinguishing Native American from African ancestry but not so good for distinguishing Native American from European ancestry. DNAPrint marker 1116, otherwise known as WI-17163*3, is found in 69.0% of Native Americans sampled but in only 17.5% and 5.4% of European Americans and African Americans sampled, respectively.[2] Such markers are helpful but not nearly as clear-cut as the Duffy Null allele for distinguishing BGA.

Other methodological questions come to mind. Of the 175 markers surveyed, what is the proportion that is informative for Native American ancestry? Do the 175 markers fairly evenly cover different ancestry categories, or are there many more that help distinguish African or European BGA? Second, does DNAPrint sample randomly across the genome in choosing markers to survey? We don't have any evidence other than the fact that they say they do. But we do know that there are actually a limited number of known AIMS (Shriver et al., 2003, p. 397). Therefore, we must ask if there actually are known AIMs "across the genome." Furthermore, does DNAPrint use skin pigmentation markers because they are convenient and not necessarily because they are the best markers for DNAPrint's purposes; i.e., determining "unique lineages"?

Bolnick notes additional technical problems with AncestrybyDNA™ that challenge DNAPrint's claim that the test can detect "precise ancestral proportions." She followed the company Web site for one year and notes some problems with individual test results posted to the site. First, some individuals claiming to have Native American ancestry are shown to have East Asian ancestry instead, thus undermining the notion that the categories are "genetically distinct." Second, Bolnick noticed drastic changes in particular individuals' "precise ancestral proportions" over the course of the year. There were no official changes to the test during that time, which "suggests that the test is not as precise or scientifically rigorous as it is claimed to be" (Bolnick, 2003, p. 4). Third, DNAPrint claims that most Mediterranean Europeans, Middle Easterners, Jews, and South Asian Indians "reliably and systematically show low levels of Native American admixture (even without an American Indian great-grandparent)" (TallBear & Bolnick, 2004). Bolnick (2003) also notes that some of the academic references cited on DNAPrint's Web site as supporting their conclusions actually do not; there is some misinterpretation of those references.

Other critiques of the DNAPrint test have more to do with cultural-political assumptions reflected in the test and with DNAPrint's marketing rather than with technical limitations. As such, not all critiques can be addressed by more sophisticated lab techniques, technologies, or data sets, as is often argued. While many scientists and social scientists disclaim the possibility of race purity and that race is genetically determined (although that doesn't mean we have totally given up the ideas), DNAPrint openly promotes popular racial categories as at least partly genetically determined and implies the possibility of purity. The DNAPrint Web site refers to individuals of "relatively pure BioGeographical Ancestry" as opposed to "recently admixed peoples." DNAPrint also uses interchangeably the phrases "racial mix" and "ancestral proportions" and then explicitly claims that the test "measures 'the biological or genetic component of race'" as it detects four "BioGeographical ancestries" or "four lineages or major population groups of the human population" (Bolnick, 2003; AncestryByDNA, 2005). More recently, the DNAPrint Web site offered the following definition of BGA: "BioGeographical Ancestry (BGA) is the term given to the biological or genetic component of race. BGA is a simple and objective description of the Ancestral origins of a person, in terms of the major population groups. (e.g. Native American, East Asian, European, sub-Saharan African, etc.) BGA estimates are able to represent the mixed nature of many people and populations today" (What is Biogeographical Ancestry, 2006). The company cannot help but reinforce the possibility of racial purity when it repeatedly refers to the notion of mixedness, whether in semantic forms of "mixed heritage" or the "mixed nature of many people and populations today" (Products and Services, 2005; What is Race, 2005). "Mixture" is predicated on purity.

DNAPrint also emphasizes that "race is a complex and multivariate construct that we tend to oversimplify in our analysis." Anticipating critiques about reinforcing biological notions of race, DNAPrint emphasizes that it is "measuring a person's genetic ancestry and not their race" (Products and Services, 2005; What is Race, 2005). Yet the "lineages" or "ancestral origins," as they are described, fall into to racial categories that were longstanding before we knew anything about human genetics.

In the "Frequently Asked Questions" on its Web site, DNAPrint describes recent disciplinary discussions about race as "merely a social construct" as oversimplified; DNAPrint argues that race is found biologically in "many particular instances of differences between the populations of the world" (What is Race, 2005). In support they cite the biological difference of skin color, which they characterize as dramatically different "across populations." Skin color markers are central to this test, and the example of skin color as a support for the notion of genetic race is telling. First, skin color is actually a good example in defense of the idea that races are socially constructed. Long before we knew anything about genetics, humans categorized races by skin color. Second, research indicates that the relationship between genes and skin color is not as deterministic as DNAPrint would have us believe (Jablonski, 2004). Third, suggesting that there is a "biological basis for race" confuses the logic of the relationship between biology and race. It is not that genetic differences have simply produced an entangled bundle of visible differences among people that we then build important (and sometimes discriminatory) social categories around, but which we could also objectively or apolitically label as "race." Rather, because we apply important social meanings to skin color, we single it out as a crucial biological difference among the many that we could use to categorize people. Skin color has long been used to break the biological continuum of humanity into races, and it shapes scientists' choices today about other biological characteristics for which to look. DNAPrint's 175 ancestry informative markers do not present us with objective scientific evidence that our race boundaries are genetically valid. DNAPrint's markers are chosen with those racial boundaries in mind and not at random.

GeneTree

GeneTree is a visible choice for "Native American DNA" testing. It registers domain names composed of strings of search terms and links those sites directly to GeneTree's main Web site. Unsurprisingly, GeneTree appears most often on Native American–theme Web sites. GeneTree describes the science behind their techno-scientific products geared toward answering questions about Native American and other ethnic ancestries: "DNA research on full-blooded indigenous populations, such as *Native American Indian populations*, from all around the world has led to the discovery of *'genetic markers that are unique to populations, ethnicity and/or deep ancestral migration patterns.'* The markers that have very specific modes of inheritance, and which are relatively unique to specific populations, are used to assess probabilities of ancestral relatedness" (emphasis GeneTree's) (Native American Genetic Ancestry, 2006). Aside from dubiously claiming that markers are unique to populations, GeneTree touts its products as providing scientifically definitive answers to what are nonscientific identity questions. Categories such as "Native American" are not genetically definitive but politically, historically, and socially negotiated. Genetics, history, culture, and law may overlap at certain points, but they also conflict. Biological relatedness and "phenotype" matter, but the degrees to which they matter and the ways in which they matter fluctuate. Particular bodies may be ethnically and racially regarded in quite different ways across time and from place to place. Genetic markers offer only weak evidence for making meaningful personal claims about heritage and identity. That is, unless we invest DNA markers with a symbolic power to indicate ethnic and racial identity. If so, we make a "fetish" (Haraway, 1997) of a genetic link to ancestors who may or may not have been seen (by themselves and society) as Native American. Molecules named for populational frequencies stand in for convoluted processes, both genetic and not genetic, of being Native American.

Genelex and Niagen

Genelex warrants attention because it markets its services in forums that are the domain of federally recognized tribes. Genelex regularly advertises in the prominent weekly *Indian Country Today* and it has advertised in the *Navajo Times*. The ad is both technically problematic and at odds with history and contemporary governance practice. In the September 22, 2004, *Indian Country Today*, Genelex proposed to readers the following: "Do you need to confirm that you are of Native American descent? Recent advances in genetic testing have put the answer to this question at your fingertips. Whether your goal is to assist in validating your eligibility for government entitlements such as Native American Rights or just to satisfy your curiosity, our Ancestry DNA test is the only scientifically rigorous method available for this purpose in existence today." The ad claims that Genelex's test reveals genetic markers that are "unique to Native Americans" (Ancestry DNA Testing, 2005).[3]

Niagen is another company that makes strong and controversial claims about the applicability of Native American DNA testing to Native American and tribal-specific identity. Their Web page, under "native Indian heritage testing," states the following:

> Native Indian heritage testing are [*sic*] for those individuals who would like to determine the presence of genetic native Indian ancestry. Utilizing the latest in DNA research individuals are able to determine the percentage of native Indian ancestry within their genetic line. Such testing is essential for documented proof of nativeness and use in government or other benefits.
>
> Native Indian heritage DNA testing determines the ancestry from most tribal origins and is a highly accurate means of determine [*sic*] percentage or full native Indian ancestry. Utilizing researched gene specific loci, specific allelic markers common within Native Indian genotypes are measured to determine the degree of Native Indian Ancestry. (Native Indian Heritage, 2006)

Several of Genelex's and Niagen's claims demonstrate how companies scientifically mischaracterize Native American DNA and then suggest its application in controversial ways. Genelex claims that ancestry DNA tests, whether mtDNA or Y chromosome (it sells both), identify uniquely Native American markers. The claim is not supported by the science. "Native American mtDNA" markers, the mutations that characterize haplogroups A, B, C, D, and X, are not found uniquely in Native Americans. They are found more often in Native Americans, but due to migration histories they are also found in other populations in smaller percentages.

Y chromosome markers such as M19 and M3 are not uniquely Native American either. Not all Native American–identified males have them and there are certainly non–Native American–identified males who do have them, which is not surprising given how populations admix. At any rate, there has not been widespread sampling for Native American markers of any type. In terms of the AncestrybyDNA™ test (which Genelex and Niagen market respectively under the headers "Native American Ancestry DNA Testing" and "Native Indian Heritage Testing"), so-called Native American markers surveyed in that test are also not unique to Native Americans. To recap, while some of the genetic markers used in mtDNA, Y chromosome, and DNAPrint's autosomal marker test have to date been found only in individuals already identified as Native American, neither Natives nor non–Native Americans—no matter how one delineates such samples—have been systematically sampled. We do not have an in-depth understanding, based on a defensible sample size, of how such markers are distributed globally.

Thus it makes little sense for Genelex to speak of such markers as found uniquely in Native Americans. Given what I have said about the impossibility of determining tribe—in the past or the present—via genetic markers, it is also not accurate for Niagen to speak of "Native Indian heritage DNA testing" as determining "ancestry from most tribal origins" (Native Indian Heritage, 2006). While some non–Native American test takers certainly find Native American markers in their genomes, Genelex's and Niagen's language implies that so-called Native American markers are definitive of racial and/or tribal identity. Despite company claims about "scientifically rigorous methods," these are not scientifically rigorous ideas. In addition, they have important political implications.

Both Genelex and Niagen exaggerate what DNA tells us about ancestry and its correspondence with racial, ethnic, and tribal identity. That said, I do not charge the companies with willfully misleading the public. Their claims demonstrate that science does not exist in a cultural vacuum. Genelex's and Niagen's language reveals an idea that persists unevenly within science and without science—that genetics relates to ethnic/racial group identity in a deterministic way. Thus tests for genetic markers can be sold as proof or validation of that identity.

DNAToday and Orchid Cellmark

At least two companies, DNAToday and Orchid Cellmark, have marketed the "DNA fingerprint" parentage test specifically to U.S. tribes and Canadian

First Nations for purposes of enrollment.[4] The parentage test is both more and less informative than a "Native American DNA" test because it gauges relatedness at a different biological level. DNA fingerprinting can be used to confirm close relations (e.g., mother, father, siblings) with very high degrees of probability. It does not analyze markers judged to inform one's "ethnic" ancestry. On the other hand, Native American ancestry DNA tracing shows that one shares relatively few, often non-coding markers and thus probable but more distant relatedness with historical Native American populations.

This technology is commonly used as a paternity test, although it can also be used to examine maternal lineages. The DNA fingerprint is also commonly used in criminal cases—to prove, for example, that a strand of hair or skin cells found on a crime victim belongs to an individual suspect. The technology examines repeated sequences of nucleotides such as "GAGAGA," called short tandem repeats (STRs). We inherit STRs from both parents, but our total individual STR pattern is, in practical terms, unique. DNA fingerprinting examines very specific patterns; i.e., only 1 in 60 million individuals would exhibit such a pattern.

While many companies sell this technology to the public, most do not expressly target tribal enrollment. Inaccurate knowledge of biological parentage is not one of the principal problems in enrollment today. But it is exactly because the companies market technical information that is often redundant that they make an interesting case for analysis. The cultural capital of "gene talk" and the zeal in the marketplace for personalized DNA services suggest that these companies market gene talk as much as the technical information to be had from a test. DNAToday markets its test in combination with a "smart [identification] card" (Whitehead, 2003) that technologizes tribal identity. Gene talk—and modernity itself—are on sale to tribes.

One ancestry DNA company scientist has expressed dismay at the marketing of parentage tests to those interested in Native American ancestry. He felt that selling paternity tests in this market was to mislead consumers by not distinguishing between the "superior" technical aspects of Native American marker tests. However, U.S. tribes and Canadian First Nations are a unique subset of the DNA-testing market. Genetic genealogists who are interested in (deep) genealogical histories find ancestry DNA tests obviously more useful for that activity. But U.S. tribes and Canadian First Nations are concerned with who gets enrolled. The parentage test precisely demonstrates recent biological ancestry that must sometimes be documented for membership.

Native American ancestry DNA should never inform enrollment policy. To do so would be to attack the very historical and political foundation upon which contemporary tribal governance and land rights are based. I do not think tribal governments would knowingly do that. On the other hand, parentage tests might be useful in certain individual enrollment cases. They are already used occasionally when an applicant's biological parentage is in question and that parent's tribal status is essential for the applicant to be enrolled. However, in the majority of tribal enrollment cases parentage tests, too, are irrelevant. For most enrollment applicants, as I have noted, biological parentage is not in question and thus should not complicate one's enrollment status. In addition, my own tribe will use affidavits from three enrolled relatives testifying as to their relationships with an applicant for the purpose of using parental lineage for enrollment. The tribe takes the stand that familial testimony about kinship is not trumped by a genetic test.

For these reasons, the marketing of parentage tests as a definitive solution to the troublesome politics of enrollment is disingenuous and demonstrates little consideration of the contemporary problems with tribal enrollment. When company representatives advocate tribal-wide paternity testing so that "only Native Americans that deserve to be members of your tribe will be" (DCI America National Tribal Enrollment Conference, 2003), they miss perhaps the most crucial and divisive problem in contemporary enrollment debates: in the majority of cases parentage is not in question, yet increasing numbers of tribal members' offspring cannot meet "blood-quantum" requirements. Individuals simply do not have a sufficient number of "blooded" parents and grandparents to meet the required blood quantum. For most tribes, most of the time, blanket parentage testing of the type promoted by DNAToday would provide already-known information at a great financial and ideological cost. The tests cannot directly address what are actually philosophical and political disagreements within tribes about who should count as Dakota or Pequot or Cree.

However, the (cultural) work of companies such as DNAToday and Orchid Cellmark and their customers may help change tribal responses by promoting the view that lineal biological descent should translate into tribal belonging. A small proportion of tribes already use blanket parentage testing for tribal enrollment. Will DNA testing arise more frequently in Indian country to mediate tribal enrollment disputes—particularly those surrounding the disbursement of casino revenues—and in so doing will it affect our understanding of what it is to be Native American?

An Indian in silhouette faces the setting sun (see fig. 12-1). The image recalls the classic "End of the Trail" depiction, the broken 19th-century Indian on horseback. But Orchid's Indian sits upright. Does he envision a more hopeful future for his progeny than did his 19th-century counterpart? After centuries of predicting their demise, can scientists now testify to the American Indians' survival? Orchid's ad announces that genomic technologies reveal "the truth." Are we to assume it refers simply to DNA markers,

FIGURE 12-1. Which truth is in the DNA?

Advertisement in *American Indian Report* (May–November 2005 issues).

or are we meant to think that Orchid's science can reveal—via those markers—an ethnic essence, the stereotypical nobility and tradition portrayed in the image?

Orchid Cellmark recently entered the Native American identity market. In a unique move, it sells a more complete array of DNA-testing services than do other companies. Like DNAToday, it promotes the standard parentage test for tribal enrollment. But it also markets tests for Y chromosome and mtDNA Native American markers and the DNAPrint test that supposedly determines one's "percentage of Native American–associated DNA." By offering all four types of tests to the Native American identity market, including to tribes and First Nations, Orchid Cellmark presents a more complex set of implications for tribes. Like DNAToday, and in reference to the paternity test, Orchid Cellmark stresses that its service "confirm[s] the familial relationship of specific individuals to existing tribal members" ("Orchid Cellmark Launches New DNA," 2005).

I worry about Orchid Cellmark marketing this test while also actively building a tribal and First Nation government (not simply Native American) customer base. It would not be such a stretch to embrace the notion of biogeographical ancestry (BGA) percentages that underlies AncestrybyDNA™, especially for those tribes that gauge relatedness to group by counting the number of one's specifically named and enrolled ancestors in order to calculate a blood-quantum fraction. If tribes begin to think in terms of combining the parentage test with AncestrybyDNA™, which "geneticizes" the old metaphorical blood-quantum fraction, Orchid Cellmark has a package that, while questionable in its technical applicability to current enrollment standards, has big symbolic value. It just might appeal to tribes, especially those with contentious enrollment processes (i.e., wealthy casino tribes flooded with enrollment applications from would-be beneficiaries). For better and for worse, Orchid Cellmark demonstrates a more complicated take-up of the cultural politics of contemporary Native American identity. Where other companies target either "race" or "tribe" more strongly, Orchid Cellmark's array of products accounts for Native American identity as both racial and tribal.

The terminology of "tribe" (or First Nation) and "race" contradict and overlap each other as they help construct the different Native American identities to which DNA-testing companies respond. Advocates for tribal government sovereignty typically talk about the fundamental difference between the concept of race and that of the tribe/First Nation in justifying self-governance: They commonly argue that tribes are nations (also implying culture), and not races. And nations decide who gets citizenship and who does not. Yet tribe and race share common ground. Part of that common ground is symbolic blood (increasingly DNA), although blood and gene metaphors take different forms when used to support different positions (i.e., reckoning race versus reckoning tribal membership). Orchid Cellmark simultaneously appeals to DNA as a metaphor for race and as representative of one's relatedness to tribe. Because tribes operate in a world that is both tribalized and racialized, they struggle to mediate a Native American identity according to those sometimes contradictory, sometimes overlapping worldviews.

More than any other company discussed, the work of Orchid Cellmark reflects the breadth of ways in which DNA testing and genetic metaphors can expand into territory previously claimed by blood-quantum regulations and blood metaphors. Unlike blood, DNA testing has the advantage of claims to scientific precision and objectivity. Orchid Cellmark's director of North American Marketing claims that in using DNA fingerprint analysis, there is "no possibility of incorporating a subjective decision into whether someone becomes a member or not" (DNA Testing for Tribal Enrollment, 2005). Of course, whether or not someone is verifiable biological kin of the type indicated by a parentage test is not an "objective" enrollment criterion. Allowing a parentage test to trump other ways of reckoning kin (e.g., blood quantum as a proxy for cultural affiliation by counting relatives, or a signed affidavit of family relatedness) for purposes of enrollment prioritizes techno-scientific knowledge of certain kinship relations over other types of knowledge and relationships. Nonetheless, the idea of scientific definitiveness attached to genetic testing is influential, even if it is not realized. Thus, parentage testing may increasingly look like a good replacement for blood (quantum) and other non-genetic criteria—especially if traditional documentation of named relations is difficult to obtain, if blood-quantum standards are unattained by most applicants, or if enrollment applications are politically and economically contentious. Some will see such a move as advantageous, as scientifically objective and less open to political discrimination or racism, but I argue that such a move can also be interpreted as politically regressive and irresponsible in terms of scientific ethics.

Conclusion

Substituting DNA for blood does not technically solve the blood-quantum problem. But increased DNA testing may encourage a switch from emphasizing community through tribal relatedness (with blood quantum as an imperfect proxy for those relations) to emphasizing race (with genetic testing as a proxy). That switch from emphasizing tribe to emphasizing race promotes the notion that race is biogenetically determined (despite perfunctory scientific disclaimers to the contrary) and thus scientifically detectable. If race or biogeographical ancestry comes to have greater influence in how we understand Native American identity, federal government historical and legal relationships and obligations to tribes may fade further from view. In turn, claims to tribal land and self-governance may be denied and justified by the absence or presence of Native American DNA in individual claimants. This may have two effects: Anti-tribal interests will have strong ammunition to use against tribes whom they already view as beneficiaries not of treaty payments, but of special race-based rights. Second, groups without historical-colonial relationships, heretofore racially identified as other than Native American, may increasingly claim tribal authority and land based on DNA. The marketing of DNA tests to tribes and others clearly has much broader implications than revisions in enrollment policy. Given the commercialization of genetic-ethnic identities and capitalism's ubiquitous reach, we must consider that the scientific object of Native American DNA (or the definitive absence of such markers) will be important in re-making Native American identity in the 21st century. DNA markers quite simply will be very valuable in certain quarters for making or refuting claims of Native American identity that otherwise might be very difficult to make or refute. Thus, DNA markers—when there is something tangible to gain—may be used to legitimate claims that contradict and potentially contravene prior tribal claims based in historical treaties, law, and policy, even if the groups that use Native American DNA analysis do not intend to undermine existing tribal claims and law.

NOTES

This material is based upon work supported under a National Science Foundation (NSF) Graduate Research Fellowship and grants from the Sisseton Wahpeton Oyate (SWO). The views expressed do not necessarily reflect the views of NSF or SWO.

1. I do not treat at length the interests of DNA-testing customers in this chapter. I did ethnographic research in one online community of DNA test users. However, the types of technologies that particular community is concerned with (ancestry DNA) and the uses to which they put them (extensive genealogy research) do not present direct risks to tribal governance as do the activities, images, and claims of companies.

2. Ann Turner, RootsWeb Genealogy-DNA-L list serve founder, cross-referenced academic papers and genetic-marker databases in order to determine several markers used by DNAPrint to calculate Native American and other BGA percentages. See Genealogy-DNA-L archives, January 11–12, 2004, at http://archiver.rootsweb.com/th/index/GENEALOGY-DNA/2004-01.

3. As of June 2006 the Genelex Web site featured revised language that more accurately reflected testing limitations.

4. Between the completion of this chapter in 2005 and publication, DNAToday declared bankruptcy. Orchid Cellmark continues to market its parentage test to tribes.

REFERENCES

Ancestry DNA Testing. (2005). *Health & DNA*. Retrieved May 2, 2005, from http://www.healthanddna.com/genealogy.html.

AncestryByDNA. (2005). Retrieved June 20, 2005, from http://www.ancestrybydna.com/products.asp.

Barker, J. (2000). *"Indian-made" sovereignty and the work of identification*. PhD dissertation. University of California, Santa Cruz.

Barker, J. (2003). "Indian™ U.S.A." *The Wicazo Sa Review, 18* (1), 25–79.

Bolnick, D. A. (2003). *"Showing who they really are": Commercial ventures in genetic genealogy*. Paper presented at the 102nd Annual Meeting of the American Anthropological Association, Chicago, Illinois.

Churchill, W. (1994). *Indians are us? Culture and genocide in Native North America*. Monroe, ME: Common Courage Press.

Churchill, W. (1999). The crucible of American Indian identity: Native tradition versus colonial imposition in postconquest North America. *American Indian Culture and Research Journal, 23* (1), 39–67.

DCI America National Tribal Enrollment Conference. (2003, October 28). New Orleans.

DNA Testing for Tribal Enrollment. (2005, July 8). In H. McKosato, *Native American Calling*. Anchorage: Koahnic Broadcast Corporation.

Edmunds, R. D. (1995, June). Native Americans, new voices. *American Historical Review, 100*, 733–734.

GENEALOGY-DNA-L Archives. (2004, January). *Roots web.* Retrieved January 11–12, 2004, from http://archiver.rootsweb.com/th/index/GENEALOGY-DNA/2004–01.

Hadley, T. J., & Peiper, S. C. (1997). From malaria to chemokine receptor: The emerging physiologic role of the Duffy Blood Group antigen. *Blood, 89,* 3088–3091.

Hamblin, M. T., & Di Rienzo, A. (2000). Detection of the signature of natural selection in humans: Evidence from the Duffy Blood Group locus. *American Journal of Human Genetics, 66,* 1669–1679.

Hamblin, M. T., Thompson, E. E., & Di Rienzo, A. (2002). Complex signatures of natural selection at the Duffy Blood Group locus. *American Journal of Human Genetics, 70,* 369–383.

Haraway, D. (1997). *Modest_Witness@Second_Millenium.FemaleMan©_Meets_OncoMouse™: Feminism and technoscience.* New York and London: Routledge.

Harmon, A. (1998). *Indians in the making: Ethnic relations and Indian identities around Puget Sound.* Berkeley: University of California Press.

Harmon, A. (2001). Tribal enrollment councils: Lessons on law and Indian identity. *Western Historical Quarterly, 32* (Summer), 175–200.

Jablonski, N. G. (2004). The evolution of human skin and skin color. *Annual Review of Anthropology, 33,* 585–623.

Jaimes, M. A. (1992). Federal Indian identification policy: A usurpation of indigenous sovereignty in North America. In M. A. Jaimes (Ed.), *The state of Native America: Genocide, colonization, and resistance* (pp. 123–138). Boston: South End Press.

Livingston, F. B. (1984). The Duffy Blood Groups, Vivax malaria, and malaria selection in human populations: A review. *Human Biology, 56,* 413–425.

Medicine, B. (2001). *Learning to be an anthropologist and remaining "Native."* Urbana: University of Illinois Press.

Meyer, M. L. (1998). American Indian blood quantum requirements: Blood is thicker than family. In V. J. Matsumoto & B. Allmendinger (Eds.), *Over the edge: Remapping the American West* (pp. 231–249). Berkeley: University of California Press.

Meyer, M. L. (2005). *Thicker than water: The origins of blood as symbol and ritual.* New York: Routledge.

Native American Genetic Ancestry. (2006). *Genetree.* Retrieved June 7, 2006, from http://www.genetree.com/ancestral/nativeAmerican.php.

Native Indian Heritage Testing. (2006). *Niagen.* Retrieved June 7, 2006, from http://www.niagen.com/dn_ntvendn_gen_data/dn_native_indian_nav.htm.

Orchid Cellmark launches new DNA testing service to confirm Native American tribal membership. (2005, June 15). Press release from Orchid Cellmark at http://www.orchid.com/news/view_pr.asp?ID=376, retrieved December 10, 2005.

Products and Services. *AncestryByDNA.* Retrieved September 15, 2005, from http://www.dnaprint.com/welcome/productsandservices/anestrybydna/.

Schneider, D. (1980). *American kinship: A cultural account.* Englewood Cliffs, NJ: Prentice-Hall.

Shriver, M. D., Parra, E. J., Dios, S., Bonilla, C., Norton, H., Jovel, C., et al. (2003). Skin pigmentation, biogeographical ancestry, and admixture mapping. *Human Genetics, 112,* 387–399.

TallBear, K., & Bolnick, D. A. (2004, December 3–17). "Native American DNA" tests: What are the risks to tribes? *The Native Voice.*

Thornton, R. (1997). Tribal membership requirements and the demography of "old" and "new" Native Americans. *Population and Research Policy Review, 16,* 33–42.

Tournamille, C., Colin, Y., Cartro, J. P., & Le Van Kim, C. (1995). Disruption of a GATA motif in the Duffy Gene Promoter abolishes erythroid gene expression in Duffy-negative individuals. *Nature Genetics, 10*, 224–228.

What is Biogeographical Ancestry (BGA)? (2006). *AncestryByDNA—FAQs*. Retrieved June 7, 2006, from http://www.ancestrybydna.com/welcome/faq/#q2.

What is Race? (2005). *AncestryByDNA—FAQs*. Retrieved September 15, 2005, from http://www.ancestrybydna.com/welcome/faq/#q1.

What is Race?—2. (2006). Retrieved June 7, 2006, from http://www.ancestrybydna.com/faqs.asp#q1.

Whitehead, S. (2003, October 27). The future of tribal enrollment software. Powerpoint presentation by Steven Whitehead, president of DNAToday, at "DCI American Tribal Enrollment Conference," New Orleans, LA.

13

The Factness of Diaspora

The Social Sources of Genetic Genealogy

ALONDRA NELSON

Family history research is a popular pastime for those seeking to discover unknown ancestors. For many, this pursuit has taken the form of genealogical journeys modeled on *Roots: The Saga of an American Family*, Alex Haley's famous (and infamously embellished) account of his successful efforts to trace his familial lineage to Africa (Haley, 1976). The book and subsequent award-winning miniseries of the same name were the result of the author's efforts to uncover the mystery of his ancestral origins with clues garnered from Gambian griots, deciphered linguistic retentions, archival research, and his own genealogical imagination. Haley's account became an ur-text of African diasporic reconciliation for a generation of Americans. Despite this example, few African Americans are able to fill in the contours of their past as Haley did, owing to the decimation of families that was a hallmark of the era of racial slavery and to the dearth of records that remain from this period. As a consequence, genetic genealogy testing, which is broadly available and also less taxing and seemingly more authoritative than conventional genealogical research, holds considerable appeal for some root-seekers of African descent.

Drawing on sampling techniques and statistical models developed in human population genetics (Cavalli-Sforza, Menozzi, & Piazza, 1994; Hammer, 1995; Jobling & Tyler-Smith, 1995; Shriver & Kittles, this volume), direct-to-consumer commercial genetic genealogy services analyze DNA in order to infer information about an individual's family history, ethnic affiliation, or "biogeographic ancestry" (see Fullwiley, this volume). These services are becoming widely used. Some are niche marketed to specific social groups, such as the testing sold by African Ancestry (www.africanancestry.com) that is promoted among African Americans and "matches" customers to ethnic groups in Africa. Molecular biologist Rick Kittles established African Ancestry with his business partner Gina Paige in early 2003 (Hostetler, 2003; Maganini, 2003; Sailer, 2003; Roylance, 2003; Kittles, personal communication, February 4, 2006). The company sells two forms of DNA analysis with the brand names MatriClan and PatriClan that trace matrilineage and patrilineage, respectively: MatriClan analyzes genetic information linked to mitochondrial DNA that is inherited by both male and female children from their mothers. The PatriClan test examines the genetic sequence of the Y-chromosome to trace lineage from son to father, to father's father, to father's father's father, and so on.

African Ancestry is an information-age business—the exchange of a fee for service takes place online and through the mail. The company mails test kits to customers that contain the implements necessary to secure a DNA sample. The customer returns the sample to African Ancestry by mail; it is amplified and sequenced by the company's lab partner, Sorenson Genomics of Salt Lake City, Utah. African Ancestry then compares the resulting data to its proprietary DNA bio-bank—the African Lineage Database (ALD)—that is said to contain more than 25,000 DNA samples from over 30 countries and 160 ethnic groups in Africa (African Ancestry, 2006). After several weeks, a customer will receive a results package that includes a printout of the customer's Y- or mt-DNA markers, a "Certificate of Ancestry," and sociohistorical information about the African continent.

A hypothetical root-seeker employing African Ancestry's analysis may learn that her mt-DNA traces to the current Mende people of Sierra Leone and that her Y-DNA test, for which she submitted her brother's DNA, traces to the Bubi group of present-day Equatorial Guinea.[1] African Ancestry analyses might thus be regarded as ethnic lineage instruments through which undifferentiated racial identity is translated into African ethnicity and kinship. By linking blacks to inferred ethnic communities and nation-states of Africa, African Ancestry's service offers root-seekers the possibility of constituting new forms of diasporic affiliation and identification.

On the surface, both the appeal and most likely outcome of ethnic lineage testing, such as that offered by African Ancestry, appear to confirm that the practice is essentialist—that genealogy testing, as a vehicle of diasporic identification, amounts to the reduction of transnational affiliation to genetics, in the process abetting the "reauthroriz[ation] of race as a biological category" (Abu El-Haj, 2007, p. 284; also Duster, 2003) and eliding the historical, political, and economic diversity of black experiences (Gilroy, 1993). As I describe here, however, the experiences of root-seekers suggest that a more complex dynamic is at play. The genetic facts rendered as the outcome of genealogy testing may provide the circumstance for reconfigured social arrangements, yet this potential transformation does not stem solely from these data. Rather, the cogency of African Ancestry's testing is derived significantly from social sources that shape how facts are anticipated, interpreted, and mobilized by root-seekers.

In this chapter, I begin to develop the concept of the *factness of diaspora* to describe the particular process of co-production (Jasanoff, 2004; Reardon, 2005) through which genetic genealogy testing attains value and validation for the root-seekers of African descent that I have encountered in the course my ethnographic fieldwork. "Factness" means possessing the state, condition, or quality of fact, yet not being only or exactly fact.[2] The "factness of diaspora" denotes the imbrication of the "biogenetic facts" (Schneider, 1968) of genealogy testing *and* aspirations for African affiliation against the backdrop of histories of forced displacement and through the subsequent enactment of "reconciliation projects"—cultural practices through which resolution of the injuries of racial slavery is sought. I elaborate the factness of diaspora through a discussion of three significant points in the interpretive trajectory of African Ancestry's genetic genealogy testing service: the "authentic expertise" of the company's chief science officer, Rick Kittles, which significantly influences root-seekers' confidence in its product; the forms of self-fashioning that may be spurred following the receipt of one's genetic genealogy test results; and, related to this, the extra-genetic forms of "kinship" that the test outcomes may foster.

Authentic Expertise

In October 2005, I attended the three-day national meeting of the Afro-American Historical and Genealogical Society (AAHGS) for the first time. The conference program consisted of social events, tours of historic sites, invited lectures by prominent figures, and concurrent panels on subjects of interest to the genealogists in attendance. Panel topics included accounts of family history research hurdles and successes, how to use the now fully digitized Freedman's Bureau records for family history research, how to participate in the Library of Congress's oral history project for veterans, and how to document one's genealogical research in compelling narrative form to share with family, friends, and local historical societies. The Saturday morning keynote address, a lecture entitled "Trace Your DNA and Find Your Roots: The Genetic Ancestries of African Americans," was delivered by Rick Kittles, African Ancestry's co-founder and chief science officer. I previously attended other public presentations by Kittles, and, as on these prior occasions, I watched with wonder as he performed his unique combination of erudition, charisma, and folksiness for a rapt audience comprising over one hundred people—women and some men who, with a few exceptions, appeared to be over age fifty, and many sixty years of age or older.

Kittles discussed how genetics could be used to help blacks trace their roots to African ethnic groups, detailing the scientific basis of African Ancestry's products. In addition to educating those in attendance about the technical aspects of MatriClan and PatriClan analysis, his presentation was also clearly intended to demonstrate how much Kittles held in common with his customer base, including their concerns about privacy and the unique historical circumstances that inspired their root-seeking pursuits. "African Ancestry is the only [genetic genealogy] company that focuses on people of African descent; it's run by black folks and it's going to stay that way," Kittles proclaimed. Genetic genealogy research should be "guarded by someone who shares the same sensitivity to the concerns of the community," he continued to applause. The many audience members who cheered in response to Kittles's assertion of community-mindedness testified to the effectiveness of the geneticist's claim to cultural authenticity.

This legitimacy was shored up as Kittles continued his pitch, changing registers slightly from man of the people to genealogist colleague. He recalled that he "caught the genealogy bug"—using a phrase common among genealogists who liken their interest in the pursuit of the past to a virus—as a doctoral student in the biological sciences at George Washington University. "AAHGS is near and dear to my heart," Kittles said. "I came to this event even before my research interests emerged in this area." Continuing to speak as a fellow traveler, with a PowerPoint slide of the iconic image of the cross section of a slave ship packed with black bodies in the background, Kittles recounted his personal frustrations with genealogical research and spoke of the challenges facing even the most diligent amateur historians seeking to trace their African roots. He discussed how genetic genealogy testing had helped him discover that his maternal mt-DNA matched to the Mandinka of Mali. His paternal Y-chromosome line traced to Germany, a result that he attributed to what he described as "the Thomas Jefferson effect"—a characterization that doubly signaled the sexual violence of racial slavery and the forensic DNA analysis that along with archival records established that the third U.S. president fathered the children of his slave Sally Hemings (Thomas Jefferson Foundation, 2000; also Bay, 2006).

At the conclusion of his formal presentation, Kittles raffled a free African Ancestry test and then spoke with the 30 or so audience members who stood in a queue that wound through the aisles of the auditorium to await their turn to offer questions, comments, and compliments. One African American woman attired in red, white, and blue clothing adorned with rhinestone American flags and a lanyard for her ID cards, embroidered with the acronym DAR (for Daughters of the American Revolution), introduced herself as a member of the "underground railroad of the DAR, called the Daughters of Color," before proceeding to ask Kittles for further interpretation of the genetic genealogy results she had recently received from his company.[3] The great majority of Kittles's audience, however, gathered in the lobby just outside of the auditorium where the lecture had taken place. Here vendors had set up tables arrayed with items for sale ranging from African-themed knickknacks to genealogical research primers. Most merchants sat noticeably idle. But one vendor's table was surrounded by enthusiastic customers. In the center of this crowd was African Ancestry co-founder Gina Paige, who struggled to stay on top of the many orders being placed for her company's testing service. Her business partner's presentation, during which Kittles had stressed his shared experience with the audience, had succeeded in persuading many of the AAHGS membership to purchase African Ancestry's genetic genealogy testing service.

The audience's indisputably positive response to African Ancestry, evidenced both by their reception to the lecture and their purchase of its product, was perhaps preconditioned by the fact that Rick Kittles is among the most well-known molecular biologists in the United States. The authenticity Kittles displayed at the AAHGS gathering is bolstered by scientific authority established through press coverage, scholarly publications, and institutional associations. He has made frequent media appearances over the last several years in his capacity as chief scientist of African Ancestry. For example, he appeared in *Motherland: A Genetic Journey*, a 2003 British Broadcasting Company documentary that aired in the United States on the Sundance Channel, as well as in the PBS documentary *African American Lives* in 2006, in which his company's services were employed to trace the roots of black celebrities. He has also been featured on ABC's *Good Morning, America* and *The Morning Show* and *60 Minutes* on the CBS network. Since 2002, scores of newspaper and magazine articles, including those in the *New York Times, Time, New York Daily News, Black Enterprise, Wired, Fortune*, and the *Los Angeles Times* have included commentary from Kittles, solidifying his position as an expert on genetic genealogy testing.

Kittles's professional ascent has included the publication of scholarly papers in leading science journals as well as stints at prestigious institutions. He has authored numerous articles in the area of human variation and genetics in notable journals, including the *American Journal of Human Genetics, Science*, the *Annals of Epidemiology*, and the *American Journal of Public Health*. He has also held positions at the National Human Genome Center at Howard University, Ohio State University, and, at present, the Cancer Research Center at the University of Chicago. Kittles's "hard" scientific research at prominent institutional settings on the genetic determinants of prostate cancer—a disease that disproportionately afflicts African American men—is concomitant with his investigations into what might be regarded as the "soft" science of the genetic genealogy testing.

As the public face of African Ancestry, Rick Kittles draws together cultural and scientific legitimacy into a complex I term "authentic expertise." In his guise as a genealogist colleague who shares his customers' desires for ancestral reckoning, Kittles establishes genetic genealogy testing as a legitimate and safe practice for African American root-seekers. At the same time, his renown as a scientist and his involvement with cutting-edge medical genetics research lend authority to his commercial genetics enterprise. Thus, many root-seekers are as compelled by Kittles as they are convinced by genetic science. "I trust Dr. Kittles," a root-seeker named Pat explained to me when I asked if she had any apprehensions prior to purchasing the MatriClan test that linked her to the Akan of Ghana and Cote D'Ivoire.[4] Although Alicia, another informant, had reservations about the genetic genealogy results she received from African Ancestry because they were inconsistent with those from another company, she took great pride in telling me that she had been in contact with Kittles by both telephone and email. Her misgivings were subsequently assuaged through her interactions with him. Kittles's authentic expertise is an unmistakably important aspect of the appeal of African Ancestry's genetic genealogy testing and of consumers' faith in the genetic facts it supplies to its customers. This symbolic capital—authentic expertise—produces value around genetic genealogy tests that extends beyond the presumed capacity of DNA to assign identity.

Affiliative Self-Fashioning

The AAHGS conference "sharing dinner" took place in a large ballroom of the on-campus hotel of Gallaudet University in Washington, D.C. During the dinner, genealogists were invited to stand and share highlights of their experiences as family history researchers, if they were so inclined. Although none at my table availed themselves of this opportunity, we spoke amongst ourselves about our respective genealogical research projects. I struck up a conversation with Bess, an African American woman in her fifties who lives near Baltimore, Maryland, and was seated next to me. I told her about my ethnographic study of conventional and genetic root-seeking—including my preliminary foray into my own family's history—that had brought me to the AAHGS meeting. Bess shared that she had been conducting genealogical research on her family for about a decade and also had recently received genetic genealogy test results from African Ancestry.

The next morning, I ran into Bess in the hotel lobby where merchants, including Gina Paige, were setting up their tables for the day. Bess said to me, "I have something for you." We sat together on the edge of a fountain in the hotel atrium and she showed me the results of her recent genetic genealogy test that she had arranged in a binder. A letter from African Ancestry indicated that mt-DNA analysis linked Bess to the Kru of Liberia "plus/or Mende-Temne of Sierra Leone." Her result package also contained a "Certificate of Ancestry" signed by Rick Kittles, a printout of the genetic markers from which Bess's African ethnicity was inferred, a map of the African continent with Liberia foregrounded, and a flier advertising Encarta Africana, a CD-ROM encyclopedia, at a discounted rate.

Bess explained to me that she wants to "do something" with her results, like perhaps "travel to Africa." Curious as to which of the two possible ethnicities suggested by African Ancestry was most compelling to Bess, I asked whether she planned to visit Liberia, neighboring Sierra Leone, or both countries in the future. "My sister was married to a man from Sierra Leone; his name was Abdul," she replied obliquely, intimating that she would likely travel to the natal home of her deceased brother-in-law. "When will you be ready to travel to Africa?" I asked. "After I get back further in time [with my genealogical research]," she responded. As is common with other root-seekers who make use of genetic ancestry tracing, Bess assumed a role in determining her test's significance and its potential import to her life. Her intention to engage in practices motivated by the findings she received from African Ancestry *after* she advanced with conventional genealogy underscores how the interpretative work that commences following the receipt of genetic genealogy results can involve consumers' efforts to "align" genetic DNA analysis with other evidence of their ancestry as well as with their genealogical aspirations, prior experience, or extant relationships (Nelson, 2008).

The conduct by which root-seekers decide to accept or jettison genetic genealogy test results in the constitution of African diasporic connection and identity can be described as "affiliative self-fashioning" (Nelson, 2008). Writing about the interface of brain imaging techniques and social identity, Joseph Dumit employs the phrase "objective self-fashioning" to explain how subjectivity can be "fashion[ed] and refashion[ed]" from the "received-facts of science and medicine" (2003, p. 39). By extending this useful analytic from objective self-fashioning to affiliative self-fashioning, I seek to emphasize that root-seekers' aspirations to be oriented on the African continent and/or within its diaspora mediate how technoscience becomes incorporated into self-making. Affiliative self-fashioning accounts for how identities culled from genetic genealogy are shaped not only by "received-facts" but also by desires for diasporic connection—a confluence that impacts root-seekers' evaluations of genetic genealogy testing and, in turn, the way that the data it provides is incorporated into their lives. Affiliative self-fashioning thus reflects, on subjective and interpersonal levels, an aspect of the interpretive arc of ethnic lineage testing that I term the factness of diaspora.

Diaspora and Relatedness

Purveyors of genetic genealogy testing claim that their services trace or reveal otherwise unavailable information about ancestry and ethnicity. However, at present, matching a consumer's DNA against proprietary genetic databases comprised of samples from contemporary populations, as African Ancestry and other genetic ancestry tracing companies do, cannot establish kinship with any certainty; ethnic lineage analysis does not associate a root-seeker with specific persons at precise spatial-temporal locations. Also, owing to both technical limitations (e.g., mt-DNA and Y-DNA tests compare a consumer's genetic sample to a selective proprietary sample database and analyze less than 1% of a test-taker's DNA, providing probabilistic outcomes) and historical dynamics (e.g., racial and ethnic identities are sociocultural phenomenon and human migration patterns suggest that contemporary social groups cannot be easily correlated with earlier ones), the associations inferred through the use of genetic genealogy are necessarily provisional (Bolnick et al., 2007; Bolnick, this volume).

In providing associations that are under-specified, genetic genealogy tracing presents consumers with the paradox of *imprecise pedigree*. Root-seekers' awareness of this paradox is indicated by their use of ostensibly redundant phrases such as "DNA cousins" and "genetic kin" (Nash, 2004). These composite descriptors, of course, acknowledge DNA analysis as the medium of affiliation. However, because the words "cousin" and "kin" are already commonly understood to connote "biogenetic ties" (Schneider, 1968), the placement of the adjectives "DNA" and "genetic" before them should therefore be unnecessary.[5] Thus the circulation of these phrases also seems to suggest that the associations supplied through genetic genealogy are qualified and, therefore, must be rhetorically set apart from "natural" kinship or, in other words, that genetic genealogy testing is categorical yet imprecise. It is in this space of indeterminancy that the factness of diaspora unfolds. Root-seekers forge "cultures of relatedness" (Carsten, 2000)—relationships, experiences, and narratives—that have some basis in molecular-level analysis but are also extra-genetic.

The recent family reunion of Marvin, a genealogist from the southern United States, featured an appearance by someone he described as a "genetic kinswoman." Some months prior, Marvin had purchased a genetic genealogy test from African Ancestry that associated him with the Mbundu people, the second largest ethnic group in the south-central African country of Angola. Marvin shared his results with a friend who subsequently put him in touch with Gertrudes, an Angolan immigrant neighbor of Mbundu ethnicity. At their first meeting, Marvin recalled Gertrudes as being "very accepting."

He continued, "She said that one of her passions is to connect with African Americans and tell them about their history in Africa and to let them know that, as she always says, 'We are one.' [She believes that] there is a disconnect between African Americans and Africans, and she's trying to bridge the gap. One of her missions is to connect with more African Americans [and] teach them about Africa."

Gertrudes subsequently invited Marvin to attend a celebration of the 30th anniversary of Angola's independence from Portugal, hosted by the voluntary association that she helms for immigrants from the African country. Here, Marvin and a cousin who attended the party with him felt accepted by the larger Angolan expatriate community as well. "Once we told everyone there that our family came from Angola, they all said, 'Welcome home. You're home now.' They even made me and my cousin get up on the dance floor. You know, they do a ring dance? . . . They told us, 'You gotta come dance. Dance for your homeland!' "

In turn, Gertrudes would attend Marvin's family reunion some months later. "Her presence was powerful," Marvin recollected. "[She talked] about the importance of us coming together as a group of Africans. She expressed that we are all Africans and that Europeans try to divide us but now we must come together. And she also told our family some very interesting facts about the Mbundu people. And that was awesome, just for the family to hear about the people we descend from . . . directly from an Mbundu person. It was very powerful. She had the full attention of the whole family. Everybody was just sitting there in awe of her presence. . . . It was uplifting and powerful just to hear her tell us something about our African roots."

The social exchange carried out between Marvin and Gertrudes points to how genetic genealogy testing circulates as a "diasporic resource" (Brown, 1998, p. 298; 2005, p. 53). As anthropologist Jacqueline Nassy Brown explains, diasporic resources can include "cultural productions such as music, but also people and places . . . [and the] iconography, ideas, and ideologies" of one black community (1998, p. 298) that are employed by another as formative schema for political consciousness, collective empowerment, and identity formation. In Brown's work, the concept describes, for example, how knowledge of a historic event such as the civil rights movement of the mid-20th-century United States circulated globally via the media, popular culture, and social networks to become an important touchstone of self-determination for blacks in Liverpool, England, in the 1990s. In the context of genetic genealogy testing, the concept of diasporic resources elucidates how genetic information occasions "biosociality" (Rabinow, 1996) between African communities and their diasporas, even in the absence of evidence of specific kinship ties. An imprecise pedigree connects Marvin and Gertrudes as "genetic kin" and "Africans." The diasporic relatedness resulting from ethnic lineage testing is genetic inference inspired by genealogical aspiration and enacted through social interaction.

The Ethics of Diaspora

In recent years, there has been increased scholarly interest in the study of "deterritorialized" (Appadurai, 1996) or diasporic communities. Rogers Brubaker recently described this ideational proliferation as a " 'diaspora' diaspora": "a dispersion of the meanings of the term in semantic, conceptual, and disciplinary space" (2005, p. 1). While efforts to refine the concept of diaspora persist, many scholars are in agreement that its hallmarks include dispersal from long-held geographic homes, the constitution of a collective identity or consciousness in response to the experience of dispersal, connection to a place of geographic origin forged through practices such as communication, travel/tourism, philanthropy, and political engagement, and the circulation of collective memories, myths, or imaginaries about the homeland. Diverse diasporas, born of distinct historical, political, and economic push-and-pull factors, share these general contours.

Some theorists suggest that the African diaspora that began in the 16th century is "exceptional" (Tölölyan, 1996, p. 13) among human dispersals of the past and present because it was a forced migration set in motion by the demand for slave labor and one which spurred the process of ethnogenesis—the substitution of specific African identities for more general collectivities, such as Pan-African, African American, and Afro-Caribbean (Gomez, 1998; Kelley & Patterson, 2000).[6] For, as Safran maintains, a "*specific* homeland cannot be restored" to slave descendents (1991, p. 90, emphasis added). Because an African homeland cannot be restored, it has been imagined or "rememoried" (Morrison, 1988).[7]

How "Africa" has been envisioned by its slave descendent diaspora is a topic of debate among theorists. At issue is the *ethics* of imagining Africa and diasporic connection: What is the substance of diaspora? Who in the diaspora gets to imagine "home"? How is it imagined and to what ends? While some scholars maintain that the conceptions of Africa that underlie diasporic consciousness may have many foundational bases, including political ideology, cultural production, desire, common experiences of oppression or redemption, and communication practices (e.g., Appadurai, 1996; Clifford, 1994, p. 304–305; Edwards, 2003; Kelley & Patterson, 2000, p. 13–15; Safran, 1991, p. 83–84; Walcott, 2005), others contend that diasporic claims to and about Africa—particularly those of African Americans—can be essentialist, homogenizing, and instrumental (e.g., Appiah, 1993; Clarke, 2004; Gilroy, 1993; White, 1990).

Paul Gilroy (1993), arguably the most influential critic of originary imaginings of Africa, argues in *The Black Atlantic* that "Africa" has been inaccurately conceived of as a trans-historical umbilicus linking blacks globally to a regal, pre-lapsarian past. Indeed, a principal theme of the theorist's body of work is anti-essentialism, in particular, the contention that notions of black transnationalism should not be based upon "the stern discipline of primordial kinship and rooted belonging" (Gilroy, 2000, p. 123). Gilroy's discomfort with the idealization of roots stems from insights gained from his valuable inquiries into "raciology"—the constellation of discourses, many drawn from the biological sciences, which sustain and justify epistemologies of race and racism and, in turn, social inequality. As an antidote to racial essentialism, Gilroy alternately advances an understanding of diaspora as network, interchange, and circulation. "Primordial kinship" and the search for roots are thus contrasted with an ethico-cultural conception of diaspora.

Notions of diapora rooted in kinship may be better conceptualized as "cultures of relatedness" (Carsten, 2000) in which biological facts are not the necessary conditions of possibility for social ones. In Carol Stack's classic ethnography, *All Our Kin* (1974), for example, kinship among urban blacks in "The Flats" is based on the exchange of economic resources and caring labor between residents. As Stack shows, kinship terms such as "aunt" and "brother" are used by members of the community, but these categories do not connote nature or blood; rather, these terms are engaged *despite* lack of demonstrable genetic links. Similarly, Judith Butler offers "the social organization of need" as one example of kinship based on "consensual affiliation" rather than "blood ties" (2002, p. 74). Recent scholarship on new reproductive technologies has shown that "biology" and "family" can be de-coupled through egg donation, surrogacy, and adoption (e.g., Carsten, 2000; Franklin & McKinnon, 2000; Grayson, 1998; Hayden, 1995).

Kinship can thus have many bases. Viewed through the prism of this recent scholarship, the discourses and practices of kinship facilitated by genetic genealogy testing can be understood to scale up to diaspora without necessarily scaling down to human biological essences. Indeed, the forms of sociality fostered by genealogy testing—both the aspirations for affiliation that inspire its use and the relationships it may occasion—are conduits through which the networked conception of diaspora that Gilroy advocates may take shape. Affiliations that incorporate biogenetic facts may nonetheless be the "families we choose" (Weston, 1997).

Conclusion

I have advanced the concept of the factness of diaspora to describe how phenomena seemingly extrinsic to genetic genealogy testing like that offered by

African Ancestry facilitate its legitimacy. The legitimacy of genetic genealogy testing is built on cultural scaffolding, including the "authentic expertise" of scientist-entrepreneur Rick Kittles; the process of affiliative self-fashioning that may be embarked upon following the receipt of test results; and the diasporic relatedness that this information may support. Mistrust of scientific authority and concerns about privacy have a non-unique but significant history in black communities. A shared background between a scientist and consumers, then, becomes crucial to making the facts of genetic genealogy testing efficacious and meaningful. As Greely (this volume) points out, the forms of association on offer through genetic genealogy testing, while tracing lines of matrilineage and patrilineage, do not and cannot establish direct lines of descent, and thus in practice are necessarily flexible and "fictive." As a consequence, root-seekers also become root-makers, taking up those elements of the testing that facilitate associations that are important to them. The factness of diaspora provides a window on how diasporic resources are put to the purpose of constructing individual and collective identity, helping us to understand why root-seekers selectively invest in genetic genealogy testing.

NOTES

I wish to thank Sandra Soo-Jin Lee, Barbara Koenig, and Sarah Richardson for their editorial guidance and Ashley Hicks for her assistance. I am appreciative of the other contributors to this volume, especially Duana Fullwiley, Rick Kittles, Kim TallBear, and two anonymous reviewers for their valuable feedback. During the course of my research, a number of individuals have been interlocutors, critics, and mentors, including Lundy Braun, Adele Clarke, Troy Duster, Anne Fausto-Sterling, Paul Gilroy, Evelynn Hammonds, and Stefan Helmreich. I thank Mary Barr, Lindsay Greene-Upshaw, and Elizabeth Olson for research assistance. An earlier version of this chapter was presented at the International Center for Advanced Studies at New York University (ICAS); thank you to the ICAS community for their insightful suggestions and the collegial manner in which they were delivered. My deepest debt of gratitude is owed to the genealogists who generously welcomed me into their root-seeking communities.

1. African Ancestry reports that approximately 25% to 30% of root-seekers using its PatriClan (Y-chromosome) test will not match any of the paternal lines in the African Lineage Database (ALD). In such instances, the customer may be advised to have his or her sample (a male relative's sample in the case of a female root-seeker) matched against a "European database" (Langley, 2003). Because the ALD is extensive but not exhaustive, however, there is some chance that matching "African" genetic markers are not yet contained in African Ancestry's database.

2. I distinguish "the factness of diaspora" from Latour's (1999) "factish" (fact and fetish). While both concepts are concerned with and reflect the dual constitution of scientific and cultural knowledge—the combination of the artifacts of reason with a field of value—with the factness of diaspora, I am specifically concerned with the historical experience of displacement and oppression and the subsequent

aspirations for reconciliation that animate a particular orientation to the cultures of science. In *Pandora's Hope*, Latour asks whether an idea is "constructed well enough to become an autonomous fact" (1999, p. 274). Here, I am interested in the values that shape the adjudication and interpretation of facts, such that we must understand them as "factness." My thanks to Natasha Dow Schull for encouraging me to draw this distinction explicitly.

3. Daughters of the American Revolution (DAR) is a genealogy- and membership-based organization for women who can trace their lineage to an individual who aided in the cause of the United States' independence from England. The DAR has a checkered history with regard to race relations. In 1939, it prohibited renowned African American singer Marion Anderson from performing at the DAR-owned venue Constitution Hall in Washington, D.C., that only allowed whites on its stage (Anderson, 1956). As recently as 1984, a woman who fulfilled all DAR membership requirements was not initially allowed to join the group because she was African American (Kessler, 1984).

4. In order to protect the privacy of my informants, all names used here are pseudonyms unless otherwise indicated. The names of the purveyors of African Ancestry are used because they are public figures. I interviewed subjects who attempted to trace their family genealogy by conventional means prior to purchasing genetic genealogy testing services as well as subjects whose first foray into genealogy was the purchase of a test kit.

5. As I address below, this is the common usage of these kinship terms, but not their exclusive meaning. See also contributions by Dolgin, McKinnon, Rapp, and Weston in *Naturalizing Power* (Yanagisako & Delany, 1995).

6. There are, of course, many African diasporas, including post-1965 immigration from the Caribbean to the United States (Bryce-Laporte & Mortimer, 1983), the "windrush" of the mid-20th century that brought blacks from the Caribbean to the United Kingdom (Phillips & Phillips, 1999), and the more recent migration of Africans to the United States in greater numbers than during the slave trade (Roberts, 2005). This chapter is concerned primarily with the older migrations that were spurred by the slave trade and dispersed Africans to the Americas, although its insights are, hopefully, more broadly applicable.

7. As developed in Morrison's *Beloved*, "rememory" is the continuous existence in the present of something lost or forgotten in the past.

REFERENCES

Abu El-Haj, N. (2007). The genetic reinscription of race. *Annual Review of Anthropology, 36*, 283–300.

African Ancestry. (2006). Retrieved July 10, 2006, from http://www.africanancestry. com.

Anderson, M. (1956). *My Lord, what a morning: An autobiography.* New York: Viking Press.

Appadurai, A. (1996). *Modernity at large: Cultural dimensions of globalization.* Minneapolis: University of Minnesota Press.

Appiah, A. (1993). *In my father's house: Africa in the philosophy of culture.* New York: Oxford University Press.

Bay, M. (2006). In search of Sally Hemings in the post-DNA era. *Reviews in American History, 34,* 407–426.

Bolnick, D., Fullwiley, D., Duster, T., Cooper, R., Fujimura, J. H., Kahn, J., et al. (2007). The business and science of ancestry testing. *Science, 318*, 399–400.

Bolnick, D. A. (2008 [this volume]). Individual ancestry inference and the reification of race as a biological phenomenon. In B. A. Koenig, S. S.-J. Lee, & S. Richardson (Eds.), *Revisiting race in a genomic age* (pp. 70–85). New Brunswick, NJ: Rutgers University Press.

Brown, J. N. (1998). Black Liverpool, black America, and the gendering of diasporic space. *Cultural Anthropology, 13*, 291–325.

Brown, J. N. (2005). *Dropping anchor, setting sail: Geographies of race in black Liverpool.* Princeton, NJ: Princeton University Press.

Brubaker, R. (2005). The "diaspora" diaspora. *Ethnic and Racial Studies, 28* (1), 1–19.

Bryce-Laporte, R. S., & Mortimer, D. M. (Eds.). (1983). *Caribbean immigration to the United States.* Washington, DC: Research Institute on Immigration and Ethnic Studies, Smithsonian Institute.

Butler, J. (2002). *Antigone's claim: Kinship between life and death.* New York: Columbia University Press.

Carsten, J. (2000). *Cultures of relatedness: New approaches to the study of kinship.* Cambridge: Cambridge University Press.

Cavalli-Sforza, L., Menozzi, P., & Piazza, A. (1994). *The history and geography of human genes.* Princeton, NJ: Princeton University Press.

Clarke, K. (2004). *Mapping Yoruba networks: Power and agency in the making of transnational communities.* Durham: Duke University Press.

Clifford, J. (1994). Diasporas. *Cultural Anthropology, 9*, 302–338.

Dumit, J. (2003). Is it me or my brain? Depression and neuroscientific facts. *The Journal of Medical Humanities, 24* (1–2), 35–47.

Duster, T. (2003 [1990]). *Backdoor to Eugenics* (2nd ed.). New York: Routledge.

Edwards, B. H. (2003). *The practice of diaspora: Literature, translation, and the rise of black internationalism.* Cambridge, MA: Harvard University Press.

Franklin, S., & McKinnon, S. (2000). New directions in kinship study: A core concept revisited. *Current Anthropology, 41*, 275–279.

Fullwiley, D. (2008 [this volume]). The molecularization of race: U.S. health institutions, pharmacogenetics practice, and public science after the genome. In B. A. Koenig, S. S.-J. Lee, & S. S. Richardson (Eds.), *Revisiting race in a genomic age* (pp. 149–171). New Brunswick, NJ: Rutgers University Press.

Fulwood, S. (2000, May 29). His DNA promise doesn't deliver. *Los Angeles Times*, p. A12.

Gates, H. L., Jr. (Producer). (2006). African American lives. Television broadcast. Public Broadcasting Station.

Gilroy, P. (1993). *The black Atlantic: Modernity and double consciousness.* Cambridge, MA: Harvard University Press.

Gilroy, P. (2000). *Against race: Imagining political culture beyond the color line.* Cambridge: Harvard University Press.

Gomez, M. A. (1998). *Exchanging our country marks: The transformation of African identities in the colonial and antebellum South.* Chapel Hill: University of North Carolina Press.

Grayson, D. (1998). Mediating intimacy: Black surrogate mothers and the law. *Critical Inquiry, 24*, 525–546.

Greely, H. T. (2008 [this volume]). Genetic genealogy: Genetics meets the marketplace. In B. A. Koenig, S. S.-J. Lee, & S. S. Richardson (Eds.). *Revisiting race in a genomic age* (pp. 215–234). New Brunswick, NJ: Rutgers University Press.

Haley, A. (1976). *Roots: The saga of an American family.* New York: Doubleday.

Hammer, M. F. (1995). A recent common ancestry for human Y chromosomes. *Nature, 378* (6555), 376–378.

Hayden, C. (1995). Gender, genetics, and generation: Reformulating biology in lesbian kinship. *Cultural Anthropology, 10,* 41–63.

Hostetler, A. J. (2003, April 24). Who's your daddy? Genealogists look inside their cells for clues to their ancestors. *Times-Dispatch.* Retrieved May 17, 2003, from http://www.timesdispatch.com/news/health/MGBZJG4MVED.html.

Jackson, T., & Baron, A. (Producers). (2003, February). Motherland: A genetic journey. Television broadcast. British Broadcasting Company.

Jasanoff, S. (2004). *States of knowledge: The co-production of science and the social order.* New York: Routledge.

Jobling, M. A., & Tyler-Smith, C. (1995). Fathers and sons: The Y chromosome and human evolution. *Trends in Genetics, 11* (11), 449–455.

Kelley, D. G., & Patterson, T. (2000). Unfinished migrations: Reflections on the African diaspora and the making of the modern world. *African Studies Review, 43* (1), 11–45.

Kessler, R. (1984, March 12) Black unable to join local DAR: Race is a stumbling block. *Washington Post,* p. A1.

Langley, G. (2003, July 20). Genealogy and genomes: DNA technology helping people learn more about who they are and where they come from. *Baton Rouge Advocate,* p. 16.

Latour, B. (1999). *Pandora's hope: Essays on the reality of science studies.* Cambridge: Harvard University Press.

Maganini, S. (2003, August 3). DNA helps unscramble the puzzles of ancestry. *Sacramento Bee.* Retrieved October 6, 2003, from http://www.sacbee.com/content/news/v-print/story/7155827p-8103039c.html.

Morrison, T. (1988). *Beloved: A novel.* New York: Plume.

Nash, C. (2004). Genetic kinship. *Cultural Studies, 18,* 1–33.

Nelson, A. (forthcoming, 2008). Bio science: Genetic genealogy testing and the pursuit of African ancestry. *Social Studies of Science.*

Phillips, T., & Phillips, M. (1999). *Windrush: The irresistible rise of multi-racial Britain.* London: Harper Collins.

Rabinow, P. (1996). *Essays on the anthropology of reason.* Princeton, NJ: Princeton University Press.

Reardon, J. (2005). *Race to the finish: Identity and governance in an age of genomics.* Princeton, NJ: Princeton University Press.

Roberts, S. (2005, February 21). More Africans enter U.S. than in days of slavery. *New York Times.* Retrieved November 6, 2005, from http://www.nytimes.com/2005/02/21/nyregion/21africa.html.

Roylance, F. D. Reclaiming heritage lost to slavery. *Baltimore Sun.* Retrieved on May 6, 2003, from http://www.sunspot.net/news/health/bal-te.african17apr17,0,2436405.story?coll=bal-health-headlines.

Safran, W. (1991). Diasporas in modern societies: Myths of homeland and return. *Diaspora, 1* (1), 83–99.

Sailer, S. (2003, April 28). African Ancestry, Inc. traces DNA roots. *Washington Times.* Retrieved May 1, 2003, from http://washingtontimes.com/upi-breaking/20030428-074922-074922-7714r.htm.

Schneider, D. M. (1968). *American kinship: A cultural account.* Englewood Cliffs, NJ: Prentice-Hall, Inc.

Shriver, M. D., & Kittles, R. A. (2008 [this volume]). Genetic ancestry and the search for personalized genetic histories. In B. A. Koenig, S. S.-J. Lee, & S. S. Richardson (Eds.), *Revisiting race in a genomic age* (pp. 201–214). New Brunswick, NJ: Rutgers University Press.

Stack, C. B. (1974). *All our kin: Strategies for survival in a black community.* New York: Harper & Row.

Thomas Jefferson Foundation. (2000). Report of the Research Committee on Thomas Jefferson and Sally Hemings. http://www.monticello.org/plantation/hemingscontro /hemings_report.html.

Tölölyan, K. (1996). Rethinking diaspora(s): Stateless power in the transitional moment. *Diaspora, 5* (1), 3–35.

Walcott, R. (2005). Outside in black studies: Reading from a queer place in the diaspora. In E. P. Johnson and M. G. Henderson (Eds.), *Black queer studies: A critical anthology* (pp. 90–105). Durham: Duke University Press.

Weston, K. (1997). *Families we choose.* New York: Columbia University Press.

White, E. F. (1990). Africa on my mind: Gender, counter discourse, and African-American nationalism. *Journal of Women's History, 2* (1), 57–73.

Yanagisako, S., & Delany, C. (Eds.). (1995). *Naturalizing power: Essays in feminist cultural analysis.* New York: Routledge.

PART FOUR

Race and Genetics in Public Discourse

14

Moving beyond the Two-Race Mantra

PAMELA SANKAR

Use of race and ethnicity as variables in genetic research has expanded rapidly over the last decade. As a result, the perennial controversy over the use of race in research has expanded beyond the narrow confines of journal pages and academic conferences into news stories, documentaries, and museum exhibits. Discussion has been vigorous and productive and has fostered much needed interdisciplinary exchange. Predictably, because divergent concepts and terminology limit participants' grasp of one another's positions, some exchanges have faltered. While common in interdisciplinary work, such misunderstandings are likely to be particularly troublesome when they concern concepts, such as race, with meanings that not only differ across academic disciplines but also have well-developed and divergent popular meanings. My interest here is with misunderstandings that have emerged between social scientists and genetics researchers, in particular when participants, typically social scientists, frame the controversy as a debate over two (and, at least implicitly, *only* two) concepts of race: biological (or genetic) race, which is wrong, and socially constructed race, which is right.[1] An article advising public school teachers how to discuss race illustrates the two-race theory: "Multicultural education has helped us to understand racism. . . . But it has not helped us understand the two concepts of race: the biological one and the social one. . . . Anthropologists . . . say that 'races don't exist' (in other words, . . . they reject the concept of race as a scientifically valid biological category) and . . . they argue instead that 'race' is a socially constructed category" (Mukhopadhyay & Henze, 2003).

Scholarly works endorse this duality as well. *Anthropology and Race: The Explanation of Differences*, a respected book that remains in print 15 years after its publication, describes one of its primary aims as putting "together two ideas of race—as a folk (sociological) concept and as a failed and dis-carded analytical (biological) concept" (Shanklin, 1993). Further, while the two-race theory might have originated in anthropology (UNESCO, 1952), a passage from a prominent legal scholar demonstrates its acceptance in other fields as well: "When some people use the [sic] 'race' they attach a biological meaning, still others use 'race' as a socially constructed concept. It is clear that even though race does not have a biological meaning, it does have a social meaning that has been legally constructed" (Haney-Lopez, 1994).

The claim that there are only two ways of looking at race has become so widely accepted that it has taken on the status of common sense. And while in some instances, such as the classroom, the two-race mantra func-tions effectively as a rhetorical device (Hartigan, 2006), in other settings it impedes rather than facilitates communication. This is especially true when people from different disciplines are discussing race. Presuming the theory is self-evident, proponents overlook the need to explain themselves, in particular, to explain how they interpret statements about race that seem to imply a biological or genetic basis to the concept. Most often they inter-pret such statements as being based on typological race, a long repudiated, essentialist notion that races are fundamentally important, hierarchically arranged, static, natural groupings. Failing to explain this assumption when debating the meaning of race hinders exchange by attributing a far broader set of assumptions about race to speakers who are largely unaware of hav-ing been assigned to the wrong side of the argument by the strict division imposed by the two-race theory. While the resulting confusion raises ques-tions enough about persisting with this theory, there is yet another unfortu-nate consequence of relying on it. Insisting that there are two and only two concepts of race turns social constructionism into one side of an argument. As such it becomes a stance to defend rather than a theory to apply. This positioning restricts the range and type of knowledge about scientific inter-est in human variation that social constructionism can help produce and stymies progress in the debate over using race.

Three points about what this chapter does *not* concern are important to emphasize before elaborating its argument. First, this chapter does not ask whether race is a social, biological, or genetic construct. Rather it explains why framing the current debate solely in these terms is a naïve mistake. Second, this chapter does not address reliance on the two-race theory, gen-erally, but focuses on its implications in discussions about recent genetic research. And third, the intersection of race and biology evident in the racial character of health disparities in the United States, while extremely important to understand (Sankar et al., 2004), is not specifically examined here. That race has profound and pervasive biological effects because of people's beliefs about race, irrespective of any biological (or genetic) basis, provides testimony to the unrelenting and pernicious presence of racism in contemporary society. Nonetheless this chapter instead addresses another intersection of race, genetics, and biology. It is something we might call statistical race, not to add to the growing list of race terms, but to help distinguish it from other uses of race and to provide a way to discuss and analyze it.

Anthropologist, Observe Thyself

The incident that prompted this chapter occurred during a 2004 conference sponsored by the American Anthropological Association (AAA) as part of its work to develop a traveling exhibit for science museums about race and human variation (American Anthropological Association, 2007). To help plan the exhibit, the AAA convened an "interdisciplinary conference" to bring "together scholars from a wide spectrum of disciplines to share their expertise and develop an agenda for future research and education on race and human variation" (American Anthropological Association, 2004). The first day of the conference featured speakers discussing human diversity from various perspectives, including law, history, culture, and genetics. Audience response to speakers who challenged or rejected determinist or biological explanations of human diversity was generally positive. Response to the one presentation endorsing current genetic research on human variation was not. The speaker for that session, "Genetics and Human Variation," was Lynn Jorde, a professor of human genetics from the University of Utah.

Jorde opened his presentation by acknowledging the controversy surrounding the use of race in medical practice and research and by citing quotes from prominent researchers. The first quote expressed unequivocal support for the concept of race while the second rejected it as biologically meaningless. Jorde began the main part of his talk by explaining that humans, as a young species, differed very little genetically and that most of the variation among humans that did exist would be found within rather than between populations and that between population differences were largely artifacts of population migration history. These differences, he continued, might account for some diseases such as hypertension occurring at different rates among different populations, such as Europeans, Africans, or Asians. He cautioned, however, that taking categories or ideas developed for describing populations and applying them to individual patients is a mistake. Thus, he stated that despite research that suggests that African Americans might not respond as well as whites to certain heart medications "a particular African American . . . could respond better than a particular white." He reiterated this point, "[One] cannot think of people as the

averages of their populations." Thus, "looking at race is not a good indicator" of the likelihood that someone has a predisposition to a particular disease, and it is "better to look at the gene."

"Why then," Jorde asked, "do researchers use race?" "Because," he
answered, "they do not have access to the gene." Thus, despite its drawbacks,
Jorde concluded that race—race as an imperfect proxy or a holding place for
genetic variation—will be used until individual genotyping eliminates the
need for it. In his closing remarks, Jorde stated, "Race is not meaningless,
but it is biologically very imprecise."

Jorde's talk was deliberate and measured. Still, there were any number
of points that anthropologists and other conference participants *might*
have taken issue with, starting with asking if his alternation among the
terms "population," "race," "European," "Asian," and "African" signaled a
conceptual distinction, a conceptual confusion, a habit of speech, or some
combination of these. Or, whether the recognized imprecision of race introduced inaccuracies into research that challenged the validity of results. Or,
considering how long it might take for individual genotyping to be available to replace race, was the scientific research in which race was merely a
placeholder sufficiently important to justify reinforcing a concept like race,
known to be ubiquitously, some would argue, inevitably, associated with
stereotyping and discrimination?

For the most part, however, audience response took little account of
Jorde's presentation. Instead, conference participants raised broad challenges, making statements such as "race is just a folk belief"; "genetic
variation and race have nothing to do with each other"; and race is not
biological or coherent but is meaningful only as a "basis for subordination
and as a basis for combating subordination." In other words, the response
cast Jorde's work as supporting a particular notion of biological race (i.e.,
typological race) and then dismissed it, citing anthropology's long-standing,
preferred theory that race is a social construction, an insight that Jorde's
talk did not dismiss so much as ignore.

In failing to respond to Jorde's specific points, social scientists missed an
opportunity to hear about the current debate in genetics over the meaning
of race or to learn about the various functions of race in genetic research.
Anthropologists in particular shortchanged themselves and their discipline.
Instead of listening to, querying, observing, and learning from Jorde as a
friendly and knowledgeable informant, they treated him as a foe.

Perhaps the novelty of witnessing a geneticist declaim about race in a
setting that they controlled explains social scientists' failure to engage in a
scholarly exchange. Few genetics researchers have taken the time to seriously respond to the challenges raised by social scientists to genetic research
that uses race. If they acknowledge such challenges at all, it is usually to

characterize them as ethical issues and so outside their expertise, or to reduce the issue to one of terminology by suggesting that if only social scientists would tell them which terms to use the problem would be solved. Or perhaps, audience members reacted as they did because they were annoyed that a prominent scientist not only would use race in his research but would discuss it openly on their home turf, thus, apparently dismissing the fact (or ignorant of it—which would be worse is hard to say) that decades ago anthropology declared race a myth. However, while not minimizing the influence of disciplinary rivalry on academic conduct, there are other factors that might account for what happened.

The audience response seemed to presume that Jorde embraces race as a basic, broadly relevant, biological feature of humans that separates them into enduring, easily identifiable groups. Since Jorde not only did not say this, but also explained in some detail his actual premise for using race as a variable (a premise relating to the history of human migration, which, notably, no one in the audience sought to challenge and which more or less matches the standard treatment of race in undergraduate anthropology textbooks today [Bailey & Peoples, 2001], we have to ask why the audience apparently attributed these assumptions to Jorde.[2]

One possible explanation is that this mistake is a predictable outcome of the two-race theory. If there are only two concepts of race—one socially constructed and the other biological or genetic—it follows that there is little reason to examine what exactly a speaker means by race. If not framed as socially constructed, there is only one other option: it must be biological or genetic. This, in turn, apparently for those committed to the rhetoric of social constructionism, also means it must be roughly equivalent to typological race. This final leap in particular demands attention.

Types of Race: Typological and Statistical

Typological race refers to the theory popular until the late 1800s and early 1900s that humans are divided into natural, discrete groups that are easily distinguished by their intrinsic properties revealed in appearance, temperament, morality, and intelligence. These properties vary across races but are consistent among members of the same race. This is the notion of race on which race science and racist political programs and beliefs are based. A question of interest to early typological race theorists was how racial types were maintained across generations, that is, how racial features were transmitted (Gannett, 2003; Sauer, 1993). Proposed explanations presumed that the features that constituted a race were passed as a bundle from parents to children, thus maintaining the races as separate groups and producing a new generation largely the same as the last. But the observation that successive generations of offspring are not identical challenged the typological race theory of transmission.

Over the final decades of the 1800s, interest in accounting for human differences coupled with growing attention to Darwin's theory of evolution intensified the drive already underway to develop a new model of heredity. The eventual rediscovery in 1900 of Gregor Mendel's pioneering 1860s genetics research on pea plants led scientists to propose that as with features of plants, so too with humans: that genes for characteristics such as those associated with racial types are not transmitted together as a bundle from parents to offspring (Gannett, 2003). No hereditary essence passes from generation to generation. Instead, in each offspring's creation alleles recombine randomly. This recombination provides the source of variation needed to support Darwin's theories of how change occurs and explains the scarcity of static traits that proponents of typological race unsuccessfully sought in support of their theory.

Although some observers had challenged the existence of human races long before this time, as a consequence of the evolutionary synthesis (bringing together Darwin and Mendel), typological race fell from favor along with scientific support of races as naturally existing, discrete reproductive groups. An alternative idea of open populations, defined on the basis of sharing a common gene pool, emerged as the model of a reproductive community. This theory holds that all humans belong to the same species. Population groups exist, but they do not have natural or fixed limits. They are not distinguished qualitatively, as typological race presumes, but quantitatively, based on differences in gene frequencies (Lewontin, Rose, & Kamin, 1984).

Whether the new theories jettisoned all of the assumptions of typological race is a topic of some dispute. Lisa Gannett's (2003) useful account argues that population genetics did not provide the complete and radical departure from typological race that standard accounts of the emergence of population genetics have suggested. A more accurate account, she asserts, would treat population genetics as having incorporated rather than wholly discarded typological race.

In genetic research today there is a use of race that cannot be easily categorized as either socially constructed or typological race. We might call it "statistical race" because of its reliance on numerical data to represent population differences. The differences it identifies and represents as racial derive from the fact that people of common ancestry are more likely to share certain genes or alleles (versions of genes) than those who do not share ancestry. This is because to some extent both ancestry and genetic variation are geographically distributed.

Genetic mutations—a source of variation—happen randomly and can be passed down and spread in human reproduction. Through natural selection,

mutations that favor adaptation to a certain environment increase in frequency while those that do not decrease or disappear altogether. As human
populations began to migrate out of Africa and journey to other continents,
groups came to inhabit markedly divergent environments. Natural selection
acted differentially on the random mutations that appeared, and, as a result,
human populations began to differ genetically from one another, albeit in
very minor ways. Neutral mutations, mutations that also occur randomly
and are shared through reproduction but that have no phenotypic effect,
added to these differences.

Groups that traveled the greatest distances from one another had less
contact and were less likely to share such mutations, while groups that
lived closer to one another and had more contact were more likely to have
these mutations in common. Over time, these genetic differences took on a
geographic pattern, and today people whose ancestors came from the same
continent or geographic region often share a set of distinctive differences,
or alleles. These differences are discernible as population-level frequencies.
They are seldom if ever universal in groups where they are likely to appear
and seldom completely absent in groups where they are rare. It is rather that,
statistically, they appear more often in some groups than in others. Thus,
while they can be used to characterize a population, they are not predictably
present or absent in any one particular member of a population.

The connection of these observations to race is that popularly recognized categories of ancestry, especially those associated with continents,
sometimes overlap with popularly recognized categories of race, such as
European, African, or Asian. For example, people who socially identify as
Asian are likely to have ancestors from Asia. Any person so identifying is
more likely to have a certain set of alleles in common with another person
so identifying than they are to share these particular alleles with someone
of European or African ancestry (Bamshad, Wooding, Salisbury, & Stephens,
2004; Jorde & Wooding, 2004). For these reasons, a person's self-identified
race might correspond to a predictable set of alleles. This is not always or
necessarily the case, but this correspondence has led some genetics researchers to assert that there is a genetic basis to race or that self-identified race
can act as a proxy for genetic variation.

As Jorde conceded, this concept of race might be "biologically very
imprecise." For that matter, one could also argue that this concept of race
is unimportant to health (relative, for example, to many well-documented
environmental stressors), or one might want to argue that it is a mistake to
promote research that focuses on it (if, for example, such research encourages racial stereotyping, which itself contributes to ill health) (Burgess, van
Ryn, & Crowley-Matoka, 2006; Fazio & Dunton, 1997). Either of these arguments or a number of other arguments *might* be made against this notion

of race. But without recognizing it as conceptually distinct from typological race, critics cannot accurately and effectively assess whether it is imprecise, unimportant, or likely to encourage stereotyping. Nor, apparently, will they be able to communicate successfully with scientists who insist on using it but who often themselves are not entirely clear about its meaning, as our recent research on the use of race as a variable in genetic research has shown (Sankar, Cho, & Mountain, 2007).

Typological Race Focus

Critics who rely on the two-race theory typically do not recognize the emergence of statistical race. Instead, they equate the characterization of race as genetic or biological with typological race. This reaction was evident in the audience response to Jorde and appears in public statements by anthropological organizations: "Pure races in the sense of *genetically homogenous populations* do not exist" (American Association of Physical Anthropologists, 1996, emphasis added). "In the United States both scholars and the general public have been conditioned to viewing *human races as natural and separate divisions* within the human species *based on visible physical differences. . . .* [H]owever, it has become clear that human populations are not *unambiguous, clearly demarcated, biologically distinct groups*" (American Anthropological Association, 1998, emphasis added). Scholarly pieces also echo this sentiment: "By . . . 'biological race' I mean the view . . . that there exist natural, physical divisions among humans that are hereditary, reflected in morphology, and roughly but correctly captured by terms like Black, White, and Asian (or Negroid, Caucasoid, and Mongoloid)" (Haney-Lopez, 1994).

Critics do not explain why they assume that assertions about possible biological or genetic features of race inevitably signal an endorsement of typological race, but statements that come close to furnishing reasons fall into two related categories. Some fall back on the logic of the two-race theory: "There is no half way between seeing race as biologically valid or not. Any reformation of race as biology will simply be interpreted as race in the older typological paradigm" (Goodman, 1998). Others, unaware that they are boxed in by the two-race theory, deflect blame to the deterministic power of language: "Just to talk about them as 'racial problems' is to make it sound as if biological variation is at the root, which is to stack the deck against a modern comprehension of, and ultimate solution to, those problems" (Marks, 2002).

Race does carry a tightly bundled set of meanings that are generally, although not always, negative, as discourse in minority communities has demonstrated. But whether the word's history has determined that its meanings cannot be unbundled and allowed to circulate and evolve with the rest of language is another argument altogether. It is not one with which all experts on race concur. K. Anthony Appiah, for one, has maintained that he has "no problem with people who want to use the word race in population genetics" (1996, p. 73). Considering the pervasive confusion over the term's meaning, Appiah's confidence is not one I share. But the point here is less to say which position might be right than to stress the need to fully examine the motivation and consequences of all positions.

Insistence that any hint of a link between race and genetics is tantamount to endorsing typological race and, as such, must be vigorously resisted might be fueled also by the worry that typological race is the underlying concept that informs public beliefs about race. This at least seems to be the position taken by the AAA's "Statement on 'Race,' " which maintains that "the general public has been conditioned to viewing human races as natural and separate divisions within the human species based on visible physical differences" (American Anthropological Association, 1998).

Unfortunately, despite the considerable concern that academics and the government regularly voice about race and racism, notably little research has been conducted that asks members of the public how they define race and what they think its relation might be to culture, history, biology, or genetics. It is not unreasonable to infer from the regular incidence of overt racist acts and of continuing, pervasive, and fundamental institutionalized racism that many people, knowingly or unknowingly, accept the tenets of typological race. Such an inference, however, is only that and suffers the same limits of similar conclusions based on what from a distance seems reasonable. A 1979 study conducted in the United States showed that most subjects agreed with statements that rejected typological race (Stark, Reynolds, & Lieberman, 1979), while a 2004 study that examined beliefs about race and genetics found considerable support for the belief that racialized groupings were discrete, a finding that seems to indicate acceptance of typological race. However, the 2004 research also suggested that people did not necessarily attribute the existence of these racially "discrete" groups to genetics (Condit, Parrott, Harris, Lynch, & Dubriwny, 2004).

Drawbacks of Focusing on Typological Race

Nonetheless, regardless of the reason for critics' focus on typological race, the result is to limit the arguments they can muster in discussions about current genetic research. For example, many critics cite Richard Lewontin's 1972 finding that there is more genetic variation within than between major racial groupings (Mukhopadhyay & Henze, 2003; Mukhopadhyay & Moses, 1997; Root, 2001; Teo, 2003). Others highlight the instability of race labels across time and place as evidence that the groups referred to cannot be natural and enduring, citing as examples people who, referred to as "white" on the East Coast of the United States, become "Anglo" in New Mexico (Lee, 1993) or who, referred to as "white" or "other" in Great Britain in 1980s, have recently become "Arab" (Aspinall, 1998).

These challenges are not wrong. There is more genetic variation within groups than between them, and race labels do vary. The problem is that they are incomplete. The fact that there is more genetic variation within groups than between them does not *in itself* rule out research about differences between groups. Perhaps, regardless of relative magnitude, differences between groups are considered too important to discount (National Institute of Arthritis and Musculoskeletal and Skin Diseases, 2006; Risch, Buchard, Ziv, & Tang, 2002). Should prostate cancer researchers completely ignore the possibility of a genetic contribution to the high rate of the disease among African American men, a rate that is twice that of white men in the United States *and* that is higher than any other group in the world? (Rebbeck, 2005).

And, while race labels are notably inconsistent and imprecise, presumably, so is any sign with respect to its signifier, whether the phenomenon of interest is framed as "natural" or as "social." The fact that labels assigned to a phenomenon vary does not in itself nullify assumptions that the phenomenon remains stable. The finding that people once labeled East Indian might at another time be labeled Caucasian constitutes a challenge to the legitimacy of genetic research that uses these categories only if one presumes that the research in question is based on a notion of race as fixed and unchanging, i.e., on typological race. To be effective, challenges raised against research that treats race as biological or genetic need to stop relying on rhetorical or formulaic arguments and instead examine such research empirically.

Socially Constructing Race in Genetic Research

Failing to recognize statistical race as conceptually distinct from typological race short-circuits effective analysis of current genetic research and, as the exchange with Jorde demonstrated, hinders communication. Until recently (Fausto-Sterling, 2004; Reardon, 2004), this failure also had the effect of deflecting attention away from a primary site for understanding how race is being socially constructed today. Ironically, this site is the very genetic research many are so eager to repudiate. Focused on defending social constructionism as a position, critics have been slow to apply it as a framework. But the current explosion in genetic research that uses race calls for a forceful, deliberate, critical response, one that could be helped along by de-nominalizing race as a social construction and focusing instead on how this controversy is constructing race.

This change shifts attention from drawing and policing the line between social constructionist and other uses of race to asking how genetics researchers are using race and how these uses reproduce and modify the range of meanings attributed to the term. This raises the question of how the meaning and function of race change across different genetic research projects and how experiences and practices in research laboratories are both cause and consequence of practices and beliefs elsewhere. It also brings activities such as this volume into the fray as yet another source of inventing, revising, and reproducing race.

Observing current genetic research through this lens does several things: First, it removes genetic research from isolation as the polar opposite of social constructionism and brings it into view as one way that race is constructed. Second, it opens the way to recognizing and incorporating contradiction into analysis. It makes clear that people hold inconsistent, shifting, and ambiguous ideas about race, including genetics researchers, many of whom readily accept race as a social construct while using it freely in genetic research. Recognizing these inconsistencies enables more nuanced analyses, highlighting, for example, how beliefs about the various roles of culture, genes, history, fate, divine intervention, and biology all shape beliefs about the nurture-nature divide. And finally, it opens up a new set of inquiries, such as asking how scientists move between popular and scientific ideas about genetic variation, and what, if anything, makes them aware of that movement. Events within the lab? Criticism from peer review? Attending a conference on race and genetic variation?

All of these inquiries mean learning more about what genetics researchers are doing. This requires paying close attention to communication and building a common vocabulary. Without greater clarity about what we, as social scientists, mean when we talk about race and about how meaning changes across contexts, we will be unable to sort out and effectively analyze the significance of evolving concepts and terminology related to race. Moreover, ignoring language blocks a likely avenue to improving how scientists frame the genetic variation that they now describe as "race." This avenue, which requires pointing out to scientists the vagaries of language and the salience of context, will remain closed to us if we ourselves continue to ignore it.

This is not to say that by paying attention to language the controversy will be solved. Language is particularly important in cross-disciplinary exchanges which progress or falter on the success participants have building a common ground from which to view one another's interests and premises.

But even here language is more means than end, if that end is to include full lives for all people unrestrained by arbitrary categories such as race.

NOTES

1. Race terms are used inconsistently across and within disciplines. Some authors distinguish biological race from genetic race; others use them interchangeably. The same holds true for race and ethnicity. Typically, genetic race is used to convey a narrow meaning referring to claims that there is a genetic basis to race categories or that there are genetic analogues to race categories that allow socially assigned race to act as a proxy for genetic differences. Biological race is sometimes used in this sense, but also is used to signal a very different concept, which is that race categories, because of racism, can have biological or health effects. While it may be awkward, I have chosen here to use both terms together, "biological race" and "genetic race," not because I do not recognize that the phrases might carry different meanings but because neither phrase carries a consistent, self-evident meaning.

2. This is an opportune moment to explain that I am not working from a tape recording or independent transcript of these events. Rather I am relying on notes I made of Jorde's presentation because of a deep interest in this topic. Only after presenting an earlier draft of this chapter at the authors' conference for this volume and being encouraged to include in it an account of Jorde's talk did I return to my notes. All of this is to say that Jorde might have said things that encouraged the audience's reaction and that I failed to note. Nonetheless, my notes are extensive and they include many statements he made that ran counter to audience reaction. Thus, even absent a complete transcript, it behooves us to ask why the audience reacted as it did and not otherwise.

REFERENCES

American Anthropological Association. (1998). Statement on race. Retrieved November 1, 2005, from http://www.aaanet.org/stmts/racepp.htm.

American Anthropological Association. (2004). Sponsored conference. Race and human variation: Setting an agenda for future research and education. Old Town Alexandria, Virginia.

American Anthropological Association. (2007). Retrieved March 25, 2007, from www.understandinggrace.org.

American Association of Physical Anthropologists. (1996). Statement on biological aspects of race. *American Journal of Physical Anthropology, 101*, 569–590.

Appiah, K. A., & A. Gutmann. (1996). *Color consciousness: The political morality of race.* Princeton, NJ: Princeton University Press.

Aspinall, P. J. (1998). Describing the "white" ethnic group and its composition in medical research. *Social Science and Medicine, 47* (11), 1797–1808.

Bailey, G. A., & Peoples, J. (2001). *Essentials of cultural anthropology* (1st ed.). Boston: Wadsworth Publishing.

Bamshad, M., Wooding, S., Salisbury, B., & Stephens, J. (2004). Deconstructing the relationship between genetics and race. *Nature Reviews Genetics, 5*, 598–609.

Burgess, D. J., van Ryn, M., & Crowley-Matoka, M. (2006). Understanding the provider contribution to race/ethnicity disparities in pain treatment: Insights from dual process models of stereotyping. *American Academy of Pain Medicine, 7* (2), 119–134.

Condit, C., Parrott, R., Harris, T., Lynch, J., & Dubriwny, T. (2004). The role of "genetics" in popular understandings of race in the United States. *Public Understanding of Science, 13*, 249–272.

Fausto-Sterling, A. (2004). Refashioning race: DNA and the politics of health care. *differences: A Journal of Feminist Cultural Studies, 15* (3), 1–37.

Fazio, R. H., & Dunton, B. C. (1997). Categorization by race: The impact of automatic and controlled components of racial prejudice. *Journal of Experimental Social Psychology, 33*, 451–470.

Gannett, L. (2003). Making populations: Bounding genes in space and in time. *Philosophy of Science, 70*, 989–1001.

Goodman, A. H. (1998). The race pit. *Anthropology Newsletter, 39*, 50–52.

Haney-Lopez, I. F. (1994). The social construction of race: Some observations on illusion, fabrication, and choice. *Harvard Civil Rights–Civil Liberties Law Review, 29* (1), 1–62, 66–67, 11–17.

Hartigan, J. J. (2006). Saying "socially constructed" is not enough. *Anthropology News, 47* (2), 8.

Jorde, L., & Wooding, S. (2004). Genetic variation, classification and "race." *Nature Genetics, 36* (11 suppl), S28–S33.

Lee, S. M. (1993). Racial classifications in the U.S. Census: 1890–1990. *Ethnic and Racial Studies, 16* (1), 75–94.

Lewontin, R. C. (1972). The apportionment of human diversity. *Evolution Biology, 6*, 381–398.

Lewontin, R., Rose, S., & Kamin, L. (1984). *Not in our genes: Biology, ideology, and human nature.* New York: Pantheon Books.

Marks, J. (2002). *What it means to be 98% chimpanzee.* Berkeley: University of California.

Mukhopadhyay, C. C., & Henze, R. C. (2003). Using anthropology to make sense of human diversity. Retrieved November 1, 2005, from http://www.pdkintl.org/kappan/k0305muk.htm.

Mukhopadhyay, C. C., & Moses, Y. T. (1997). Reestablishing "race" in anthropological discourse. *American Anthropologist, 99* (3), 517–533.

National Institute of Arthritis and Musculoskeletal and Skin Diseases. (2006). Strategic plan for reducing health disparities. *National Institute of Health*, from http://www.niams.nih.gov/an/stratplan/strategicplanhd/strategicplanhd.htm.

Reardon, J. (2005). *Race to the finish: Identity and governance in an age of genetics.* Princeton, NJ: Princeton University Press.

Rebbeck, T. R. (2005). Genetics, disparities, and prostate cancer. *LDI Issue Brief, 10* (7), 1–4.

Risch, N., Burchard, E., Ziv, E., & Tang, H. (2002). Categorization of humans in biomedical research: Genes, race, and disease. *Genome Biology, 3* (7), 1–12.

Root, M. (2001). The problem of race in medicine. *Philosophy of Social Sciences, 31* (1), 20–39.

Sankar, P., Cho, M. K., Condit, C. M., Hunt, L. M., Koenig, B., Marshall, P., et al. (2004). Genetic research and health disparities. *Journal of the American Medical Association, 291* (24), 2985–2989.

Sankar, P., Cho, M. K., & Mountain, J. (2007). Race and ethnicity in genetic research. *American Journal of Medical Genetics Part A, 143A* (9), 961–970.

Sauer, N. J. (1993). Applied anthropology and the concept of race: A legacy of Linnaeus. *National Association for the Practice of Anthropology Bulletin, 13* (1), 79–84.

Shanklin, E. (1993). *Anthropology and race: The explanation of differences.* Boston: Wadsworth Publishing.

Stark, J. A., Reynolds, L. T., & Lieberman, L. (1979). The social basis of conceptual diversity: A case study of the concept of "race" in physical anthropology. *Research in Sociology of Knowledge, Sciences, and Art, 2,* 87–99.

Teo, T. (2003). Ethnocentrism as a form of intuition in psychology. *Theory & Psychology, 13* (5), 673–694.

UNESCO. (1952). *The race concept: The results of an inquiry.* Paris: UNESCO.

15

Cops, Sports, and Schools

How the News Media Frames
Coverage of Genetics and Race

SALLY LEHRMAN

At a workshop at Harvard University on covering racial justice and health disparities, a group of journalists listened to science historian Evelynn Hammonds outline the most recent scientific findings on human genetic variation and race (Institute for Justice and Journalism, 2005; Shields et al., 2005). In the discussion afterward, one broadcaster told her colleagues that the news media should drop the word "race" entirely. Journalists were misleading the public by holding onto distinctions that have little if any biological validity, she said. Worse yet, they were reinforcing social categories that had been used to protect privilege. Perhaps, but no editor would allow reporters to excise "race" from the language, one newspaperman responded wryly. What sense would it make anyway, another wondered aloud: How then would they write about race relations and race-based disparities, the issues they were there to discuss?

Most of this multicultural group of journalists covered government, education, the environment, and other areas, not health or medicine. They were quite familiar with race-based inequities in education, racial profiling in the criminal justice system, and environmental racism. They had researched the institutions and history that contribute to social stratification today. If anything, they might be more skeptical than most people about claims that tightly link the fate of racial groups to biology. When faced with the complexity of exploring social disparities, genetics, and race, some jumped quickly to what might seem the easy way out—let's avoid the term "race" altogether.

Of course, this would be next to impossible in news coverage today. Race is an ongoing theme in many of the topic areas that news media emphasize and in which reporters specialize: beats from sports to cops to schools. The reporters and producers at the workshop expressed the fundamental challenge that faces journalism each time it touches on the subject. Whether in the context of breaking news, social issues, or science, how can coverage involving race reach beyond deeply bound social assumptions and shared public understandings?

As journalists attempt to bring the public up to date on the subject of human genetic variation, they can find themselves facing this question without many tools. Science and medical reporters may have little knowledge about the history of race in the United States, while those who report regularly on racial or ethnic groups are likely to have sparse background in science. Whatever their specialty, they are likely to have spent little time analyzing the ways that an American racial philosophy unconsciously permeates their work. When journalists miss the mark, they can deeply influence their audiences' perceptions about both genetics and race. The general public has struggled to keep up with genetic research as it progresses at dramatic speed. A large proportion apparently harbors serious misunderstandings: In one poll, for instance, 26% of respondents thought genetic testing could detect intelligence or strength (Kalfoglou, Suthers, Scott, & Hudson, 2004, p. 9). In a separate, in-depth telephone survey, researchers at the University of Michigan found great confusion and limited understandings of basic genetics (Lanie et al., 2004). The press can take at least some responsibility.

The news media also convey influential messages about race, although often less consciously and directly. The preponderance of black people as central characters in crime, sports, and entertainment stories, for instance, contributes to stereotyping and racial animosity (Entman & Rojecki, 2000). The framing of a news story as a conflict, an economic question, or a moral choice can call up stereotypes and beliefs that audiences already hold (Domke, 2001). Even word choices that bring to mind racial imagery can influence audiences' political views—for instance, whether to support punitive approaches to crime (Hurwitz & Peffley, 1997; Gilliam & Iyengar, 2000). News about genetics and race enters this loaded arena. Messages that link race, genes, and health can dramatically increase racist attitudes, recent research at the University of Georgia suggests (Condit & Bates, 2005).

Journalists attempting to report accurately on both the scientific exploration of race and problems such as race-based disparities know that they navigate difficult terrain. They must cover science with skepticism and intelligence. They must cover race with sensitivity and awareness. Most journalists would claim this isn't so hard—they pride themselves on confronting society with uncomfortable ideas and carefully questioning what most others assume is true (Society of Professional Journalists, 1926). Yet, in a number of ways, the tested, reliable tools of journalism fail them. The values that underpin news judgment can push reporters toward leaps of logic and interpretation when covering both science and ideas about race. Traditions such as maintaining balance, developing beats, and identifying authoritative sources, while important for producing credible journalism, can actually limit accuracy if not applied with care. And beyond the craft itself, buried assumptions and understandings held by individual journalists can lead the news media astray.

Breaking the Biology Frame

The news reports on the role of SLC24A5 in skin pigmentation offer a window into the often-conflicting values, pressures, and assumptions that can seem to push journalists over the edge of reason. The December 16, 2005, paper in *Science* by Keith C. Cheng and his colleagues describes a zebrafish pigmentation gene and its human counterpart (Lamason et al., 2005). The researchers identified a single base change that lightens the coloring of pigment granules in the golden zebrafish. They compared the fish sequence to human versions of the gene in West African, East Asian, and European populations and concluded that the mutation may help to explain the difference in skin tone between black and white people. WKYC-Cleveland opened its report on the finding with a dramatic proposal: "Changing the color of your skin may be as easy as taking a pill" (WYKC-Cleveland, 2005). At the other end of caution, *Newsday* declared that the discovery highlighted "what scientists say is the power of nonhuman genetic stand-ins to provide new information about human traits and disease" (Nelson, 2005).

In a sampling of television, newspaper, and magazine coverage of the finding, assumptions about its implications for the biology of race generally seemed to dominate. Reporters for at least two major newspapers went further, connecting biology—this single mutation, in fact—to broad statements about the history of race relations in the United States. In a *Washington Post* story reprinted across the country, reporter Rick Weiss reminded readers that "race is a vaguely defined biological, social and political concept," but emphasized that the "finding helps solve one of biology's most enduring mysteries and illuminates one of humanity's greatest sources of strife" (2005, p. A1). Writing for the *Boston Globe*, Gareth Cook agreed: "For something that has divided America for centuries and laid the foundation for the worst war in our history, it has a quite unremarkable name: SLC24A5" (2005).

Jumping to Conclusions

Why do journalists so often quickly fix on a biological frame for race, even as they seem to dismiss it? Why did they so easily leap from a variant in an allele that influences color in a fish to sweeping generalizations about human relations and, in many stories, evolutionary history as well? It may be helpful first to consider who these journalists are and what they see as their jobs. How do they define news and craft its presentation? What strictures limit their ability to convey the nuances of genetics as they apply to research on race? Standard practices, demographics, and unquestioned assumptions all play a role.

Seeking Truth

The Society of Professional Journalists says that journalists' mission is "to seek truth and report it" in the service of a free and democratic society (Society of Professional Journalists, 1926). Thoroughness, fairness, and accuracy make up the conceptual cornerstones of this effort. The ideal is admirable, certainly, and one that most journalists take quite seriously. They seek truth and, in particular, truth that is relevant to social discourse and democratic decision-making. "News" must meet one or more of these conditions: it must be timely, have impact, involve prominent people or institutions, involve issues that generally concern the public, be surprising or unusual, or have physical or emotional proximity to the audience. As one media analyst put it, "It must be important" (Stoll, 2006).

Reports on research in zebrafish thus would naturally emphasize the implications for news audiences: that is, humans. Then, depending on whom journalists choose as sources, their background knowledge about the subject, and even their own personality or that of their medium, they may carefully define the specific importance of new research or decide to stretch much further:

> A little striped fish has helped scientists begin to solve one of the biggest mysteries in biology—which genes are responsible for differences in human skin, eye and hair color. (Fox, 2005)
> The team of 25 geneticists, molecular biologists and anthropologists . . . says the work could have implications for skin cancer treatment, crime-scene analysis, and even cosmetics. (Avril, 2005, p. A01)
> An Indiana University Northwest associate professor cloned a human gene related to variations in human skin color—an action that one day could improve how different races are treated for disease. (O'Shaughnessy, 2005)

Despite what might seem to be extreme leaps in the second two examples, both reporters back up their assertions with supporting quotes from the scientists themselves. The WKYC report cites a researcher who "even suggests there might be ways perhaps to modify color without damaging the skin in tanning parlors" (WKYC-TV, 2005). This concept as well as the emphasis on human skin color as "one of biology's enduring mysteries" can also be found in the original press release from Pennsylvania State University

Hershey Medical Center (Manlove, 2005). Journalists often fail to greet university press releases with the same skepticism as they apply to missives from city hall. Even so, a reference to tanning parlors or potential research on skin lighteners hardly justifies the claim that "changing the color of your skin" could be accomplished with a pill. More than one reporter took a rhetorical leap and ran with this line of thinking, never seeming to have asked historians, sociologists, psychologists, or even cosmeticians what the level of interest in changing skin color might be. In the effort to show importance, journalists who are inexperienced in covering science, race, or both subjects can easily slip into the sort of drama and shorthand more commonly found in crime and human-interest stories.

Racial Ideology

Aided by dramatic press releases and evocative journal covers, journalists also may be tempted to overemphasize a particular finding or leap from basic science right to medicine, society, and culture. *Science* magazine highlighted the skin pigmentation discovery—and what could be interpreted as its fundamental link to race and human ancestry—with an illustration on its cover. The image featured two hands, one white, one black, lifting a globe containing the two types of zebrafish swimming opposite directions. Most people think race and racism doesn't take much time or depth to understand (Dovidio, 2003), and journalists are no exception. The declarations that reporters make about this gene's implications for race relations may have popped out reflexively from their personal understandings about race, rather than any reporting or background research. In the *Washington Post* and *Boston Globe* stories, skin color in black and white people is linked seamlessly to "race" and in turn to racial stratification and strife.

In both cases the writers used a sentence structure that presents racism as the result of activity by a gene. Whether intentional or not, this presentation suggests that race and racism exist as an effect with no other cause than skin color. And even though the study authors specifically point out that the mutation helps explain skin color only in white and black people, these journalists are happy to let it underpin race in its entirety. This is consistent with much of racial discourse in the United States today. Furthermore, there is a stunning absence of any mention of politics, economics, and other contributions to "race."

By linking race to a single base-pair change, these and other reporters may have hoped to minimize the social category that they had described as producing such harm. Taking a stance is generally considered unacceptable for journalists, but not when it comes to advocating for the downtrodden or, as the Society of Professional Journalists (1926) code says, "giv[ing] voice

to the voiceless." Under the headline "Black? White? Not So Different,"[1] Tom
Avril wrote in the *Philadelphia Inquirer*, "So a significant part of the difference
we perceive between the races is caused by just one rung on the twisted lad-
der of our DNA" (Avril, 2005, p. A01).

Many of the news frames used for the zebrafish study are consistent with
a coverage ideology dubbed "white-centered" news by University of Maryland broadcast researcher Don Heider (2000). He and other media analysts
refer to traditions and implicit belief systems that have grown up in newsrooms historically dominated by white (and male) journalists. Story angles,
word choices, and the emphasis of certain ideas can unconsciously reflect a
white-centered point of view. The *Washington Post*, for instance, highlights
the importance of "the skin whitening mutation" that helped "give rise to
the lightest of the world's races" and raises questions concerning "why white
skin caught on so thoroughly in northern climes." "Unlike most mutations,
this one quickly overwhelmed its ancestral version," the story asserts, then
goes on to characterize skin color evolution as "white skin's rapid rise to
genetic dominance" (Weiss, 2005).

Media analysts have identified structural and institutional race-based
bias in news judgment, and this story may be no exception. Geneticists have
found several other genes associated with skin color, including at least one
that may approach the influence of SLC24A5. But "the newly found glitch is
the first found to play a role in the formation of 'normal' white skin," the
Post says (Weiss, 2005), or as Reuters puts it, "the distinctive light European
coloring" (Fox, 2005). Along with more mundane factors, such as the overall amount of news activity that day, could the mutation's association with
whiteness help explain the high level of attention it received?

Terminology

Workers in the news media often find themselves both translating technical
subjects into "plain English" and struggling to respect the special nuance
each field attributes to its own terms. In most subject areas, they arrive at
a compromise that keeps material understandable, honest, and true to its
original meaning. But when it comes to the language of genetics and race,
journalists seem mired in confusion. Reporters often resort to broad, gener-
alized terms like "African" and then mix and match terms haphazardly. Even
a single story may not manage internal consistency in terms. Considering
the various worlds that the news media attempt to connect and straddle
when covering genetics and race, this may not be so surprising. Anthropologists, clinical geneticists, population geneticists, sociologists, philosophers, and the general public can all mean different things by words such
as "race," "population," and "ancestry." Already, as community composition

and preferences change, news organizations are rethinking the usefulness of terms such as "Asian" or "Asian American" across all of their beats. Who do they mean by "white"? When should they use "Latino," "Hispanic," "black," "African American," and "Native American" or "American Indian"?

In the skin-color stories, reporters generally defined race as three groupings: Africans, Asians, and Caucasians or people of European descent, with one story mentioning American Indians. These choices likely stemmed from the *Science* article authors' description of alleles found in people they described as Africans, East Asians, and Europeans (Lamason et al., 2005). Reporters, apparently applying those assumptions that race is easy to understand, don't seem to have asked the researchers how they defined these groups and who they represented. The journalists who went on to make general conclusions about race did not clarify the populations involved. Their oversight, of course, makes it easy for audiences to assume that race is a clear scientific concept and that declarations about black and white relations based on skin color can be extrapolated to explain a broad spectrum of social phenomena.

In writing about this area of genetics and medicine, journalists often use the terms "race," "ethnicity," "ancestry," and "nationality" interchangeably. A 2002 article in the *New York Times*, for instance, focused on the University of California at San Francisco's population geneticist Neil Risch's proposition that race is biologically meaningful. Under the headline, "Race Is Seen as Real Guide to Track Roots of Disease," reporter Nicholas Wade explains "the apparent correlation between race, genetic data, and disease" (Wade, 2002, p. F1). But in offering an example, he points to nationality, not to what are commonly understood as racial groupings: "For example, a mutation that causes hemochromatosis, a disorder of iron metabolism, is rare or absent among Indians and Chinese but occurs in 7.5 percent of Swedes." In a December 20, 2002, article, Wade uses sickle cell anemia among "Africans" and hemochromatosis among "Swedes" as examples of diseases that are more common among some "ethnic groups" than others, and then segues right into race. "Researchers seeking the genetic variants that cause such diseases," he continues, "must take race into account" (Wade, 2002, p. 37).

Reporters may be taking their cue from the inconsistent use of terms not only within genetics and population genetics research, but also in clinical medicine and the social sciences (Fullwiley, Kahn, and others in this volume). Researchers studying human genetic variation are in the midst of one debate about terms; clinical researchers have a separate set of disagreements. As Condit asserts, statistical population clustering by continent leaves great ambiguity regarding not only the designated geographical areas, but also the people within them (Condit, Parrott, Harris, Templeton, et al., 2004). Judith Kaplan and Trude Bennett (2003) detail the problems encountered when attempting to include race and ethnicity in medical research in a meaningful way; among other challenges, they name broad, imprecise, and overlapping categories, inconsistency across studies and data systems, and the complexities of self-identification.

Outside of the academy, popular notions of race merge identity, culture, and physical features (Condit, Parrott, Harris, Lynch, et al., 2004). Whether "race" can be a useful term to describe and apply scientific knowledge about genetic variation remains an open question. Certainly, the word has long been discarded as anything approaching a precise biological term. Geneticists continue to debate the reliability and meaning of defining groups of people according to geographic origin based on genetic variation. Faced by disagreement among scientists, daunting subject matter, and unfamiliar terrain, many journalists seek safe ground. The complexity of a social category concedes to what reporters seem to understand as simple "reality" and practicality. Yet all too often, they merge race and biology in news stories without critique or explanation.

Journalists cover a great variety of news in some very technical areas under severe time pressure. Advance preparation and greater awareness of potential pitfalls can make it easier to cover the field of genetics as it intersects with race more accurately. This might include reading about the sociology and history of race in the United States, studying the statistics that underlie claims about population groupings, and analyzing the terminology commonly used by scientists. Journalists cannot assume that a scientist in one discipline is knowledgeable about another area of inquiry. Thus, when writing about race, sourcing must expand beyond those involved in molecular biology, genetics, or clinical research. Scientists can assist by offering background interviews and short educational sessions, suggesting a text or reading list, or inviting journalists to sit in on their courses. They should avoid the temptation to make a finding seem especially exciting or to "dumb things down," as one geneticist reported he was advised to do. Clarification is helpful; over-simplification often leads to confusing and misleading stories.

Newsroom Demographics

Some of the underpinnings to confusion, awkward framing, and lack of analysis lie in the demographics of the newsroom itself. In 1968, the Kerner Commission chastised the news media for its tendency to "report and write from the standpoint of a white man's world." Since then, print and broadcast outlets have pledged to change their habits, first of all by building a workforce that reflects the population they serve. More representative news decisions and coverage will follow, they reason. More than three decades later, however, both print and broadcast media have remained stubbornly dominated by whites. According to the most recent surveys, journalists of color make up just over 13% of editorial staff in newspaper newsrooms and 22% of broadcasters (American Society of Newspaper Editors, 2007; Papper, 2007, p. 21). Predictably, the proportions shrink when it comes to managers and editors in decision-making positions.

With newsrooms skewing heavily toward the white population, coverage of race and genetics faces foreseeable limitations. In the following description of tribal origin studies, not only is genetic "evidence" set against American Indian "stories," but the Indians themselves fade into history. The author refers to a "we" that clearly does not include Native Americans:

> According to Indian folklore, the Cherokee are distant relatives of the ancient Hopewell Indians, the mysterious race of Mound Builders who constructed huge geometric earthen monuments in central and southern Ohio.
>
> Genetic evidence, however, doesn't back up the stories, a researcher says. . . . Her study, part of a relatively new area of research known as genetic anthropology, could help fill gaps in what we know of Ohio's prehistoric Indians. (Lafferty, 2003)

The perspective so plainly delineated in this article shows up with more subtlety in other coverage. Journalists were quick to report a mutation associated with alcoholism that was identified among American Indians, for instance. They described the problem as rampant among Native Americans, without noting the significant differences in rates of alcoholism among tribes or other research that has identified a correlation between adverse childhood experiences, including boarding school, and alcoholism (Koss et al., 2003).

Unconscious Assumptions

Like everyone else, journalists who live and work in environments populated mainly by people like themselves can easily fall prey to assumptions about groups other than their own (Nosek, Banaji, & Greenwald, 2002; Olsson, Ebert, Banaji, & Phelps, 2005). White journalists may venture beyond familiar, homogeneous environments only to cover sports, poverty, and crime. Their awareness about people different from themselves may be structured on these themes, based mainly on reporting about the ills of society (Entman & Rojecki, 2000). Journalists also share the unconscious human tendency to associate negative behavior in out-groups with innate characteristics and to choose language that promotes this view (Maass, Milesi, Zabbini, & Stahlberg, 1995; Maass, Ceccarelli, & Rudin, 1996). The troubling patterns identified in content analyses of news reporting on non-white populations can be assumed to transfer into reporting on genetics associated with those populations.

The unconscious belief that whiteness is the norm, or as social psychologist Mahzarin Banaji puts it, "American = White" (Devos & Banaji, 2005), leads to reporting practices that in turn can limit journalists' understanding of the issues they cover. By consulting only social scientists, geneticists, and other experts from a narrow demographic band, journalists are less likely to uncover the context that might bring needed complexity into genetics stories involving race. Tasha Dubriwny, Benjamin Bates, and Jennifer Bevens's (2004) focus group studies on differing perceptions about race support this possibility. While people of European descent relied on physical appearance to define race, African Americans described a more fluid concept that included culture and self-definition. As discussed above, the stories about skin pigmentation thus reflect what might be described as a "white" point of view.

Who Gets a Say

While news outlets have made efforts to include more of the population in coverage, content analyses indicate that journalists generally favor white men as both subjects and sources (Media Tenor, 2003). In one recent study by an international media analysis group, only about 0.5% of individuals appearing on the network news or newsweeklies could be identified as black or African; 1.6% of those appearing on the network news and 1.2% in the newsweeklies were Asian. In 170,212 reports in six major outlets over three years, only 322 featured Native Americans, Latinos, African Americans, Arab Americans, or Asian Americans as main characters (Media Tenor International, 2004).[2]

News media analyses tend to rely on popular definitions of racial or ethnic groups, often going by last name, skin color as interpreted through a photograph, or references to the subject's self-described identity. Studies of local news media have found higher representation of people of color than in national and international coverage, but within limited news categories. In a random sample of 72 newscasts by each of six news stations in 1999, researchers found that Latinos and African Americans featured most often in sports or crime stories; and when the team studied 52 issues of six newspapers, almost half of the people of color in stories were sports figures or entertainers (Pease, Smith, & Subervi, 2001). The severe imbalance is obvious, despite a probable lack of consistency in definitions and terms. Why might the news media tilt toward white male source and subject be important to science reporting? First, it leads the public to connect certain types of activity and social roles with particular racial groups (Entman & Rojecki, 2000). Audiences then interpret news about race and genetics within this context. Journalists themselves, of course, are not immune to such influences, especially when they experience and repeat them day after day on the job. And without much experience covering all types of people in all types of stories, journalists become less capable of asking good questions and constructing coverage that involves race in an intelligent way.

Solutions

Journalists can correct these tendencies by following some of the strategies for reporting fairly and accurately on race that are recommended by organizations such as the Poynter Institute and the Maynard Institute for Journalism Education as well as resources such as *News in a New America* (Lehrman, 2006). These include interacting more with people different from themselves, questioning their own assumptions, distinguishing race from class and other social factors, and adding complexity and context to their pieces. Scientists can point the way by acknowledging the uncertainty in their work and pointing to colleagues who can share a different perspective. They can challenge and encourage journalists to do a more complete and nuanced job.

Journalistic Truisms

As stated earlier, journalists see it as their mission to "seek truth." Most recognize that they are not likely to discover "the Truth" or even "the truth." But news-gathering practices are founded upon the belief that dogged pursuit of the facts will get them close. This commitment to the idea of some existing truth, even if it must be patched together from different perspectives, can lead to an important disconnect between journalism and science. Scientists also may be said to seek truth, but there is a different underlying assumption. In a story about science in the courtroom, *Science News* uses an interview with Douglas L. Weed of the National Cancer Institute to explain the scientific method to general readers: "Scientists question, test, evaluate, and retest various hypotheses, looking for an explanation that best fits the observations. . . . They're not expecting 'truth,' he notes, because they know that 'uncertainty flows through science like a river' " (Weed quoted in Raloff, 2005).

Complexity and uncertainty don't fit very well into the journalism culture, which relies on identifying facts and consulting experts until truth emerges. And when faced with lack of agreement, journalists are more likely to shift to another way of approaching truth—balancing two or more points of view.

Balance

Early American newspapers prided themselves on unabashed partisanship, but modern journalists seek balance in order to cover the news fairly and accurately. They eschew taking sides, aiming to present a complete picture so that audiences can make their own decisions. And while committed to revealing injustice, they avoid straying into advocacy. The concept is straightforward when it comes to covering a lawsuit, say, or civic controversy. But balance can get science journalists into trouble. Instead of reporting about the pursuit of truth, they end up looking for the "sides" in research or even in data interpretation and analysis. The result is summaries such as this, which do little to help readers understand the underlying biology or social factors referenced: "A view widespread among many social scientists, endorsed in statements by the American Sociological Association, is that race is not a valid biological concept. But biologists . . . have found that there is a structure in the human population" (Wade, 2003, p. 30).

The search for balance in genetics, as in science coverage generally, often entails identifying groups of people as distinct in their views and then rhetorically pitting them against one another. In the above example, sociology is confronted by biology. Because the biologists are placed second, essentially as a retort, their view is conferred additional validity. Here, race becomes the dividing factor, with the presumably non–African American geneticists given the final word:

> A difference of opinion about the genetic basis of race has emerged between scientists at the National Human Genome Center at Howard University and some other geneticists. . . .
>
> . . . [S]cientists at Howard, a center of African-American scholarship, generally favor the view that there is no biological or genetic basis for race.
>
> . . . But several other geneticists writing in the same issue of the journal say the human family tree is divided into branches that . . . coincide with common racial classification. (Wade, 2004)

Conflict

As these two examples illustrate, balance tends to work hand in hand with one particular journalistic frame: conflict. Writers know they generally can rely on this approach to create interest and draw audiences into a subject. But as a means to engage people with scientific ideas, presenting a difference of interpretation as a dichotomy can be quite misleading. Instead of educating readers or viewers about the underlying research and its significance, journalists depict a catfight. As in the following excerpt, reporters often pit "hard" science against "soft" science in order to explain differences of opinion over race-based genetics: "The new interest in racial genetics comes at a time when the softer sciences, like anthropology and sociology, have declared that race is a cultural construct, without any biological significance" (Henig, 2004).

Solutions

Journalists can provide balance in coverage of a topic such as race-based genetics, but they must venture further than simplistic dichotomies. Controversies should be presented, but not as two-sided arguments based on distinctions such as hard and soft science or African American scholars versus everybody else. The news media can better approach true fairness and balance by relating the data underlying various interpretations, making uncertainties clear, exploring what is missing, and dissecting the assumptions behind interpretations. Journalists also must bring a greater degree of skepticism to reporting in this area. Scientists must be questioned and their data and documentation examined, just as reporters do with every other type of source. Geneticists and social scientists can help journalists identify competing interpretations, weaknesses in a particular study, and questions that remain to be answered.

Beats

News organizations assign their reporters to beats, thus allowing reporters to follow a subject closely, understand its complexities, and develop reliable, trustworthy sources. While this organizational structure does encourage reporters to develop knowledge in an area, it also can box them into a type of reductionism. Common beats include crime and cops; government; schools; transportation; sports; business; arts and culture; health, medicine, or fitness; and science and technology. These are the frames through which the newsroom sees the world. The news pages are laid out and broadcast programs are designed using this paradigm, making it difficult for individual journalists to stray from the central themes.

Depending on where the news originated (that is, a journal article, an advocacy group, a biotechnology company, or a university), journalists reporting on genetics and race are likely to have a background in one of the following beats: science and medicine, social issues, business, or, perhaps, the local university. They will usually approach the story from the angle most relevant to their beat without venturing much further. This alone can skew coverage—for instance, the crime reporter sees DNA evidence and data banks as an important tool, but the business reporter may ask questions about privacy, insurance, and liability. The community reporter may cover DNA testing for race as a fascinating family project, while the science reporter is likely to ask more questions about its validity. Science writers who have covered anthropology, sociology, and biology may explore popular concepts of race and ethnicity and how these may or may not relate to genetics. Those who tend toward a science and technology focus may explore the social sciences a little less. Medical reporters may be drawn to genetic explanations for racial differences, especially if they have covered race-based disparities in health.

In cultivating expertise, specialty reporters also can develop myopia of both vision and thinking. Most reporters who cover science regularly have become familiar with interpreting and probing disagreement in a field. But those who cover medicine more commonly bring a different set of tools. They may question the science less and look more closely—and often more enthusiastically—at its potential applications.

When Science and Business Don't Mix

As science, politics, and business increasingly intermingle in stories about genetics and race, few reporters address this new level of complexity. Hit by a flood of press releases, science journalists who do not have a business background may find it difficult to discern news from promotion and marketing efforts. They may not easily sort out those simply meant to serve full disclosure requirements under Securities and Exchange Commission rules.[3] Business journalists with no science background face a comparable problem.

Many science reporters don't check the commercial affiliations of scientists they feature as story subjects. They simply don't think of it and they have little background in reading proxies and prospectuses. Business journalists may bring less skepticism to the science, but they have far better radars for conflicts of interest. In the following stories about an ancestry test by DNAPrint Genomics, the term "consultant" in one, compared to "creator" and "partner" in the other, conveys something quite different about the test's inventor. The second description, of course, more accurately alerts the reader to the scientist's interest in the company's success. Both are features about a test developed by DNAPrint Genomics to trace ancestry or, as the articles call it, "race": "The case also illustrates the limits of DNA testing, says Mark Shriver, a professor of anthropology at Penn State and a consultant for DNA Print" (Lamb, 2005, p. 14). "The test . . . was created by Professor Shriver and his commercial partners at DNA Print Genomics Inc." (Daly, 2005, p. 8).

Even beat reporters must jump from one area to another, working rapidly on stories in a pressure-cooker environment. They must quickly spot important news developments, report and research them, identify their meaning in a broader context, and distill technical aspects for public consumption. Faced by these constraints, reporters can take risky shortcuts. The beat system encourages reductionism because a reporter is likely to turn to the area of science he or she knows best in order to explain all aspects of a phenomenon. Journalists come to believe that science itself is the same as truth. Reporters too are vulnerable to determinism, or the view that our gene complement more or less dictates and predicts an individual's future. With so much scientific focus on the Human Genome Project and corporate effort going into biotechnology, it is easy to ascribe central authority to DNA. And because journalists move quickly from one story to another, often not turning back to check on new ways of thinking in an area they already covered, paradigms stick. As recently as 2003, for instance, the *Times-Picayune* was relying on the "blueprint" metaphor, which has been largely discarded by geneticists. The metaphor conveys a determinist, essentialist picture of DNA, as this example goes on to describe explicitly: "Scientists often describe DNA as a blueprint for who we are, determining what we look like as well as predicting certain inherited traits" (Stanley, 2003, p. 2).

Solutions

Journalists who want to cover genetics accurately must keep up with the field. They must follow the science as closely as a business reporter might keep tabs on the 10ks and proxy reports filed by companies in the industries they track. Good resources are review papers in *Science* and *Nature* as well as the medical and public health journals. Journalists also can watch funding trends and track citations. Those who have developed a familiarity with genetics or biotechnology can make an effort to broaden their vision beyond the lab and the clinic, checking into the business side of the science, too. Business writers can study up on the underlying science.

Journalists in both areas of specialty can turn back to one of the most important precepts of good journalism: context and background. By weaving in the social, historical, and economic fabric of a story, not only will they report more accurately, but the result will become more interesting. Geneticists and social scientists can help by suggesting reading materials and sources that will shorten the news media's learning curve.

Conclusion

Journalists serve as information connectors and mediators in a democratic society. They link the general public to political developments, social trends, advances in science, and other news that can help people of all backgrounds become informed participants in public life and policy-making. All too often, however, news reports fall into predictable patterns when they feature complex new developments in race-related genetics research. Journalism's failures in this area neglect to inform the public adequately about a powerful new area of science. Furthermore, they have been documented to increase racial hostility or discrimination (Condit & Bates, 2005).

Journalists must recognize that coverage of race and genetics research requires a high level of thoughtfulness and expertise. They cannot rely on experts within just one discipline to help them understand discoveries, evaluate scientists' interpretations or identify social meanings and implications. They must learn more about the structural tendencies within the news business itself and work on ways to counteract those that might distort coverage of race. They must educate themselves on an ongoing basis about scientific methods and analysis, research trends and discoveries, and the social system in which new findings are interpreted and used. They must ask better questions—of the research and of their own reporting.

This may sound daunting, but with the help of the research community and the learning that takes place through the daily reporting process, it can be done. The rewards will help journalists bring audiences closer to understanding the search for "truth" in its fullest meaning in science and society.

NOTES

1. Others in the Knight Ridder chain made quite different headline choices. The *Myrtle Beach Sun News* in South Carolina announced, "Scientists discover gene that causes difference in skin color," while the *Arizona Daily Star* wrote, "Part of gene is big player in deciding skin color" (both December 16, 2005).

2. These categories, of course, are just as difficult to decipher as those used in other disciplines. Reflecting its international point of view, the research organization describes international coverage with one set of broad terms ("Asian, African, black, European") and U.S. domestic coverage with another ("Asian American," "African American," "white American"). Reflecting the debate about terms in the United States, earlier reports used the category "Hispanic," while more recent ones employ "Latino."

3. These regulations require companies to report any activities or events that might materially affect their financial health or stock. The result has been an overflow of press releases about conference presentations, preliminary animal trials, and other less-than-newsworthy developments.

REFERENCES

American Society of Newspaper Editors. (2007). *Newsroom employment census.*

Avril, T. (2005, December 15). Black? White? Not so different. *Philadelphia Inquirer*, p. A01.

Condit, C. M., & Bates, B. (2005). Lay understandings of the relationship between race and genetics, public understanding of science. *Clinical Genetics, 68* (2): 97–105.

Condit, C. M., Parrott, R. L., Harris, T. M., Lynch, J., & Dubriway, T. (2004). The role of "genetics" in popular understandings of race in the United States. *Public Understanding of Science, 13* (3): 249–272.

Condit, C. M., Parrott, R., Harris, T., Templeton, A., Dubrivny, T., Lynch, J., et al. (2004). *Psycho-social, clinical, and scientific barriers to race-based genetic medicine? A query by the health and heritage team, University of Georgia.* Retrieved February 14, 2006, from http://www.biomed.uga.edu/RaceBasedMedicine2.pdf.

Cook, G. (2005, December 19). A fish, a gene, and a source of skin color. *Boston Globe.*

Daly, E. (2005, April 13). DNA tells students they aren't who they thought. *New York Times,* p. 8.

Devos, T., & Banaji, M. R. (2005). American = white? *Journal of Personality and Social Psychology, 88,* 447–466.

Domke, D. (2001). Racial cues and political ideology: An examination of associative priming. *Communication Research, 28* (6): 772–801.

Dovidio, J. (2003). Phone interview with Dovidio, professor of Social Psychology, Yale University.

Dubriwny, T. N., Bates, B. R., & Beven, J. L. (2004). Lay understandings of race: Cultural and genetic definitions. *Community Genetics, 7,* 185–195.

Entman, R., & Rojecki, A. (2000). *The black image in the white mind: Media and race in America.* Chicago: University of Chicago Press.

Fox, M. (2005, December 16). Stripy fish helps pinpoint human skin color gene. Reuters. Reprinted by ABC News from International News Service, London.

Fullwiley, D. (2008 [this volume]). The molecularization of race: U.S. health institutions, pharmacogenetics practice, and public science after the genome. In B. A. Koenig, S. S.-J. Lee, & S. S. Richardson (Eds.), *Revisiting race in a genomic age* (pp. 149–171). New Brunswick, NJ: Rutgers University Press.

Gilliam, F. D., Jr., & Iyengar, S. (2000). Prime suspects: The influence of local television news on the viewing public. *American Journal of Political Science, 4* (3), 560–573.

Heider, D. (2000). *White news: Why local news programs don't cover people of color.* Mahwah, NJ: Lawrence Erlbaum Associates.

Henig, R. M. (2004, October 10). The genome in black and white (and gray). *New York Times.*

Hurwitz, J., & Peffley, M. (1997). Public perceptions of race and crime: The role of racial stereotypes. *American Journal of Political Science, 41* (2), April 1997.

Institute for Justice and Journalism (2005, September 19–24). Covering news of race in a new world. Racial Justice Fellowship Conference conducted at Harvard University, Cambridge, MA.

Kalfoglou, A., Suthers, K., Scott, J., & Hudson, K. (2004). *Reproductive genetic testing: What America thinks.* Washington, DC: Genetics and Public Policy Center.

Kahn, J. (2008 [this volume]). Patenting race in a genomic age. In B. A. Koenig, S. S.-J. Lee, & S. S. Richardson (Eds.), *Revisiting race in a genomic age* (pp. 129–148). New Brunswick, NJ: Rutgers University Press.

Kaplan, J. B., & Bennett, T. (2003). Use of race and ethnicity in biomedical publication. *Journal of the American Medical Association, 289* (20), 2709–2716.

Kerner Commission. (1968). *Report of the National Advisory Commission on Civil Disorders.* Washington, DC: U.S. Government Printing Office.

Koss, M. P., Yuan, N. P., Dightman, D., Prince, R. J., Polacca, M., Sanderson, B., et al. (2003). Adverse childhood exposures and alcohol dependence among seven Native American tribes. *American Journal of Preventive Medicine, 25* (3): 238–244.

Lafferty, M. (2003, September 23). 2,000-year-old DNA connects Hopewell Indians, Asian races, researcher says. *The Columbus Dispatch,* p. A4.

Lamason, R. L., Mohideen, M.-A.P.K., Mest, J. R., Wong, A. C., Norton, H. L., Aros, M. C., et al. (2005). SLC24A5, a putative cation exchanger, affects pigmentation in zebrafish and humans. *Science, 310* (5755), 1782–1786.

Lamb, G. (2005, April 28). Mixed roots: Science looks at family trees. *Christian Science Monitor*, p. 14.

Lanie, A. D., Jayaratne, T. E., Sheldon, J. P., Kardia, S.L.R., Anderson, E. S., Feldbaum, M., et al. (2004). Exploring the public understanding of basic genetic concepts. *Journal of Genetic Counseling, 13* (4), 305–320.

Lehrman, S. (2006). *News in a new America.* Miami: John S. and James M. Knight Foundation.

Maass, A., Ceccarelli, R., & Rudin, S. (1996). Linguistic intergroup bias: Evidence for in-group-protective motivation. *Journal of Personality and Social Psychology, 71* (3), 512–526.

Maass, A., Milesi, A., Zabbini, S., & Stahlberg, D. (1995). Linguistic intergroup bias: Differential expectancies or in-group protection? *Journal of Personality and Social Psychology, 68* (1), 116–126.

Manlove, M. W. (2005, December 15). Fish gene sheds light on human skin color variation. Press release from Pennsylvania State University Hershey Medical Center, EurekAlert!

Media Tenor. (2003). It's a man's world—even more in the media. *Media Tenor Quarterly Journal, 4*, 36–37.

Media Tenor International. (2004). *Coverage of ethnic and racial groups in U.S. Media: 01/01/2002—5/31/04.* New York, Media Tenor report, p. 2.

Nelson, B. (2005, December 16). Skin color gene found: Protein variants seen in zebrafish are shown to account for pigment differences among humans. *(Long Island) Newsday*, p. 62.

Nosek, B. A., Banaji, M. R., & Greenwald, A. G. (2002). Harvesting implicit group attitudes and beliefs from a demonstration web site. *Group Dynamics, 6* (1), 101–115.

O'Shaughnessy, T. L. (2005, December 16). Is skin color only gene deep? *Northwest Indiana Times.*

Olsson, A., Ebert, J. P., Banaji, M. R., & Phelps, E. A. (2005). The role of social groups in the persistence of learned fear. *Science, 309* (5735), 785–787.

Papper, B. (2007, July/August). Women and minorities in the newsroom. *Communicator*, p. 21.

Pease, E., Smith, E., & Subervi, F. (2001). *The News and Race Models of Excellence Project—overview connecting newsroom attitudes toward ethnicity and news content.* p. 46. Retrieved February 15, 2006, from Poynter Online: http://www.poynter.org/content/content_view.asp?id=5045.

PR Newswire. (2005, May 11). DNAPrint genomics forensics technology plays key role in probe of California's Mammoth Lakes murder case.

Raloff, J. (2005). Benched science: Increasingly, judges decide what science—if any—a jury hears. *Science News, 168* (15), 232.

Shields, A. E., Fortun, M., Hammonds, E. M., King, P. A., Lerman, C., Rapp, R., et al. (2005, January). The use of race variables in genetic studies of complex traits and the goal of reducing health disparities: A transdisciplinary perspective. *American Psychologist, 60* (1).

Society of Professional Journalists. (1926). *Code of ethics, 1996.* Retrieved February 14, 2006, from http://www.spj.org/ethics_code.asp.

Stanley, S. A. (2003, March 29). Innovative test of serial killer DNA sought: Work may paint clearer portrait. *New Orleans Times-Picayune*, p. 2.

Stoll, M. (2006, January). Interview by Stoll with Lehrman for "Grade the News" story: What top-10 lists say about the people who produce the news: "Purely unscientifically," journalists pick stories based on a mix of values.

Wade, N. (2002, July 30). Race is seen as real guide to track roots of disease. *New York Times*, p. F1.

Wade, N. (2002, December 20). Gene study identifies 5 main human populations, linking them to geography. *New York Times*, p. 37.

Wade, N. (2003, March 20). 2 scholarly articles diverge on role of race in medicine. *New York Times*, p. 30.

Wade, N. (2004, October 27). Articles highlight different views on genetic basis of race. *New York Times*.

Weiss, R. (2005, December 16). Scientists find a DNA change that accounts for white skin. *Washington Post*, p. A1.

WKYC-TV. (2005, December 23). Changing the color of your skin with a pill? WKYC-TV, Cleveland, OH, Web site at http://www.wkyc.com/news/news_article.aspx?storyid =45206.

16

Race without Salvation

Beyond the Science/Society Divide
in Genomic Studies of Human Diversity

JENNY REARDON

Science has long been associated with projects of social salvation. It is the perennial hope of the post-Enlightenment West that humans will recover from their flaws—their biases, prejudices—by employing the rational tools of the scientific method. One of the most recent public manifestations of this has been in the domain of race and genomics. Beginning in the mid-1990s, articles began to appear in the popular and scientific press promising that genomic findings would lead to the final demise of socially divisive ideas of race. In 1995, several major U.S. newspapers, including the *Los Angeles Times* and the *Boston Globe*, reported that "race has no biological basis" (Flint, 1995; Hotz, 1995; Venter, 2000). In the years that directly followed, the American public would be regaled with continuing reports of science's demonstration that race had no biological meaning, culminating with the announcement of the completion of the first draft of the human genome sequence on June 26, 2000. On that day, Craig Venter, head of Celera Genomics' private sequencing effort, proclaimed that his company had sequenced "the genome of three females and two males who have identified themselves as Hispanic, Asian, Caucasian or African-American" and found that "there's no way to tell one ethnicity from another" (2000, p. D8).

Rather than a world constrained by race, advocates of genomics promise that this novel form of technoscience will make good on the liberal democratic promise of producing a world that promotes the well-being of every individual—in this case, through tending to the biochemical and biological needs of every citizen through personalized genomic medicine. "This Drug's For You," announced the front cover of *US News and World Report* in January of 2003 (Fischman, 2003). In April 2000 the National Institute of General Medical Sciences published a pamphlet called *Medicines for YOU*, explaining how understanding genetics could help doctors prescribe each individual the right drug, in the right dose.

Yet, despite these claims, recently a more sobering reality has emerged. Rather than therapies tailored to individuals, we are witnessing the emergence of therapies such as the heart failure drug BiDil. BiDil is, indeed, marketed as a drug that will help particular individuals. However, these "individuals" all have one thing in common: they are all said to share a common race. In June 2005, the U.S. Federal Drug Administration approved BiDil as a drug specifically for "African-Americans" (Sankar & Kahn, 2005).

In addition to racially coded drugs, in recent years it has also become clear that racial categories are playing key roles in forming the medical diagnostics emerging from genomics. Envisioning a future of gene chips designed to scan for heart disease, one human geneticist at an East Coast university recently explained to me: "The decision as to which SNPs [single nucleotide polymorphisms, or genetic variants] you should be screening for is dependent upon ethnicity. . . . We might be able to have a cardiovascular chip that's got the Japanese and the Korean and the African American and the Nordic—you know, all of these different variants will be on that chip."[1]

So certain are many that different races will have different medically relevant genetic variants that Howard University, a historically black college in the United States, has taken the unprecedented step of creating its own biobank. Some Howard University researchers argue that without a collection of their own DNA, those of African descent may be left out of the genomic revolution (Pollack, 2003). Their fear is not that researchers will study African Americans, but that they *won't* study African Americans.

These are just a few of the many indications that far from the promised transcendence of race and uplift out of the scourge of racism, genomics only poses the problematic paradox of difference that has long endured at the heart of liberal democratic practice. As the historian and social theorist Joan Scott concisely states in her book *Only Paradoxes to Offer*, by positing a citizen who was at once universal, but also white, Western, and male, the conceptual structures underlying the "universal rights of man" positioned claimants to occupy "the very difference that the idea of the prototypical human individual was meant to deny" (1996, p. 1). Similarly, many genome scientists promise that genomics will lead us out of socially divisive group categories—namely, race—and produce data tailored to individuals. Paradoxically, however, they then proceed to use these very categories of race that they seek to dismantle.

This dilemma is not new. Indeed, it has long been recognized to inhere in affirmative action policies. Nor has the predicament escaped the notice of proponents of genomics. Advocates of personalized genomic medicine increasingly recognize that to enjoy new freedoms from disease, human beings must first occupy categories that might subject them to new forms of control. In an effort to respond to this dilemma, and to reconstruct genomics in a manner that will accentuate its democratic potential, genome scientists and policy makers have in recent years proposed that groups "identified" as the proper objects of research should also be treated as subjects who have the right to regulate, and even design, genomic studies. In so doing, these reformers of genomic practice draw upon a model of interacting with research subjects that emerged in the wake of National Institutes of Health's (NIH) response to AIDS activists in the 1980s and early 1990s (Epstein, 1996). Like the AIDS activists who eventually came to sit at the table with NIH officials and researchers, these administrators of genome research projects envision that members of "identified populations" asked to participate in genomic studies will now also participate in research design and regulation.

Yet, unlike the experience with AIDS activists, this offer to partner with research subjects has not always been met with open arms (Reardon, 2006). In several instances, "the people" asked to participate in genomic research design and regulation resist taking on these new, purportedly democratic roles (Reardon, 2006). Further, more recently, many of the bioethicists and social scientists who originally had been so enthusiastic, and who were charged with the task of soliciting the input of people, have grown cautious.

In this chapter, I seek to begin to illustrate why. Drawing upon case studies from the Human Genome Diversity Project (HGDP) and the International Haplotype Map Project, I argue that far from ameliorating concerns generated by genomics' paradoxical relationship to race, these initiatives have only generated their own novel paradoxes. Namely, rather than merely *including* new participants (i.e., groups) in research and its regulation, these novel procedures *create* new participants as they seek to include them—at once altering the constitution of the objects that can be the focus of scientific inquiry and the subjects who can make claims against governmental institutions and their research programs. Given the importance of these acts—constituting what can be known and who can be a citizen—not surprisingly, they have generated concerns, particularly since they have proceeded with little or no oversight and deliberation.

To draw this novel and consequential paradox of participation into view, and to address the dilemmas of difference within which genomics confronts us, in this chapter I argue that rather than simply extending liberal democratic practices—such as participation—to new collectivities, we must ask more fundamentally about the conceptual structures that undergird these practices. As I demonstrate below, these structures presume both the pre-existence of human groups and the demarcation of science from society. Thus, they facilitate the belief that vexing questions about how to demarcate groups (for purposes of their participation in both science and democratic governance) can be solved by turning to either "precise" scientific criteria, on the one hand, or "precise" social criteria, on the other. But as the episodes explored in this chapter make clear, such efforts falter for they enact the mistake that the philosopher Alfred North Whitehead called "misplaced concreteness"—that is, they posit the existence of clear and distinct categories, "the social" and "the scientific," where none exist (Whitehead, 1925, pp. 74–77).

The chapter argues that more promising efforts to address concerns generated by genomic studies of human differences must be guided by a different conceptual framework. Such a framework would not demarcate science from society, but rather call into view the ways in which particular orderings of the "social" world affect the kinds of categories scientists can use to characterize human diversity and the ways in which these "scientific" categories in turn "loop back" (à la the philosopher of science, Ian Hacking) to shape conceptions of the "individual" and "group" that undergird liberal democracies (Hacking, 1999). Such a framework would also help us not to presume, but rather to ask who the people are and what they can know in this paradoxical world where identities and interests and natural and social order never precede technoscience but rather form in tandem with it. This framework, I suggest, might provide the requisite conceptual training wheels needed to resist the alluring promise that precise, expert criteria can resolve enormously consequential questions about who can count as a citizen, with a voice and rights, in a world shaped by novel forms of technoscience, such as genomics. In so doing it would facilitate not salvation, but rather a chance at reflective choices.

The Human Genome Diversity Project

Claims that genomic studies of human diversity might save human beings from their racial prejudices emerged in the early 1990s as population geneticists and evolutionary biologists in the United States attempted to gain support for a worldwide survey of human genetic diversity. Proponents of the survey—which became known as the Human Genome Diversity Project—proposed sampling "isolated" indigenous groups, for they believed that the DNA of these groups would reveal the structure of human genetic diversity from which all human beings evolved. This sampling, they urged, should commence immediately before these isolated groups "vanished" (Cavalli-Sforza, Wilson, King, Cook-Deegan, & Cantor, 1991).

In addition to answering fundamental biological questions, organizers contended that the project would also have great social value as it would help "combat the scourge of racism" through demonstrating that "there is no absolute 'purity' " and no "documented biological superiority of any race, however defined" (Cavalli-Sforza, 1994). Rather than promote racial division, project organizers promised that the initiative would demonstrate "humanity's diversity and its deep and underlying unity" (Cavalli-Sforza, 1994, p. 1). As one step toward these goals, one of the project's main scientific leaders, Luca Cavalli-Sforza, advocated abandoning the category of "race" in favor of the categories "group" and "population" (1994, p. 11).

Yet despite these early promises to promote human unity and harmony, the HGDP would go on to become infamous for the animosity it sparked. Of all the critiques the project provoked, those issued by indigenous rights organizations proved most severe. After decades working to recognize the existence of indigenous people, indigenous rights leaders found Diversity Project organizers' descriptions of indigenous people as "vanishing" or "disappearing"—groups that were of historical interest that needed to be "preserved" for study before they went extinct—maddening. This was *exactly* the representation of indigenous people they had spent decades opposing. They argued precisely that they were not disappearing, that they were here to stay and deserved representation in international governing bodies just like any other free peoples.

In an age of electronic communication, it took very little time for some indigenous rights groups to broadcast this problematic connection between the indigenous rights movement and the Diversity Project (Lock, 1994). On July 8, 1993, the Third World Network posted the following message on Native-L, an aboriginal, First Peoples news net: "URGENT! URGENT! URGENT! CALL FOR A CAMPAIGN AGAINST THE HUMAN GENOME DIVERSITY PROJECT" (Native-L, 1993). The notice that followed argued that the Diversity Project put an interest in historical curiosity above an interest in the future well-being of indigenous groups. This focus on indigenous groups as objects of historical interest that were about to go extinct, as opposed to "fully human communities with full human rights," they found particularly egregious in 1993, the U.N.-decreed year of indigenous peoples (Lock, 1994).[2] Dozens of proclamations penned by indigenous groups denouncing the Diversity Project followed soon thereafter. Most famously, in December 1993, the World Council of Indigenous Peoples labeled the Diversity Project the "Vampire Project"—a project more interested in taking the blood of indigenous groups than in their ultimate survival.[3]

In the wake of these alarming critiques by indigenous groups, project organizers created an ethics subcommittee and commenced writing what would become known as the Model Ethical Protocol (MEP). According to its drafters, the goal of the protocol was to respond to the concerns of indigenous rights activist critics. The proposed ethical practice at the centerpiece of the document—group consent—reflected this goal. According to the code of ethics developed in the wake of the Belmont Report, in order for research on humans to be considered ethical, informed consent must be obtained from individuals.[4] Group consent expanded informed consent rights from individuals to groups. Drafters of the group consent provision hoped that this new ethical provision would respond to critiques made by indigenous rights groups by assuring them that indigenous populations would not be "picked" for research, but would have the chance to choose to participate. Yet, quite to the contrary, instead of quelling controversy, group consent sparked new and grave concerns. To understand why, and to see the novel paradox of participation generated by genomic studies of human difference, I draw the conceptual assumptions that underlie group consent into view.

Group Consent

In the first instance, group consent assumed that groups are the objects of human genetic diversity research and thus should be the subjects of consent.[5] As the MEP explains, "Although this requirement [group consent] goes beyond the strictures of existing law and ethical commentary, we believe it flows necessarily from the nature of the research, which is, by definition, research aimed at understanding *human populations* and not individuals" (North American Regional Committee, 1997, emphasis added, p. 1437). To use the language of the law, the study of human populations was one of the "facts of the case" to which group consent responded.

Some scientists working to organize the Diversity Project, however, strongly disagreed. These scientists had participated in the early debates sparked by the project about what kind of phenomenon "human genetic diversity" was—an individual or group-level phenomenon? One of the project's main proponents, the prominent human population geneticist Luca Cavalli-Sforza, argued that human genetic diversity was a group phenomenon and could be sampled through collecting human tissue samples from populations. The other key project leader, University of California at Berkeley biochemist and molecular evolutionary biologist Allan C. Wilson, challenged the validity of this population-based approach. Some genetic variants, he argued, occurred only within a small geographic radius of 50 or 100 miles. Thus, if the project only sampled populations, much genetic variation would be missed. To solve this problem, Wilson proposed sampling individuals on a geographic grid. Collecting in this manner, he explained, would enable researchers "to be explorers, finding out what is there, rather than presuming we know what a populations is" (quoted in Roberts, 1991, p. 1615).

In the context of group consent, however, this prior debate about the existence and status of groups in structuring human genetic diversity shut down. Group consent merely assumed that biological groups mapped onto cultural ones. Many scientists feared that group consent would unwittingly introduce reductionism into their science and, in effect, invent biological groups where none had existed before. One European population geneticist recalled in an interview:

INT: The Europeans were against this [group consent] because they don't need it and they are afraid that it may generate enormous problems for them.

JER: To have to identify these groups?

INT: Yeah, there are no such groups. So they'll have to create them in order to talk to them [*laughter*].[6]

In other words, efforts to give the objects of research a voice, and thus render them subjects, had inadvertently placed organizers in the paradoxical position of constituting their very objects of research. As examples like this indicate, nature did not simply exist apart from the moral universe, available to be enrolled as a resource. Rather, both were at stake in the design of the project.

Likewise, social order did not exist in a domain separate from and unaffected by scientists' inquiries into natural order. This we can see if we look at the controversy that surrounded the second assumption that underlies group consent: *The social world is already ordered into groups, and these groups, and their spokespeople, can be defined by experts.*

While the MEP did not raise questions about the existence of groups, it did recognize that many questions about how to define groups might arise. As one drafter of the group consent provision explained: "Is the population this village? Is it the Plains Apache? Is it all the Apache? Is it all Nadene speakers? Is it Native Americans? What's the population?"[7] To answer these questions, the protocol settled on two definitions of groups, each sanctioned by an expert community. The first drew upon the expertise of Diversity Project researchers. The protocol states, "If the researchers intend to sample only a particular part of such a unit [town, village, or religious unit], *the group* from whom consent must be sought is *defined by the researchers'* sampling criteria. For example, if researchers wanted to sample all members of a village who spoke a particular language, that portion of the village would be the relevant community" (North American Regional Committee, 1997, p. 1444). The second drew upon the expertise of anthropologists. To determine whether, for example, Maori groups in New Zealand called *Iwi*[8] recognize a pan-Maori organization that could act as a "culturally appropriate authority," the protocol stipulates, "Researchers would need to have a very sophisticated knowledge of Maori culture and politics before proceeding" (North American Regional Committee, 1997, p. 1446). It is presumed that expertise about Maori culture and politics will resolve complex questions about who speaks for the Maori.

This assumption that groups could be defined by experts elided the broader set of concerns expressed by members of indigenous rights organizations in the summer of 1993. As I already mentioned, these groups worried that expert definitions of indigenous groups would construct the identity of indigenous groups in ways that would threaten their security and autonomy. For many, the Model Ethical Protocol (MEP) group consent provision confirmed these fears. As a member of the Pine Ridge Ogalala/Sioux tribe, also a former policy analyst for the National Indian Health Board, explained at the First North American Conference on Genetics and Native Peoples in October of 1998, "If you read this thing [the Model Ethical Protocol], it sounds like a how-to book, and it sounds like it was written for Bureau Agents back in the 1800s or something."[9] He went on to imagine the Diversity Project's ethics director heading out into Indian country with a safari hat in search of the true natives. Finally, he explained the potential consequences of such acts: "Tribal membership is one of many sovereign rights that outsiders have attempted to circumvent or deny of native people, and every time their attempts have proven divisive and very harmful if actually put into practice. Other attempts will be similarly harmful. Only native people are able to make these decisions, not scientists, lawyers, or ethicists."[10] As this commentator on the MEP makes clear, failure to understand that the Diversity Project—through its ethical practices—might participate in the enormously consequential process of defining groups prevented organizers from seeing that far from expanding the rights of groups, the initiative might inadvertently act to constrict them.

In short, like the Diversity Project more broadly, group consent failed to gain support because it assumed that prior questions about the status and definition of human groups in nature and human groups in society had already been settled and were not themselves at stake in the design of human genetic diversity research. Ethicists and lawyers simply could not turn to scientists to resolve questions about how to define human groups. Nor could scientists look to ethicists and lawyers to create a moral order that could underwrite their effort to order human diversity in nature, for both moral and natural order were coming into being together. Not to recognize this was to risk falling into covert acts of power (e.g., the power to define human groups) associated with racism, a kind of racism organizers had vowed to oppose.

Enduring Dilemmas

In the years since the Human Genome Diversity Project debates, subsequent organizers of studies of human genetic variation have made every effort to distinguish their initiatives from the Diversity Project. One principal way they have sought to do this is through designing processes for including groups in the design and regulation of research. The Diversity Project faltered, many argue, as it failed to include indigenous groups in the design and planning of the project. By the time that any indigenous groups were asked to participate in discussions about the project, several planning meetings had already taken place (Reardon, 2005). Indeed, organizers sought to include indigenous groups only after some of these groups began to critique the project.[11] Current administrators of high-profile efforts to study human genetic variation—most notably, the International Haplotype Map Project—have attempted to avoid the controversies sparked by the Diversity Project through making proactive efforts to include groups early on in the design and regulation of research (International Haplotype Consortium, 2004).

Yet far from moving beyond the troubles of the Diversity Project, these more recent initiatives have encountered similar troubles. As we will see below, these enduring dilemmas make clear that the unease provoked by studies of human genetic variation will not be relieved by expanding the reach of liberal democratic values, such as participation, to include more kinds of people—in this case, groups, not just individuals. Instead, what is required is a more fundamental re-thinking of the conceptual foundations of research and the liberal democratic practices enrolled to stabilize it. To date, these practices have assumed that people pre-exist who can simply step into the role of being participants in the activities of liberal democracies—in this case, their knowledge-making practices. Yet, as will soon become clear, to be included as participants in research design and regulation, the existence of groups must be confirmed and their contours defined. The practices of genomics—both those deemed scientific and those deemed ethical—play roles in these processes of groups-in-the-making. Thus, genomics in part constitutes and defines the very groups it seeks to study. It is this novel paradox that the International Haplotype Map (HapMap) has made explicit.

The International Haplotype Map Project (HapMap)

Difficulties faced by HapMap organizers' efforts to include groups in the design and regulation of research begin to emerge by asking the simple question: What is the HapMap Project? If one looks at the project's official Web page, a seemingly straightforward answer can be found: "The International HapMap Project is a multi-country effort to identify and catalog genetic similarities and differences in human beings. Using the information in the HapMap, researchers will be able to find genes that affect health, disease, and individual responses to medications and environmental factors" (International HapMap Project, 2005). This sounds simple enough. But from its very inception, the endeavor faced the same problematic, fundamental questions faced by the Diversity Project: What kind of phenomenon is human genetic variation? An individual or group phenomenon? If a group phenomenon, what kind of group? How should groups be defined?

At first glance, the HapMap documents appear to provide a straightforward answer to these questions. As the Web site explains, "The International HapMap Project is analyzing DNA from populations with African, Asian, and European ancestry" (International HapMap Project, 2005). Yet although presented simply enough, the meaning and implications of this statement are far from clear. The population categories African, Asian, and European would appear to resonate with the racial categories African, Mongoloid, and Caucasoid. However, project documents as well as many of the project's organizers deny that the project is at all concerned with questions of race. Consider the following exchange with a project leader:

JER: Let me ask about what these populations in fact represent. . . . I imagine that if I asked you, "Can the samples be said to be representative of race?" you would say, "No."

INT: Right.

JER: And if I said it was representative of Africans or Chinese?

INT: The answer would be no.[12]

Rather than a project hopelessly entangled in the complexities and uncertainties of race, many organizers imagine theirs to be a project that uses the "precise" tool of the scientists' sampling criteria to define its objects of study.

In addition to refusing any link to race, like HGDP organizers, proponents of the HapMap hope that scientists' sampling criteria will resolve the nettlesome questions about representation. The exchange continued:

JER: So, who could be considered a member of the population that was sampled when [the geneticist] and others went to Africa?

INT: The criteria—I'm just guessing because I can't remember. The [sampling] criteria may include residents in this local community and self-representation as . . . Yoruba.

JER: Right. And so then that sample would only represent whatever the sampling strategy was?

INT: That's what all samples represent, is just whatever the sampling strategy was.[13]

But, there was a problem with this attempt to answer nettlesome questions about group definition by retreating to the seemingly objective ground of the sampling criteria:

JER: But you were trying to represent global diversity.

INT: So then the question is how can you extrapolate from that, what's your rationale for extrapolating from that to other things. Yes. . . . Well, OK, the argument made there was that . . . if we had left African populations out altogether, we would have been ignoring not only the greatest diversity, but also the source of human diversity.

JER: Right. So in what sense would it be wrong to say that the Yoruba population is supposed to be representing the African populations? The issue was "We don't want to leave out African populations," like you said, "so the way we're going to deal with that problem is we're going to sample Yorubas." You didn't say, "Yorubas have the greatest diversity of human populations, so we're sampling Yorubas," you said—

INT: So you could say it's not *the* strategy, I guess, but *a* strategy to get at that diversity—that greater diversity. Yes, you could say that.[14]

In the wake of the admittedly painful Diversity Project debates, current organizers of human genetic variation research understandably shy away from any claim that links them to the messy world of politics, representation, and race through attempting to only make claims backed by purportedly apolitical, objective criteria, such as a sampling strategy. The word the official Web site uses for this is "precision." It explains: "From a scientific standpoint, *precision* in describing the population from which the samples were collected is an essential component of sound study design. . . . From a cultural standpoint, precision in labeling reflects acknowledgement of and respect for the local norms of the communities that have agreed to participate in the research. From an ethical standpoint, precision is part of the obligation of researchers to participants, and helps to ensure that the research findings are neither under-generalized nor over-generalized inappropriately" ("Guidelines for Referring," 2005, emphasis added). The problem with such a position, however, is that it presumes that the existence of the object, i.e., the population, is not in question; the issue is only one of precisely defining it. But exchanges like the one above make clear that calls for precision are founded upon the mistake that the process philosopher Alfred North Whitehead called "misplaced concreteness"—that is, they posit the existence of a concrete object where none exists (Whitehead, 1925, p. 77). What did researchers come home with when they went to Africa and collected blood from Yorubas in Ibadan, Nigeria? The DNA of the human species? Of a race? Of Africans? Of Yorubas? Of 90 people who lived in Ibadan, Nigeria? Answers to these questions are far from resolved. Instead, the objects of study in the HapMap are still very much emergent; their contours will not be resolved by turning to so-called precise empirical data.

This point becomes more evident if we consider the results of the community engagements conducted in Japan. As mentioned above, following the perceived missteps of Diversity Project organizers, HapMap administrators placed an emphasis on engaging "the population" they wished to sample, seeking their input on project design and regulation—in particular, on the labeling of samples. But like the case of the Diversity Project, this effort to include groups in human genetic variation research, both as objects to be known *and* subjects who could know and regulate research, caused organizers to trip as it became clear that no clarity, and certainly no precision, existed on the most basic issue of deciding the boundaries and definition of the "group" to be included. Consider the following exchange:

INT: In the Japanese case, there was an effort to make those criteria [for labeling] more specific. . . . [T]hrough the engagement, through discussion with the Japanese . . . [it became clear that] the community didn't want it more specific.

JER: Really? Why not, do you know?

INT: They thought—which is reasonable—that they didn't all come from this place in Tokyo, that they came from all these other different places, and so they shouldn't be labeled as coming from that place. . . . There's also apparently something going on about Japanese national identity and how they feel about that national identity, and how they may have mythologized it to some extent, that they didn't want to be labeled as something other than that national label.

JER: So . . . I'm wondering in what sense these folks who were in the community engagement have the power to speak for "the Japanese" on how the label "Japanese" would be used.

INT: Well, let me rephrase that—you're asking the people who were the prospective donors to speak about how they wanted *their* samples labeled. So you're not asking them to speak for all Japanese . . . but rather as the people who are being asked to make the sample donations.

JER: Right. But . . . [there] is going to be a label "Japanese" associated with genetic information. . . . And just like with Native American tribes, there's an issue of social identity here.

INT: . . . I think we're going to try very hard not to make that claim that they are representative biologically of genetic frequencies among all Japanese—rather they're representative of genetic frequencies among those Japanese sampled for the HapMap.

Here we encounter the same problems of defining the object of analysis evident in the case of sampling in Africa. But this time these problems come entangled in questions about how to define the subjects to be consulted: Who did organizers of the HapMap think they were talking to when they conducted their consultation in Tokyo, Japan? Clearly, this organizer felt uncomfortable with my suggestion that "the community" consulted could speak for "the Japanese." I empathize with his discomfort. In what sense do the people at a university in Tokyo speak for a group called "the Japanese"? Does a group, "the Japanese," even exist? Further, because the project's community engagement process was founded upon the idea that the population whose genes were sampled would be the community whose views would be consulted, to say that the project consulted with "the Japanese" was tantamount to saying that there were "Japanese genes." This raised the concern that, despite claims to the contrary, the HapMap Project would lead to a reinvigoration of genetic constructions of nation and race. Ironically, this time the biologization of race would not be the result of a science unreflective of its social dimensions, but rather the result of a scientific initiative that was explicitly trying to address those social dimensions by democratizing its practice.

From Salvation to Reflection

Despite allegations in recent decades that biology—particularly in its genomic form—is motivated by economic interests, many of the scientists involved in the planning of the Human Genome Diversity Project and the International Haplotype Map remain committed to the vision that science transcends base materialist interests in order to further the higher values of truth and justice. There is much to admire in this vision and in these scientists' commitment to it. However, as I have sought to demonstrate in this chapter, it is precisely this understanding of science as a domain apart from base social interests and prejudices—indeed, as a force that might counteract them—that prevents engagement with questions about "groupness" that lie at the heart of contemporary concerns about human genetic variation research.

Recall the cases described in this chapter. Despite some claims to the contrary, in practice both projects discussed are founded upon a distinction between the domain of science and the domain of ethics and society.[15] Each domain is positioned as the outside to the other, figuring both as external resources that can redeem or provide expertise. For example, in both cases explored in this chapter, "scientific" criteria (e.g., sampling criteria) act to answer questions about how "ethical" protocols should define groups for the purposes of engaging with them. In turn, norms of ethical behavior (e.g., researchers must obtain group consent) in effect answer nettlesome "scientific" questions about the status of groups in nature. As a result, the more fundamental questions about the status of human groups and how these groups are (re)constituted through the construction of both knowledge and ethics never formally arise. Thus, despite the extensive and well-meaning efforts to address the ethical concerns raised by human genetic variation research, to date these more fundamental questions about groupness—questions that have long been of utmost concern to critics—have, in effect, been pushed to the side, rendered invisible by the very practices created to address them.

To bring these more fundamental questions into view and not be lulled by misplaced promises of precision, democratic science, and a race-free genomic future, we are in need of new analytic tools, forged by critical scholars of science and technology, that draw into view the mutual constitution of scientific and social practice, of objects of study and subjects with rights. Such analytic tools would help make explicit the social choices embedded in any effort to order human beings into groups for the purposes of understanding their differences and commonalities. Genomics might, indeed, bring new levels of refinement to these efforts to order and know human beings, but that process of refinement cannot gain meaning apart from its ties to the social order. Our best chance at attending to the human concerns raised by this novel form of technoscience, and at making good on its promise of deeper human understanding, will follow from deliberative practices that mind these entanglements. Salvation from our social biases through precision may not be possible, but reflective choices about who we are becoming through our efforts to refine knowledge about ourselves are within our reach.

NOTES

Research for this essay was made possible by a grant from the National Science Foundation (NSF #0613026). Any opinions, findings, conclusions, or recommendations expressed in this material are those of the author and do not necessarily reflect the views of the National Science Foundation.

1. Interview with Jenny Reardon, May 6, 2004. Although this interviewee uses "ethnicity," I contend, following Omi and Winant (1986), that these uses of "ethnicity" are often invoked to obscure the obvious political problems that attend race.

2. As part of an effort to recognize and promote the rights of indigenous groups in the midst of the 500-year celebration of the "discovery" of America, in 1993 the United Nations announced the beginning of the Decade of Indigenous Peoples (General Assembly Resolution 48/163 of December 21, 1993).

3. For a description of the opposition to the Diversity Project, see Reardon (2005, p. 100, 2006)

4. The National Commission for the Protection of Human Subjects of Biomedical and Behavioral Research issued the Belmont Report in 1979. This report outlines the major conditions of informed consent: consent must be given voluntarily by

competent and informed individuals. It explains that this ethical practice follows from the conviction that "individuals should be treated as autonomous agents," a principle that is consistent with the liberal democratic tradition of protecting and preserving *individual* autonomy and rights (National Commission for the Protection of Human Subjects of Biomedical and Behavioral Research, 1979).

5. This section draws upon materials previously published in Reardon (2001).

6. Interview with Reardon, July 2, 1996. INT refers to the interviewee and JER refers to Jenny Reardon in this and all subsequent interview excerpts.

7. Interview with Reardon, June 13, 1996.

8. *Iwi* refers to subgroups within the Maori, analogous to clans or tribes (Reardon, 2005, p. 120).

9. First North American Conference on Genetic Research and Native Peoples, October 1998, Polson, MT.

10. Interview with Reardon, January 24, 2001.

11. It is important to note that after these critiques emerged, project organizers did put considerable time and energy into efforts to consult with indigenous groups. I explain the reasons for the failure of these efforts to address the concerns of indigenous groups elsewhere. See, in particular, Reardon (2005, chap. 6).

12. Interview with Reardon, December 1, 2004.

13. Interview with Reardon, December 1, 2004.

14. Interview with Reardon, December 1, 2004.

15. For claims to the contrary, see, in particular, International Haplotype Consortium (2004).

REFERENCES

Cavalli-Sforza, L. (1994). The Human Genome Diversity Project. An address delivered to a special meeting of UNESCO, Paris, France.

Cavalli-Sforza, L., Wilson, A. C., King, M., Cook-Deegan, R. M., & Cantor, C. (1991). Call for a worldwide survey of human genetic diversity: A vanishing opportunity for the human genome project. *Genomics, 11* (Summer), 490–491.

Epstein, S. (1996). *Impure science: AIDS, activism, and the politics of knowledge.* Berkeley: University of California Press.

Fischman, J. (2003, January 20). On target: A new generation of drugs offers customized cures. *U.S. News and World Report*, p. 51–56.

Flint, A. (1995, March 5). Don't classify by race, urge scientists. *Boston Globe*, B1.

Guidelines for referring to HapMap populations in publications and presentations. (2005). *International HapMap Project*. Retrieved December 22, 2005, from http://www.hapmap.org/citinghapmap.html.en.

Hacking, I. (1999). *The social construction of what?* Cambridge, MA: Harvard University Press.

Hotz, R. L. (1995, February 20). Scientists say race has no biological basis. *Los Angeles Times*, A1.

International Haplotype Consortium. (2004). Integrating ethics and science in the International HapMap Project. *Nature Reviews Genetics, 5*, 467–475.

International HapMap Project. (2005). Retrieved December 21, 2005, from www.hapmap.org.

Lock, M. (1994). Interrogating the Human Diversity Project. *Social Science and Medicine, 39* (5), 603–606.

National Commission for the Protection of Human Subjects of Biomedical and Behavioral Research. (1979, April). *The Belmont Report: Ethical principles and guidelines for the protection of the human subjects of research.* Washington, DC: Office for Protection from Research Risks, National Institutes of Health, Department of Health, Education, and Welfare.

National Institute of General Medical Sciences. (2000). *Medicines for YOU: Studying how your genes can make a difference* (NIH Publication No. 05–4562). Retrieved February 19, 2006, from NIGMS Web site: http://publications.nigms.nih.gov/medsforyou/MedsForYou.pdf.

Native-L. (1993). Call to stop Human Genome Diversity Project. Retrieved June 24, 2006, from http://nativenet.uthscsa.edu/archive/nl/9307/0036.html.

North American Regional Committeee of the Human Genome Diversity Project. (1997). Proposed model ethical protocol for collecting DNA samples. *Houston Law Review, 33* (5), 1431–1473.

Omi, M., & Winant, H. (1986). *Racial formation in the United States: From the 1960s to the 1990s.* New York: Routledge.

Pollack, A. (2003, June 27). Large DNA file to help track illness in blacks. *New York Times.* Retrieved June 5, 2003, from http://www.nytimes.com/2003/05/27/national/27GENE.html?ex=1055051936&ei=1&en=d918e2aee2c22fc2

Reardon, J. (2001). The Human Genome Diversity Project: A case study in coproduction. *Social Studies of Science, 31* (3), 357–388.

Reardon, J. (2005). *Race to the finish: Identity and governance in an age of genomics.* Princeton, NJ: Princeton University Press.

Reardon, J. (2006). Creating participatory subjects: Race, science, and democracy in a genomic age." In S. Frickel & K. Moore (Eds.), *The new political sociology of science: Institutions, networks, and power.* Madison: University of Wisconsin Press.

Roberts, L. (1991). Scientific split over sampling strategy. *Science, 252,* 1615.

Sankar, P., & Kahn, J. (2005, October). BiDil: Race medicine or race marketing. *Health Affairs,* 455–463.

Scott, J. (1996). *Only paradoxes to offer: French feminists and the rights of man.* Cambridge, MA: Harvard University Press.

Venter, C. (2000, June 27). Statement on decoding of genome. *New York Times,* p. D8.

Whitehead, A. N. (1925). *Science and the modern world.* New York: The MacMillan Company.

17

The Feasibility of Government Oversight of NIH-Funded Population Genetics Research

JACQUELINE STEVENS

Background on Two Recommendations to Regulate the Study of Population Diversity by National Institutes of Health (NIH) Grantees

Even before the invention of laboratory techniques to create genotypes of putative difference from the political categories of race and ethnicity,[1] scholars concerned about inequality rooted in these phenomenological differences offered their concerns. Troy Duster's *Backdoor to Eugenics* (1990) is the first alarm for this emerging research. Its warnings and cautions hold up with alarming accuracy over 15 years later. As most readers of this chapter know, Duster and others with similar worries had their cautions institutionalized under the auspices of a standing committee, Ethical, Legal, and Social Issues (ELSI), generated by the Human Genome Project (HGP). A congressional mandate requires researchers working on the HGP to take ELSI into account. The effect of this mandate is discouraging (see Sankar et al., 2004; Stevens, 2003). Government scientists are ignoring the evidence produced under the aegis of ELSI and further geneticizing population differences.

One recent symptom of ELSI's failure to make an impact on government health policy is the patent that has been issued for a race-specific drug for hypertension. In 2005 the Federal Drug Agency (FDA) gave the pharmaceutical firm NitroMed the exclusive right to sell BiDil to Black[2] patients until 2020. It turns out that NitroMed had a patent set to expire in 2007 on the general population's use of the two drugs composing BiDil. The new race-specific patent gives NitroMed proprietary rights to this narrow market until 2020 (Kahn, 2005). Were norms from work by ELSI panels over the last decade followed, such a patent never would have been granted. Jonathan Kahn's analysis in the Correspondence section of *Nature Genetics* states:

At work here is an appropriation of race as reified in the BiDil story to serve larger political agendas aimed at transmuting health disparities, rooted in social and economic inequality, into mere health differences, rooted in biology and genetics. Attempts to address social disparity generally implicate the power of the state or other nonmarket institutions consciously to intervene both in the allocation of resources and the sanctioning of racist practices. In contrast, attempts to address genetic difference may be located at the level of the molecule and targeted by pharmaceuticals developed and dispensed through the purportedly impersonal forces of the market.

Not only are African-Americans now going to be charged monopoly prices instead of generic fees for this medicine, but the public and even many scientists who are casual consumers of drug studies now have the false impression that drugs should be race-specific, to reflect genetic differences among races. (2005, pp. 655–656)

The BiDil approval and similar press reports (Stevens, 2002, 2003) fuel the self-fulfilling prophesy of genetic studies, many funded by the NIH, that assume but do not prove that races are genetically discrete. The BiDil patent is the logical outcome of NIH-funded research ignoring evidence-based analysis of racial categories. When the NIH itself produces specious race-specific findings of genetic differences, this encourages the FDA to support a race-specific patent. Reciprocally, the FDA's approval of BiDil makes race-based drug treatments a reality and hence buttresses the misguided intuitions of scientists and others that race is a genetically based difference.

Alas, more than 15 years of advisory warnings to scientists publishing misleading and harmful views about race and ethnicity have not worked. More stringent measures are necessary. In other areas, failed warnings lead government to take coercive measures. A number of federal agencies allow for private firms to work closely with the government to assure compliance, but if that compliance is not forthcoming, then penalties ensue. The Environmental Protection Agency may issue cautions for a limited period of time, but if a polluter shows no interest in fixing the identified problems, then legal action will be taken. Similarly, for several years after evidence on secondhand smoke's harms became available, polite requests to forego smoking were the sole tool available for citizens to prevent their exposure to this contaminant. But eventually, when it was clear that publicity alone was not sufficient to prevent smokers from harming non-smokers, states took other measures, including outlawing smoking everywhere from restaurants to beaches. Using a similar evidence-based cost-benefit approach to race research, I have suggested elsewhere that the NIH require, and not simply advise, the use of scientifically defensible research designs from grantees intending to publish results indicating a genetic basis of racial and ethnic differences (Stevens, 2003). These proposals follow principles animating similar policies followed by some science and medical journals, as well as recommendations made by Judith Kaplan and Trude Bennet (2003), though theirs were directed toward the criteria for medical and scientific journals evaluating health claims based on taxonomies of race and ethnicity, while the proposals below are for implementation by the U.S. Department of Health and Human Services (DHHS), the agency overseeing the National Institutes of Health (NIH). They also are urged for use by the independent National Science Foundation. The two recommendations are as follows:

1. The DHHS should issue a regulation prohibiting its staff or grantees, including those receiving NIH funding, from publishing in any form—including internal documents and citations to other studies—claims about genetics associated with variables of race, ethnicity, nationality, or any other category of population that is observed or imagined as heritable unless statistically significant disparities between groups exist and description of these will yield clear benefits for public health, as deemed by a standing committee to which these claims must be submitted and authorized prior to their circulation in any form beyond the committee.

2. The NIH should issue a clarification of the current Congressional requirement that "women and members of minorities and their subpopulations *are included in all human subject research*," a proviso which has been interpreted to require ethnic and racial taxonomies (*59 Federal Register 11*, p. 146 [1994, March 9]). The new regulation should specify that federally funded medical research study many populations, including those that vary by childhood residence, current residence, occupation, diet, exercise, age, wealth, income, and regularity of medical care. (Stevens, 2003, p. 1075)

The evidence motivating these recommendations—the unsubstantiated, hyperbolic claims about alleged racialized genetic etiologies for diseases that have proven environmental causes of substantial magnitudes, especially breast cancer and asthma—is discussed in detail in Stevens (2003). This article has been cited in medical journal articles on population genetics, and I have received no correspondence disputing the characterization of the studies criticized. Despite the lack of disagreement with the article's observations about shoddy, government-funded population genetics hyperbole, the article's regulatory recommendations generated a number of criticisms from researchers in the field. This chapter responds to these criticisms, shows how one public health epidemiologist exemplifies how race and ethnicity can be operationalized without making genetic assumptions, and infers from this work practical guidelines for scientists and policy makers that further explain my original proposals.[3]

Criticisms of Recommendations

The major criticisms of the approach above are these: (1) the proposals are insufficiently motivated, as the research on which they are based does not support equation of geneticized racial and ethnic classifications with harms similar to those posed by other harms the government regulates, such as gun use and smoking; (2) the proposals would violate researchers' First Amendment rights to free expression; (3) the proposals would violate norms of academic freedom; (4) the proposals are insensitive to scientists from racial and ethnic minorities conducting this research on behalf of their respective communities; and finally, (5) the proposals are too cumbersome and would prove onerous for NIH staff and grantees already burdened with numerous other regulatory requirements.

Objection 1: Harms from Racial Categorization Unproven

Stevens (2003) claimed that geneticized racial and ethnic variables caused harms that far exceeded their actual benefits. This analysis relied on studies showing that experiencing victimization consequent to racial stereotyping was correlated with increases in blood pressure that might be associated with the disease of hypertension, that the mark of hereditary race itself was proven to cause sharply graded economic disparities, and that racial difference correlated with differences in housing quality (Stevens, 2003, pp. 1067–1073). The further postulate also discussed was that those differences associated with adverse health outcomes perceived as hereditary caused harms of a magnitude considerably greater than those perceived as incidental to choice, merit, or chance after birth (Stevens, 2003, pp. 1069–1071, 2004).

The causal mechanism animating the association between racial differences per se and adverse impacts is neither easily observable nor intuitively obvious. Colleagues pointed to the absence of studies directly depicting the harms caused by a geneticized view of difference. Absent direct observation that specifically geneticized racial and ethnic classifications in scientific publications cause serious harms, they argued against any rules that would prevent circulation of these representations. Before reviewing the evidence of these harms, it is important to note that the assumption that scientific agencies make policy decisions on the basis of scientifically researched evidence is a false one.[4] Not only does the scientific method reject the possibility of an irrefutable truth, as opposed to a provisional theory that has not been falsified (Popper, [1934] 1965), but agencies frequently enact measures because policy makers have good evidence to believe they will be effective and failure to act will cause harm.

The evidence on the cumulative pernicious effects of hereditary categories on human suffering is at least as strong as that allowing planners to tell us to travel at 35 mph and not 40 mph. For information on the potential health and safety gains from eliminating certain hereditary classifications, we have world history. An infinite variety of groups and individuals compete, but only a handful do so by exhorting their members to kill and risk death: it is only those groups wrought by kinship and religion, no others, that yield such dire consequences of needless systematic violence (Stevens, 1999, 2004). In 1985 Surgeon General C. Everett Koop enlarged the mandate of health care professionals by making injuries from violence the province of medical and not just legal experts (Mercy, 1993). Early work in this area focused on gang violence (Mercy, 1993), but the leading cause of death and injuries from violence among strangers occurs between groups taking up causes associated with ethnicity, nationality, race, or religion (WHO, 2002, pp. 213–240). World history teaches us that only those groups distinct as putatively ethnic, racial, religious, or otherwise familial are those for which masses of people will kill and die; and, indeed, war itself is now an emerging concern for public health officials (Roberts, Riyadh, Garfield, Khudhairi, & Burnham, 2004). In addition to the life-threatening afflictions of violence directly attributable to the prejudices wrought by hereditary communities, the staggering differences among countries' standards of living, far greater than disparities within countries, suggest that divisions of race, nation, and ethnicity contribute significantly to life outcomes (Stevens, 1999, 2004). We do not have to guess about this but have thousands of years of recorded history as a natural experiment demonstrating the harms caused by national and ethnic myths. This is far better evidence than that of controlled studies, the latter of which seek to mimic the real life of which our history actually consists.[5]

Admittedly, racial categorization is not the same as racism, but racial categorization never happens without also producing racial hierarchies (Stevens, 1999, pp. 172–208). These hierarchies produced through race have direct health consequences (Rich-Edwards et al., 2001, p. 126, citing Krieger & Sidney [1996], James et al. [1984], Krieger [1990] [positive correlations], Dressler [1990], Broman [1996] [no correlations]) and other poor behavioral and psychological outcomes (Rich-Edwards et al., 2001, p. 126, citing Jackson et al. [1996], Ladrine & Klonoff [1996], Mays & Cochran [1997], Williams, Jackson, & Anderson [1997], Kessler, Mickelson, & Williams [1999], Ren, Amcik, & Williams [1999], Klonoff & Landrine [1999], Zierler et al. [1991]). It is not conflict but conflict associated with race that produces the high levels of stress thought to trigger adverse effects (Rich-Edwards, Krieger, et al., 2001, citing

Amstead, Lawler, Gorden, Cross, & Gibbons [1989]; Jones, Harrell, Morris-Prather, Thomas, Omowale, 1996; Morris-Prather et al., 1996). One study found that among 85 low-income, African American mothers of singletons in Chicago, women who bore very low birth-weight infants (1500 grams) were "twice as likely to report having experienced racial discrimination while pregnant than were mothers of normal birthweight infants (>2500 g) admitted to the neonatal intensive care unit for ventilator management or to the normal newborn nursery" (Rich-Edwards et al., 2001, p. 125, citing Collins et al. [2000]).

The evidence from history, current large-scale medical studies, and clinical studies shows that race causes racism and racism causes adverse health outcomes. Mythical hereditary classifications of race proceed by state action (e.g., Dominguez, 1986; Haney-Lopez, 1996), making it the government's responsibility to take corrective measures to eliminate this taxonomy of its making. The contribution to be played in the field of public health consists of government scientists carefully scrutinizing unfounded claims, not funding their thoughtless circulation.

Objection 2: Censorship

Some colleagues complained that requiring an NIH review of racial and ethnic population assertions before publication is the equivalent of censorship. This response reflects a poor understanding of censorship and was addressed in the initial publication (Stevens, 2003, pp. 1078–1079). Censorship is not the failure of the government to fund, much less represent all points of view. For the government to be guilty of censorship, it must violate the First Amendment's proviso that Congress "shall make no law . . . abridging the freedom of speech." The Supreme Court has construed this amendment to prohibit only the government's restrictions on private speech, and not to mean government must indiscriminately fund all speech. Restricting government funding of claims the government's elected and appointed officials find odious or even unsupportable is well within constitutional parameters.[6] The Constitution protects the right of Creationists to spend their own time and money disputing Darwin's theory of evolution, but the First Amendment does not require the National Science Foundation to fund this research. As long as the government is not restricting privately funded researchers from making claims about race, there is no First Amendment issue at stake.

Moreover, the success with which studies of race and ethnic difference with genetic etiologies have been able to proceed with private funding proves there is an avenue for those strongly committed to geneticized narratives. The scientists who maintain ancestral databases for the purposes of matching mitochondrial DNA have been supported by individuals; religious, ethnic, and racial organizations; and the National Geographic Foundation.[7]

Objection 3: Restricts Academic Freedom

A more considered objection is that regardless of their legality, the proposals impose unacceptable constraints on academic freedom. This raises the problem of when a scholarly community may state that certain claims have been falsified so many times that they lack credibility and hence, in light of the harms such statements cause, should not be published, even as speculative. With this in mind, the journal *Nature Genetics* published an editorial deriding the U.S. government's use of race and ethnic categories on its 2000 census and initiated a policy requiring contributors to "explain why they make use of particular ethnic groups or populations, and how classification was achieved" (Nature Genetics, 2000, p. 98). Moreover the editors wrote that reviewers would be asked to consider these explanations, and they concluded by writing that they hoped their policy restricting publications of unsubstantiated claims about group ancestry would "raise awareness and inspire more rigorous design of genetic and epidemiological studies" (Nature Genetics, 2000, p. 98). Other journals with similar requirements include the *British Medical Journal, Paediatric and Perinatal Epidemiology*, and the *Archives of Pediatrics and Adolescent Medicine* (Kaplan & Bennett, 2003, p. 2709). *JAMA* has a similar policy, though it appears not to be followed (for a *JAMA* survey of repeated violations of its editorial policies on race variables, see Winker, 2006). How many times do professional associations, peer-reviewed journals, and leading scientists need to repeat that racial classification does not reflect genetic differences before this is off-limits for government-funded debate? If scientists want the government to fund further discussion about whether the earth is flat, the sun revolves around the earth, and God built the world in seven days—all of which have been dismissed by the scientific community despite credibility among larger publics for thousands of years— then scientists may take a principled stance against the proposals above.[8]

Objection 4: Scientists from These Communities Support This Research

Some scientists and other stakeholders from socioeconomically disadvantaged racial and ethnic communities in the United States are very supportive of research seeking to ascertain population genetic differences associated with disease etiology. From the University of Massachusetts' African-American DNA Roots Project to various scientists studying DNA variation among Native Americans, these researchers believe they are doing something to help improve the health status of members of their communities (Fullwiley, this volume) and may resent any restrictions on this research.[9]

Invocation of any preferences based on scientists' group commitments is troubling. First, the NIH has a long-standing history of funding projects based on objectively ascertainable scientific merit, not the subjective affinities of its potential grantees. The NIH does set aside a preferential category of grants to "promote diversity in health-related research" (U.S. NIH, 2004), but this is to diversify the pool of research communities. This, too, arguably should be changed to reflect a range of diversity criteria, including methodologies, and not just race and ethnicity. In any case, the NIH has never committed itself to fund research projects that lack scientific merit, regardless of the urgency of investigators' community commitments. Second, the financial incentives to study race and ethnicity as genetic differences and not as environmentally situated variables may be a strong factor in these scientists' support for genetic research. Insofar as there is evidence that scientists tend to find research questions under whatever lamp the biggest dollars (usually the federal government's) pay to light up,[10] it seems likely that if the federal government were to shift the $437 million devoted to the HGP (FY 2003), and the hundreds of millions spent on DNA variation projects in other parts of the NIH, to studying the effects of environmental racism and similar non-genetic adverse outcomes, then researchers from minority and majority communities would find these questions worthy of the same intensive study presently devoted to hereditary factors, and household background registries documenting exposures to pollution, exercise, nutrition, and so forth would be as prevalent as ancestry registries. Especially because some of the racial and ethnic groups supporting research into population DNA variation have dubious priorities, e.g., to restrict membership to tribes that profit from casinos (TallBear, this volume), it seems suspect for long-term guardians of the nation's health to take their cues from any such self-interested groups. The legitimate concerns about eliminating health disparities among minority populations can be amply and perhaps more adequately addressed by the guidelines proposed here.

Objection 5: Review Panels Are Cumbersome

Scientists who participate as applicants and reviewers in the NIH funding process already are saddled with following detailed, complicated requirements that drain time and attention that would best be devoted to research. Therefore, one complaint about the proposals above is that they would burden this already fraught process.

It certainly is true that NIH grantees endure significant requirements for documentation, but if the goal is to curtail research along guidelines endorsed by the NIH and its advisory councils, especially the U.S. National Research Council (1997), then one outcome of the proposals' implementation would be a salutary null response. Researchers are not forced to make claims about heredity unsupported by scientific research. There is no need for additional paperwork: researchers may abjure publishing these claims.[11]

Moreover, as discussed in the following section, the ability to follow these guidelines is not an abstract possibility but already realized in the best research methods practiced in this field. If some public health scholars can operationalize race and ethnicity as synchronic variables, i.e., as variables indicating phenomena of the immediate environment's rules, perceptions, and prejudices, and not as inherited, diachronic variables, then there is no reason others cannot as well.

Operationalized Measures of Race and Ethnicity as Synchronic: Research by Nancy Krieger

An important consideration is the availability of research models that would comply with the proposed guidelines. Indeed, these guidelines are not an abstract possibility, but already realized in the best research methods practiced in this field. Medical researchers have at hand protocols for studying race and ethnicity as synchronic variables—that is, differences within a generation and not inherited (diachronic). Numerous studies undertaken across a variety of disciplines in the sciences and the social sciences exemplify such synchronic approaches to race and ethnicity (e.g., Anderson, 1983; Benedict, 1947; Allen, 1997; Haraway, 1989; Hey, 2001; Kaya, 2001). In the field of public health in particular, numerous scholars demonstrate the feasibility of the proposed guidelines in dozens of their studies, some of which are discussed below. One scientist who has contributed a great deal to the empirics and heuristics of race and ethnicity in the field of public health is Nancy Krieger, a professor in the Harvard School of Public Health. Since 1994 Krieger has been publishing research on race and ethnicity that self-consciously adheres to the standards above, and long before she has been urging these for her colleagues as well.[12]

Krieger is responding, with more than a little frustration, to a long history of medical experts themselves recognizing the limits of stories in which genetics assumes an undeserved leading role, while nonetheless failing to produce more accurate narratives. Even when scientists are well aware of the vast epidemiological and clinical research concluding that environmental variation has a far greater impact on health differences than genetic variation, this common knowledge is often ignored. Krieger describes early medical research on the "links between economic deprivation, bodily characteristics, and health" in the 1930s (Krieger & Smith, 2004, p. 93, citing Sydenstricker [1933] & Vernon [1939]), and she shows that, nonetheless, the "biomedical literature remains rife with studies attempting to discern whether 'race'—as an alleged biological trait—explains U.S. black/white disparities in hypertension (not to mention other health outcomes)" (Krieger, 2001a, p. 44, citing Fray & Douglas [1993], Cooper & Freeman [1999], & Mutaner, Nieto, &

O'Campo [1996]). Medical researchers seem unable to stop asking a question that already has been answered, a phenomenon found in fields from political theory to sociology, where, beginning with Aristotle's attack on the idea of autochthonous Athenians (1984 [c. 320s B.C.]), each generation must prove anew that their intergenerational communities exist because of rules and not birth from the earth, gods, or molecules.

In one publication Krieger and her colleagues point out that 80 years after public health experts called for more statistical information about socioeconomic contributions to health status, these data "typically have not been a component of published U.S. vital statistics; data instead have been stratified solely by age, sex, and what is referred to as 'race' . . . and are primarily used by researchers to 'control' for, rather than study the effects of, socioeconomic position on health" (Krieger, Williams, & Moss, 1997, p. 342, citing Liberatos, Link, & Kelsey [1988], Navarro [1986], & Syme [1992]). In a publication five years later, Krieger points out that major public health studies done by the U.S. government still did not collect data necessary for evaluating the socioeconomic contributions to health status: the 2002 edition of *Health, United States* "lacked socioeconomic data in 85.5% of its 71 tables on 'health status and determinants'" and "70% of the 467 U.S. public health objectives for the year 2010 had no socioeconomic targets" (Krieger, Chen, Waterman, Rehkopf, et al., 2003, p. 1655).[13] Krieger also has pointed out that scientific publications may not just misinform, but that their mistakes may be "outright harmful" (Krieger, 2003a, p. 194, citing Krieger [2001b] & Schwartz & Carpenter [1999]). Finally, while Krieger recognizes the harms racial classifications themselves cause, she also understands the need to research race and takes issue with those who would refuse researching race and ethnicity altogether (2003a, p. 196, disagreeing with Fullilove [1998] & Stolley [1999]).

Krieger studies "race/ethnicity" as a "social rather than a biological category, referring to social groups, often sharing cultural heritage and ancestry, that are forged by oppressive systems of race relations justified by ideology" (Krieger, 2003a, p. 195); and she shows that characterizing race and ethnicity otherwise, as simple genetic variables, impedes research. Although scientists work as though new techniques will justify viewing race and ethnicity as biological variables, Krieger points out that all the new digital hardware and software in the world cannot give scientists the names they use for analyzing what they see. To shatter old prejudices about race and ethnicity, it is new thinking and not new machines that are necessary: "One way to move to the next stage [of research on the health effects of racism] is to consider current conceptual issues in the field, given that scientific knowledge is more often spurred by clarification of our thinking than by technological breakthroughs" (Krieger, 2003a, p. 194, citing Fleck [1979] & Zaman [2000]).[14] If the government wants to effectively support health research on the causes of disease variation, then it should take this insight seriously by using its grants to encourage conceptual innovation as well as technological improvements (see also Haslanger, this volume).

Krieger approaches the study of race by attempting to understand how it is construed as a hereditary category even though it is made in society. Her research is inclusive of a wide range of subject characteristics, including stress from racist discrimination. To colleagues who carp that race discrimination is a soft and ideological variable inappropriate for serious health research she replies: "The canard that research on health consequences of racism is 'political' rather than 'scientific' is blatantly incorrect: it is in fact political and unscientific to exclude the topic [of racial discrimination] from the domain of legitimate scientific inquiry and discourse" (Krieger, 2003a, p. 197, citing Satel [2000] [an example of the misplaced criticism], Krieger [2001a, 1999], & Mutaner & Gomez [2002] [defending broad-based epidemiological studies]).

As an example of this, a jointly authored study of racial differences for preterm birth rates observes controls for education and other socioeconomic status (SES) variables, still leaving statistically significant disparities in racial variables' contribution to preterm birth weights (Rich-Edwards et al., 2001, p. 124, citing Shoendorf, Hogue, Kleinman, & Rowley [1992] & Kleinman & Kessel [1987]). Such disparities led Rich-Edwards and her colleagues to attempt to isolate the "physiological mechanism through which chronic threat could imitate labour" and hence to a corticotrophin-releasing hormone (Rich-Edwards et al., 2001, p. 125). Rich-Edwards et al. also cite similar work on other diseases, including a study showing that recent immigrants from the same respective racial and ethnic groups have better health outcomes than immigrants from the same communities who had been living in the United States for several years, thus precluding a genetic explanation of majority/minority health differences (Rich-Edwards et al., 2001, p. 124, citing David & Collins [1997]). In addition to contextualizing the imbalance of government data on SES variables and calling for reconceptualization of racial and ethnic categories, Krieger's research is consistent with the specific parameters that would be applied by the proposed NIH standing committee, hence demonstrating that this is feasible. As shown above, Krieger characterizes race and ethnicity as synchronic; and, as shown below, when studying these variables she uses numerous variables and includes explicit discussion of her data analysis.

Studies Use Many Variables, Not Only Race and Ethnicity

Krieger focuses on the goal of studying the *embodiment* of certain discriminatory life experiences, including race, "while avoiding the trap of equating 'biologic' with 'innate' " (Krieger & Smith, 2004, p. 95). To do this, she studies the "experiences and policies" that reflect inequities caused by "socioeconomic position, not the hereditary category of race"; to study "social disparities" in cancer rates, Krieger includes more than 10 variables in addition to "race/ethnicity": "occupation, income, wealth, poverty, debt, and education and discrimination [as well as] gender, sexuality, age, language, literacy, disability, immigrant status, insurance status, geographic location, housing status, and other relevant social categories" (Krieger, 2005, p. 7). Her earlier work also draws on a similarly large array of variables (Krieger, Chen, et al., 2002; Krieger, Waterman, et al., 2002). By including a number of variables that are associated with discrimination and not heredity, Krieger is able to represent race as one possible synchronic and not diachronic cause of differences in health status.

The study of early pregnancy differences in infant weight is a good example of how this method is used. Krieger and her colleagues designed questionnaires asking their subjects about "medical history, sexual orientation, household size, income and assets" and also childhood financial hardships (Rich-Edwards et al., 2001, p. 128). Additionally, the study's questionnaires sought data on experiences of discrimination, asking about "experiences of 'unfair treatment because of race or ethnicity' in eight domains, such as 'at work' or 'getting housing' " (Rich-Edwards et al., 2001, p. 131). Using a race variable as one of many variables makes it easier to see the effects of these other variables and allows, though does not require, readers to see race as one more possible synchronic contribution to health status.

Research that examines many variables, to have results with statistical significance, must have a large number of subjects, and Krieger's work does this. In a 1994 study comparing rates of breast cancer and accumulation of a possible carcinogen in one's blood, Krieger et al. (1994) used a cohort of 57,040 women. This allowed them to study numerous biological and social variables, including age of menarche, body mass index, pregnancy status (ever versus never pregnant), and menopausal status as well as education, social class, poverty level, birth place, and a body mass index (Krieger et al., 1994). Her study of the impact of racial discrimination on high blood pressure used 4,086 subjects (Krieger & Sidney, 1996), and the birth-weight study included 6,000 subjects (Rich-Edwards et al., 2001). This is an entirely different magnitude of sample size than that used by those seeking genetic etiologies, where the number of people tested and followed are typically fewer than 500 and, in pharmacological studies making use of direct SNP analysis, fewer than 50, or even none at all but simply inferences based on some molecular experiments (see, for citations, Stevens, 2003).

Krieger and her colleagues adhere to another practice that appears in the proposed criteria for the NIH Committee on Population Genetics: explicit discussion of the various techniques used in analyzing the data. This helps readers assess the usefulness of the probability estimates, that is, the likelihood that a correlation reflects an underlying relation between the observed variables and is not a result of randomness. For any particular statement of an association, researchers conventionally test for a 95% probability of it reflecting a persistent feature of the phenomena studied. The claim that the sample size allows for this degree of certainty means very little, however, if a researcher has analyzed the data using 20 different possible hypotheses or techniques, since 1/20 is 0.05 and, hence, any positive association discovered by 1/20 of the guesses being confirmed by the data falls into the range of randomness.

Another problem is "data mining"—either eyeballing the results or using software to find patterns—and then defining a hypothesis. In research invoking the standards of statistical rigor for supporting its findings, this approach is a way of cheating. These approaches to analyzing data in a single study are difficult to detect, but the best evidence that numerous guesses and data mining are widely used by population geneticists is the many published studies about racialized genetic disease variations with statistically significant results that later have been shown to be non-reproducible (Stevens, 2003).

While data mining is common in population genetics research, there are numerous other possibilities for misleading results in the range of statistical significance. In her devastating critique of racializing population research, Deborah Bolnick reviews the *structure* software documentation used by two prominent publications claiming genetic taxonomies corresponding with idiomatic racial categories (Bolnick, this volume). After the software user inputs a number for the clusters of potential populations, the program responds by providing an estimate of the likelihood of that number providing a good fit with the data's variation. In one case there was no better fit between the number they published (one corresponding with the dominant phenomenology of racial populations) and those of 19 other numbers for which the researchers tested their cluster hypothesis. Yet the authors only reported the results of the one fit that corresponded with their racializing intuitions, failing to mention the 19 other similar results.

Krieger's statistical work is a model for being up front about its assumptions and techniques, consistently mentioning possible biases in sample selection (Krieger, Chen, Waterman, Rehkopf, et al., 2003, p. 1660), possible effects of misclassification of race/ethnicity (ibid., p. 1660), potential flaws in the heuristics used (ibid., p. 1669), and alternative hypotheses that would account for correlations other than the data points directly analyzed. Especially noteworthy is her transparency about the different statistical techniques that she uses (e.g., Krieger et al., 1994, p. 591), including mention that while hers was the largest study to date of possible carcinogenic contributions to breast cancer, the research still would have benefited from including even more subjects (Krieger et al., 1994, p. 593).[15] This is a standard good practice in any statistical analysis, and hence the failure of high-profile studies of race and ethnicity to document their work in the manner Krieger and others document theirs (Hey, 2001) is a key reason that a special committee is needed to assess hereditary claims made about race and ethnicity.

Scientists may not believe there is a need for a committee to review the presentation of their ideas and may find the proposals somewhat confusing. Generally perplexed by the idea that language is anything other than a simple medium for pointing to hard truths, it may be helpful to scientists to be specific about the exact language and omissions that would prompt their publications' scrutiny from a standing committee, so that they may decide if they have evidence sufficient to pass the review.

Criteria for a Population Group Triggering a Review

What are the criteria for a population group triggering a review? In deference to the official consensus that race and ethnicity are folk categories, the words "race" and "ethnicity" are infrequently used in population genetics articles. "Population" and "ancestry," referring to genetically constituted groups with all the characteristics of idiomatic races and ethnicities, are often used instead, e.g., the "African population." Thus it seems useful to be clear on the sorts of references that would trigger a review: mention of disease etiologies or treatments for populations with genotypes or phenotypes associated with a geographic territory of origin (Stevens, 1999, pp. 172–208).

The list of geographical territories of origin that would prompt review is referenced by taxonomies including but not limited to the following designations: Caucasian, European, White, African, Black, Asian, Hispanic, Latino, Native American, Arab, any name of a geographical region (e.g., Pacific Southwest, West Africa), tribe, past or present nation (this includes all ethnic groups).[16]

This definition of a racial or ethnic population excludes nonprejudicial research on other medical populations with hereditary diseases, e.g., clusters of people with hereditary blindness or diabetes. Likewise, if scientists wanted to study correlations between genes associated with height and heart disease, their studies would not be subject to scrutiny. Membership in groups with phenotypes of heart condition or height per se is not taxonomized according to a specific geographic territory of origin. Researchers commonly publish associations of diseases with racial and ethnic groups, which is precisely what leads to so many spurious and harmful inferences. If researchers are convinced that they can prove the robustness of their population associations, then they are encouraged to submit their data and statistical techniques to a review panel.

If scientists did decide to publish their results with race and ethnic correlations, the review panel would have to stipulate that the article include at least 10 nonhereditary variables: e.g., geographic location of childhood residence, diet, exercise, exposure to toxins, occupation, income, wealth, poverty, debt, education, discrimination, sex, sexuality, age, weight, height, language, literacy, disability, immigrant status, insurance status, present geographic residence. The list is suggestive, not exhaustive, and some variables will be more important to some research questions and not others. Nonetheless, as Krieger's investigations show, it is important to study numerous possible contributions to adverse health experiences. Specifying a target number, and 10 is one suggestion, allows researchers to select those variables most relevant to the research questions being investigated, while ensuring hereditary variables are contextualized. The likelihood that fewer correlations will have statistical significance is not to be feared: this is an accurate reflection of the diffuse character of disease etiology.

Conclusion

The greatest threat to human thriving is the group divisions imposed by governments directly or indirectly imposing kinship rules and genealogies as well as granting official status to certain differences—e.g., Sunni and Shiite in Iraq, Catholic and Orthodox in the Balkans—or nationality confining one to relatively wealthy or poor countries. Now that the international public health body, the World Health Organization (WHO), has issued a major report indicating the responsibility of public health officials to prevent the violence endemic to such distinctions (WHO, 2002) and not to any others (Stevens, 1999, 2004), it is time for each country's public health practitioners to consider their role in implementing policies to diminish the number of deaths and injuries stemming from group differences of race, nationality, and ethnicity. In view of previous policies' failures, it seems a good time to try new measures that will directly curtail unsupported official assertions of genetic etiologies for racial and ethnic differences, publications that recent data, logic, and history demonstrate harm more people than they help.

NOTES

Many thanks to the editors Sandra Soo-Jin Lee and Sarah Richardson for their generous and detailed suggestions for revisions, and especially to Barbara Koenig for pressing me to address her excellent questions.

1. It would be a mistake to refer to the techniques of SNP (single nucleotide polymorphism) analysis for racial and ethnic correlations as a discovery. All research categories are invented, with different generations and communities interested in different observed or imagined phenomena. Some of these categories are useful and therefore do not receive heuristic scrutiny. Others pose problems and may not be useful; these warrant more careful reconsideration.

2. I capitalize names of racial groups (e.g., Black and White) because their specificity and political contingency resemble the proper nouns of nationalities and ethnicities.

3. This chapter is especially responsive to the concerns voiced by health researchers, to whom I am very grateful for their generosity in sharing their responses to the proposals above. Conversations with the other participants during the workshop convened by the editors of this volume were extremely useful.

4. To name one obvious example, there has never been an instrument that could measure the precise point at which the number of life hours saved by a speed limit exceeds those lost by lower speed limits. Presumably, the only certain speed for safe auto travel is the one infinitesimally close to zero but not zero itself. Yet thousands of city and highway planners worldwide allow us to move faster than snails on the basis of rough estimates for categorical speed limits, parameters that fundamentally affect how we move daily.

5. It is possible that future events could diverge radically from past patterns. But if policy makers believed that predictions based on past experiments should be discarded because future environments might be different, then policy makers would have to disregard all experimental results, including those from controlled settings. Insofar as large data sets and a large number of variables are the best way to measure results, history is a far better indicator of our real behaviors than even the best and largest series of laboratory experiments. It is also possible that we only invent the past as conflicts based on intergenerational attachments and we forget the others. In this case, it will be interesting to see what sorts of new memories might occur henceforth.

6. *Rust v. Sullivan*, 500 U.S. 173 (1991); *National Endowment for the Arts v. Finley*, 524 U.S. 569 (1998). A recent U.S. Supreme Court decision stated that the First Amendment does not protect government employees fired for expressing views distasteful to their superiors (*Garcetti et al. v. Ceballos*, 547 U.S. [2006]).

7. The Web site http://www.familytreedna.com/ sponsored by National Geographic offers links to tests for Native American ancestry, African ancestry, and Jewish ancestry. It also includes links for people wanting to find their Irish ancestry and to determine if they are linked to Ghengis Khan. Retrieved July 2, 2006.

8. For about two decades, scientists studied whether radium would improve vitality and relieve medical problems ranging from arthritis to schizophrenia (Rowland, 1994, esp. pp. 7–8). Enthusiasm waned after the death of a famous businessman, Eben Byers, who could afford the expensive Radithor, the trade name for bottled radioactive water. Byers's autopsy report "stated that he had comsumed about 1,400 bottles of Radithor. . . . The cause of death was stated to be necrosis of the jaw, abcess of the brain, secondary anemia, and terminal pneumonia" (Rowland, 1994, p. 8, citing Gettler & Norris [1933]). Until then, scientific journals contained numerous reports on the health benefits of radiation (Rowland, 1994; Gittinger, 2001). The government no longer funds research into whether exposure to radiation will alleviate schizophrenic symptoms. Now that the medical research community also has reached an overwhelming consensus that racial and ethnic classifications do not reflect genetic variations, including those associated with disease (for a similar claim see Braun, 2002; Fullilove, 1998; Goodman, 2000; Kaplan & Bennett, 2003; Osborne & Feit, 1992), and now that several large-scale controlled studies of health outcomes have not found a relation between race or ethnicity as hereditary variables and disease results, it would seem the government might also insist that grantees reporting earlier, faulty research would not have their work given the imprimatur of the NIH. It is harmful for scientists to cite

discredited research suggesting radioactive water will increase longevity (Gittinger, 2001) when data show it causes early death. This is no doubt a general rule of thumb at the NIH, where people who publish claims based on refuted research lose their support. When this norm consistently is overlooked because of the power the mythical views of genetics hold over most in the agency, and when the harms of these views being published is demonstrable, then these claims require special scrutiny.

9. Chapters by Kimberly TallBear, Henry Greely, and Rick Kittles and Mark Shriver in this volume provide numerous examples of such research projects, especially for African Americans, Jews, and Native Americans.

10. An old social science methods anecdote refers to an inebriated fellow looking for a quarter he dropped outside a bar at night. His companion tries to help by indicating the quarter was lost a few feet away from the spot being searched, to which his intoxicated friend replies impatiently, "Yes, I know that. But the light is better here." The government's massive support for genetic research is a very bright light indeed, enticing researchers, including those from ethnic and racial minorities, to seek answers to their questions about difference in this well-funded corner and not in the less-funded environmental and social areas that are more likely to yield answers to their questions.

11. The Office of Biotechnological Activities (OBA) is one example of how the NIH has been responsive to previous safety concerns, in this case, that of recombinant DNA research. Its regulations apply to "all recombinant DNA research within the United States (U.S.) or its territories" as well as to NIH grantees (NIH Guidelines for Research Involving Recombinant DNA Research, Section 1-C-1-a. Retrieved March 7, 2007, from http://www4.od.nih.gov/oba/rac/guidelines_02/NIH_Guidelines_Apr _02.htm).

12. To be clear, absolutely nothing that Krieger has written implies she would endorse the NIH implementation of her research practices as requirements for funding and not just good epistemological norms.

13. Krieger appears to endorse lower standards for monitoring health status compared to studies looking for causal explanations. "Although a plurality of measures may be useful for aetiological research, in the case of monitoring, such heterogeneity impedes comparing results across studies and across outcomes, let alone tracking changes over time" (Krieger, Chen, Waterman, Soobader, et al., 2003, p. 186). In her research pursuing the former, she advocates relying on the crude measure of zip code as this provides what she and her colleagues believe is a rough proxy for socioeconomic information that otherwise might go entirely unconsidered. Krieger, Chen, Waterman, Rehkopf, and Subramanian note that this study and those similar to it in state and federal agencies "obscure socioeconomic gradients in health overall and within diverse race/ethnicity-gender groups, as well as the contribution of economic deprivation to racial/ethnic and gender inequalities in health" (2003, p. 1655). Their study found that using the single-variable "percentage of persons below poverty" did as well as composite measures, such as the Townsend index (Krieger, Chen, Waterman, Rehkopf, et al., 2003, p. 1660). This research does not make any claims at all about genetic etiologies and hence would not trigger NIH review according to the proposals discussed in this chapter. Krieger's sensitivity to the different ends for which studies are conducted, however, is worthy of explicit mention.

14. Krieger does not confine her research to analyses of race and ethnic heuristics but also explains the various approaches to "gender" and "sex," as well as how these affect health policy studies (Krieger, 2003b). Consistent improvements in the technologies to count SNPs and therefore to produce subpopulations phenotypically identified by these characteristics do not mean that races and ethnicities idiomatically associated with these classifications are easier to identify or more real than the political taxonomies of race made through kinship rules and other state documents. Rather, it means that anyone, including scientists, can list variables and label them in a pattern consistent with any idiomatically existing group. These labels will perform the group's existence and therefore make it real; without that meaning the group is more or less real than any other putative group scientists or other experts with authority, especially state authority, might classify and therefore produce. For more on a theory of genetic heuristics for subpopulations, see Krieger (2001c, p. 668) and Stevens (2002, 2003). For methods to overcome stubborn intuitions, see also Kaplan and Bennett (2003).

15. Although current medical journals now require full disclosure of all grant and other funding support, researchers are rarely as forthcoming and precise as is Krieger. She reports, for instance, in her acknowledgements a "small honorarium" paid by a conference, "Theory and Action," where she presented a paper from which the published article was drawn (Krieger 2001c, p. 674).

16. Ethnic and national groups whose religious practices feature strongly in their self-identity may lead some to believe, incorrectly, that religions may be passed on through ancestry. Any time a religious group is thought ancestral, this is because the group, in addition to its religious practices, is named with reference to a past, present, or aspirational nation, e.g., the Israelites of the Hebrew Bible.

REFERENCES

Allen, T. (1997). *The origin of racial oppression in America.* London, Verso.

Amstead, C. A., Lawler, K. A., Gorden, G., Cross, J., & Gibbons, J. (1989). Relationship of racial stressors to blood pressure responses and anger expression in black college students. *Health and Psychology, 8,* 541–556.

Anderson, B. (1983). *Imagined communities: Reflections on the origin and spread of nationalism.* London: Verso.

Aristotle. (1984). *The complete works of Aristotle.* Princeton: Princeton University Press.

Benedict, R. (1947). *Race: Science and politics.* New York: Viking.

Bolnick, D. A. (2008 [this volume]). Individual ancestry inference and the reification of race as a biological phenomenon. In B. A. Koenig, S. S.-J. Lee, & S. S. Richardson (Eds.), *Revisiting race in a genomic age* (pp. 70–85). New Brunswick, NJ: Rutgers University Press.

Braun, L. (2002). Race, ethnicity, and health: Can genetics explain disparities? *Perspectives in Biology and Medicine, 45,* 159–174.

Broman, C. L. (1996). The health consequences of discrimination: A study of African Americans. *Ethnicity and Disease, 6,* 148–152.

Collins, J. W., David, R. J., Symons, R., Handler, A., Wall, S. N., & Dwyer, L. (2000). Low-income African-American mothers' perceptions of exposure to racial discrimination and infant birthweight. *Epidemiology, 11,* 337–339.

Cooper, R. S., & Freeman, V. I. (1999). Limitations in the use of race in the study of disease causation. *Journal of the National Medical Association, 91,* 379–383.

David, R., & Collins, J. W., Jr. (1997). Differing birth weight among infants of U.S.-born blacks, African-born blacks and U.S.-born whites. *New England Journal of Medicine, 337*, 1209–1214.

Dominguez, V. (1986). *White by definition: Social classification in Creole Louisiana.* New Brunswick, NJ: Rutgers University Press.

Dressler, W. W. (1990). Lifestyle, stress, and blood pressure in a southern black community. *Psychosomatic Medicine, 52*, 182–198.

Duster, T. (1990). *Backdoor to eugenics.* New York: Routledge.

Fleck, L. (1979). *Genesis and development of a scientific fact.* Chicago: University of Chicago Press.

Fray, J.C.S., & Douglas, J. G. (Eds.). (1993). *Pathophysiology of hypertension in blacks.* New York: Oxford University Press.

Fullilove, M. T. (1998). Comment: Abandoning "race" as a variable in public health research—an idea whose time has come. *American Journal of Public Health, 88*, 1297–1298.

Fullwiley, D. (2008 [this volume]). The molecularization of race: U.S. health institutions, pharmacogenetics practice, and public science after the genome. In B. A. Koenig, S. S.-J. Lee, & S. S. Richardson (Eds.), *Revisiting race in a genomic age* (pp. 149–171). New Brunswick, NJ: Rutgers University Press.

Gettler, A. O., & Norris, C. (1933). Poisoning from drinking radium water. *Journal of the American Medical Association, 100*, 400–402.

Gittinger, J. (2001). Radiation and cataracts: Cause or cure? *Archives of Ophthalmology, 119*, 112–116.

Goodman, A. H. (2000). Why genes don't count (for racial differences in health). *American Journal of Public Health, 90*, 1699–1702.

Haney-Lopez, I. (1996). *White by law: Legal constructions of race.* New York: New York University Press.

Haraway, D. (1989). *Primate visions: Gender, race, and nature in the modern world of science.* New York: Routledge.

Haslanger, S. (2008 [this volume]). A social constructionist analysis of race. In B. A. Koenig, S. S.-J. Lee, & S. S. Richardson (Eds.), *Revisiting race in a genomic age* (pp. 56–69). New Brunswick, NJ: Rutgers University Press.

Health, United States, 2002, with chartbook on trends in the health of Americans. (2002). Hyattsville, MD: National Center for Health Statistics.

Hey, J. (2001). *Genes, categories, and species.* New York: Oxford University Press.

Jackson, J. S., Brown, T. N., Williams, D. R., Torres, M., Sellers, S. L., & Brown, K. (1996). Racism and the physical and mental health status of African Americans: A thirteen-year national panel study. *Ethnicity and Disease, 6*, 132–147.

James, J. A., LaCrois, A. Z., Kleinbaum, D. G., & Strogatz, D. S. (1984). John Henryism and blood pressure differences among black men. *Journal of Behavioral Medicine, 7*, 259–275.

Jones, D. R., Harrell, J. P., Morris-Prather, C. E., Thomas, J., & Omowale, N. (1996). Affective and physiological responses to racism: The roles of Afrocentrism and mode of presentation. *Ethnicity and Disease, 6*, 109–122.

Kahn, J. (2005). Misreading race and genomics after BiDil. *Nature Genetics, 37*, 655–656.

Kaplan, J., & Bennett, T. (2003). Use of race and ethnicity in biomedical publication. *Journal of the American Medical Association, 289*, 2709–2716.

Kaya, A. (2001). *Sicher in Kreuzberg: Constructing diasporas: Turkish hip-hop in Berlin.* Piscataway, NJ: Transaction Press.

Kesler, R. C., Mickelson, K. D., & Williams, D. R. (1990). The prevalence, distribution, and mental health correlates of perceived discrimination in the United States. *Journal of Health and Social Behavior, 40*, 208–230.

Kleinman, J. C., & Kessel, S. S. (1987). Racial differences in low birth weight: Trends and risk factors. *New England Journal of Medicine, 317*, 749–753.

Klonoff, E. A. (1999). Cross-validation of the schedule of racist events. *Journal of Black Psychology, 25*, 231–254.

Krieger, N. (1990). Racial and gender discrimination: Risk factors for high blood pressure? *Social Science and Medicine, 30*, 1273–1281.

Krieger, N. (1994). Epidemiology and the web of causation: Has anyone seen the spider? *Social Science Medicine, 39*, 887–903.

Krieger, N. (1999). Questioning epidemiology: Objectivity, advocacy, and socially responsible science. *American Journal of Public Health, 89*, 1151–1153.

Krieger, N. (2000). Refiguring "race": Epidemiology, racialized biology, and biological expressions of race relations. *International Health Services, 30*, 211–216.

Krieger, N. (2001a). Commentary: Society, biology and the logic of social epidemiology. *International Journal of Epidemiology, 30*, 44–46.

Krieger, N. (2001b). The Ostrich, the albatross, and public health: An ecosocial perspective—or why an explicit focus on health consequences of discrimination and deprivation is vital for good science. *Public Health Report, 116*, 419–423.

Krieger, N. (2001c). Theories for social epidemiology in the 21st century: An ecosocial perspective. *International Journal of Epidemiology, 30*, 668–677.

Krieger, N. (2003a). Does racism harm health? Did child abuse exist before 1962? On explicit questions, critical science, and current controversies: An ecosocial perspective. *American Journal of Public Health, 93*, 194–199.

Krieger, N. (2003b). Genders, sexes, and health: What are the connections—and why does it matter? *International Journal of Epidemiology, 32*, 652–657.

Krieger, N. (2005). Defining and investigating social disparities in cancer: Critical issues. *Cancer Causes and Controls, 16*, 5 14.

Krieger, N., Chen, J. T., Waterman, P. D., Rehkopf, D. H., & Subramanian, S. V. (2003). Race/ethnicity, gender, and monitoring socioeconomic gradients in health: A comparison of area-based socioeconomic measures—the Public Health Disparities Geocoding Project. *American Journal of Public Health, 93*, 1655–1671.

Krieger, N., Chen, J. T., Waterman, P. D., Soobader, M.-J., Subramanian, S. V., & Carson, R. (2002). Geocoding and monitoring of U.S. socioeconomic inequalities in mortality and cancer incidence: Does the choice of area-based measure and geographic level matter? The Public Health Disparities Geocoding Project. *American Journal of Epidemiology, 156*, 471–482.

Krieger, N., Chen, J. T., Waterman, P. D., Soobader, M.-J., Subramanian, S. V., & Carson, R. (2003). Choosing area based socioeconomic measures to monitor social inequalities in low birth weight and childhood lead poisoning: The Public Health Disparities Geocoding Project. *Journal of Epidemiological Community Health, 57*, 186–199.

Krieger, N., & Sidney, S. (1996). Racial discrimination and blood pressure: the CARDIA study. *American Journal of Public Health, 86*, 1370–1378.

Krieger, N., & Smith, G. D. (2004). "Bodies count" and body counts: Social epidemiology and embodying inequality. *Epidemiologic Reviews, 26*, 92–103.

Krieger, N., Waterman, P. D., Chen, J. T., Soobader, M.-J., Subramanian, S. V., & Carson, R. (2002). ZIP code caveat: Bias due to spatiotemporal mismatches between ZIP codes and U.S. Census-defined areas—the Public Health Disparities Geocoding Project. *American Journal of Public Health, 92*, 1100–1102.

Krieger, N., Williams, D. R., & Moss, N. E. (1997). Measuring social class in U.S. public health research: Concepts, methodologies, and guidelines. *Annual Review of Public Health, 18*, 341–378.

Krieger, N., Wolff, M., Hiatt, R., Rivera, M., Vogelman, J., & Orentreich, N. (1994). Breast cancer and serum organochlorines: A prospective study among white, black, and Asian women. *Journal of the National Cancer Institute, 86*, 589–599.

Ladrine, H., & Klonoff, E. A. (1996). The schedule of racist events: A measure of racial discrimination and study of its negative physical and mental health consequences. *Journal of Black Psychology, 22*, 144–168.

Lee, S. S.-J., Mountain, J., & Koenig, B. A. (2001). The meaning of race in the new genomics: Implications for health disparities research. *Yale Journal of Health Policy, Law, and Ethics, 1*, 33–75.

Liberatos, P., Link, B. G., & Kelsey, J. L. (1988). The measurement of social class in epidemiology. *Epidemiological Review, 10*, 87–121.

Mays, V. M., & Cochran, S. D. (1997, July 28–31). Racial discrimination and health outcomes in African Americans. Paper presented at the National Center for Health Statistics, Joint Meeting of the Public Health Conference on Records and Statistics and the Data Users Conference, Washington, DC.

Mercy, J. A. (1993). Public health policy for preventing violence. *Health Affairs, 12*, 7–29.

Morris-Prather, C.E.E., Harrel, J. P., Collins, R., Leonard, K.L.J., Boss, M., & Lee, J. W. (1996). Gender differences in mood and cardiovascular responses to socially stressful stimuli. *Ethnicity and Disease, 6*, 123–131.

Muntaner, C., & Gomez, M. B. (2002). Anti-egalitarianism, legitimizing myths, racism, and "neo-McCarthyism" in social epidemiology and public health: A review of Sally Satel's PC, M.D.: How political correctness is corrupting medicine. *International Journal of Health Services, 32*, 1–17.

Muntaner, C., Nieto, F. J., & O'Campo, P. (1996).The bell curve: On race, social class, and epidemiological research. *American Journal of Epidemiology, 144*, 531–536.

Nature Genetics. (2000). Census, race, and science. Editorial. *Nature Genetics, 24*, 97–98.

Navarro, V. (1986). U.S. Marxist scholarship in the analysis of health and medicine. In B. Ollman and E. Vernoff (Eds.), *The left academy: Marxist scholarship on American campuses* (pp. 208–236). New York: Praeger.

Osborne, N. G., & Feit, M. D. (1992). The use of race in medical research. *Journal of the American Medical Association, 267*, 275–279.

Popper, K. ([1934] 1965). *The logic of scientific discovery*. New York: Harper and Row.

Ren, X. S., Amick, B. C., & Williams, D. R. (1999). Racial/ethnic disparities in health: The interplay between discrimination and socioeconomic status. *Ethnicity and Disease, 9*, 151–165.

Rich-Edwards, J., Kreiger, N., Majzoub, J., Zierler, S., Lieberman, E., & Gillman, M. (2001). Maternal experiences of racism and violence as predictors of preterm birth: Rationale and study design. *Paediatric and Perinatal Epidemiology, 15* (Supp. 2), 124–135.

Roberts, L., Riyadh, L., Garfield, R., Khudhairi, J., & Burnham, G. (2004, November 20). Mortality before and after the 2003 invasion of Iraq: Cluster sample survey. *Lancet, 364*, 1857–1864.

Rowland, R. E. (1994). Radium in humans: A review of U.S. studies. Argonne National Laboratory, available at http://www.ipd.anl.gov/anlpubs/1994/11/16311.pdf.

Sankar, P., Cho, M., Condit, C., Hunt, L., Koenig, B., Marshall, M., Lee, S. S.-J., & Spicer, P. (2004). Genetic research and health disparities. *Journal of the American Medical Association, 291,* 2985–2989.

Satel, S. (2002). PC, MD: How political correctness is corrupting medicine. *International Journal of Health Services, 32,* 1–17.

Schoendorf, K. C., Hogue, C.J.R., Kleinman, J. C., & Rowley, D. (1992). Mortality among infants of black as compared with white college-educated parents. *New England Journal of Medicine, 326,* 1522–1526.

Schwartz, S., & Carpenter, K. M. (1999). The right answer for the wrong question: Consequences for type III error for public health research. *American Journal of Public Health, 89,* 1175–1180.

Stevens, J. (1999). *Reproducing the state.* Princeton: Princeton University Press.

Stevens, J. (2002, Spring). Symbolic matter: DNA and other linguistic stuff. *Social Text, 20* (1), 106–140.

Stevens, J. (2003). Racial meanings and scientific methods: Policy changes for NIH-funded publications reporting human variation. *Journal of Health Policy, Politics and Law, 28,* 1033–1098.

Stevens, J. (2004). Legal aesthetics of the family and nation: AgoraXchange and notes toward re-imaging the future. *New York Law Review, 49,* 317–352.

Stolley, P. D. (1999). Race in epidemiology. *International Journal of Health Services, 29,* 905–909.

Sydenstricker, E. (1933). *Health and Environment.* New York: McGraw-Hill.

Syme, S. L. (1992). Social determinants of disease. In J. M. Last & R. B. Wallace (Eds.), *Public and health and preventive medicine* (13th ed.) (pp. 687–700). Norwalk: Appleton and Lange.

TallBear, K. (2008 [this volume]). Native-American-DNA.com: In search of Native American race and tribe. In B. A. Koenig, S. S.-J. Lee, & S. S. Richardson (Eds.), *Revisiting race in a genomic age* (pp. 235–252). New Brunswick, NJ: Rutgers University Press.

U.S. National Institutes of Health (NIH). (2002). NIH guidelines for research involving recombinant DNA molecules. Section 1C. See http://www4.od.nih.gov/oba/rac/guidelines_02/NIH_Guidelines_Apr_02.htm#_Toc7261552.

U.S. National Institutes of Health (NIH). (2004). Research supplements to promote diversity in health-related research. PA-05–015. See http://grants.nih.gov/grants/guide/pa-files/PA-05–015.html.

U.S. National Research Council/Committee on Human Genome Diversity. (1997). *Evaluating human genetic diversity.* Washington, DC: National Academy Press.

Vernon, H. M. (1939). *Health in relation to occupation.* Oxford: Oxford University Press.

Williams, D. R., Jackson, Y., & Anderson, N. (1997). Racial differences in physical and mental health: Socioeconomic status, stress, and discrimination. *Journal of Health Psychology, 2,* 335–351.

Winker, M. (2006). Race and ethnicity in medical research: Requirements meet reality. *Journal of Law, Medicine and Ethics, 34,* 520–525.

World Health Organization (WHO)/Kurg, E., Dahlberg, L., Mercy, J., Zwi, A., Lozano, R. (Eds). (2002). *World report on violence and health.* Geneva: WHO.

Zaman, J. M. (2000). *Real science: What it is and what it means.* Cambridge: Cambridge University Press.

Zierler, S., Feingold, L., Laufer, D., Velentgas, P., Kantrowitz-Gordon, I., & Mayer, K. (1991). Adult survivors of childhood sexual abuse and subsequent risk of HIV infection. *American Journal of Public Health, 81,* 572–575.

18

Racial Realism and the Discourse of Responsibility for Health Disparities in a Genomic Age

SANDRA SOO-JIN LEE

In the first formal announcement of the Human Genome Project's completion in early 2000, project leaders were widely reported as discounting any genetic basis for racial grouping.[1] In concert, they stated definitively that the only race one could speak of was of the "human race" as a whole. President Clinton, in his formal statements from the White House in the same year, proclaimed that the "great truth to emerge from this triumphant expedition inside the human genome is that in genetic terms, all human beings, regardless of race, are more than 99.9% the same. . . . The most important fact of life on this earth is our common humanity."[2]

Despite the fanfare that accompanied these pronouncements, the message rang familiar. The statement that there is more genetic variability *within* populations than *between* had been echoed among social and life scientists alike, long before the completion of the Human Genome Project. The focus on the commonality of humanity, at least in biological terms, had been harnessed to the ubiquitous 99.9% figure that served as clear evidence that differences within the human species were minimal at best (Lewontin, 1972). Although one could argue that we did not need a multimillion-dollar Human Genome Project (HGP) to reiterate this message, one would expect that the project's completion would have laid to rest once and for all that race is not genetic. However, since these momentous celebrations, a "turn towards difference" has resulted in an emboldened trajectory of studies that have shown greater difference than was once expected. Recent research on human genetic variation has challenged the 99.9% figure, shifting it downwards to between 85% and 95% depending on the type and number of markers used and the populations in question (Redon et al., 2006; Rosenberg et al., 2002). It seems increasingly true that the mantra that "we are all the same" has morphed to the slightly awkward, if not ironic, statement that "we are the same, but also different."

In the post-HGP period, ambivalence is the space in which those interested in human genetic variation must take up residence. Researchers take for granted that human groups will share the majority of their genes with each other, but in pursuing pertinent health questions of disease and drug response, the scientific literature reveals that they are doggedly focused on identifying differences. These interstitial areas, they explain, are where the clues to disease etiology and effective therapeutics lie. This turn toward difference creates the current landscape of genomic medicine that belies grand pronouncements of commonality and lays bare a stubborn, possibly more durable, resistance to the decoupling of race from genes. Despite great pains taken by some scientists, policy makers, and journalists to complicate the relationship between race and current patterns of genetic sequences, the emerging framework is strangely reminiscent of earlier, facile perspectives of race as genetically determinative (Braun, 2002; Sankar and Cho, 2002; Ossorio and Duster, 2005).

This shift from emphasizing commonality to a search for differences is critical to understanding the emerging meaning of race for genomic medicine and how these differences are interpreted through social practices in scientific research and in explanatory models of why differences exist at all. In approaching race, I focus on the beliefs, practices, and institutional arrangements that produce dominant understandings of population differences. As I have argued elsewhere, this turn toward difference has catalyzed processes of racialization (Omi & Winant, 1989) where race is reified through a convergence of trajectories, including focused research on genetic differences among racially identified groups, the use of race as a proxy for risk in clinical practice, and mounting interest in new market niches by the pharmaceutical and biotechnology industries (Lee, 2005). This convergence reveals shifting moral stakes around how to approach population differences.

In this chapter, I examine the emerging narrative on personal responsibility in this discourse and how genetic science is incorporated into public policy directed at health disparities. I argue that social practices around identifying, characterizing, and applying race in human genetic variation research inform the parameters of possible solutions for the public health campaign to eliminate health disparities among racially and ethnically identified populations. The racialization of genetics and disease emerges from underlying assumptions of who is responsible for one's illness (Harding, 2002). I argue that justice in health care in the United States is inextricable from the prism of racial difference. To illustrate the emerging framework of racial justice, its undercurrent of calls for personal responsibility, and its bearing on the struggles over how to think about race, disease, and difference, I offer accounts of three recent events: reactions among scientists to the "colorblind" DNA Polymorphism Discovery Resource created by the National Institutes of Health (NIH); endorsement of and subsequent activism around BiDil, the first racially labeled medication, approved by the Food and Drug Administration (FDA); and revisions to drafts to the National Health Disparities Report by the Department of Health and Human Services (DHHS). In discussing these three seemingly disparate events, my hope is to describe the contours of the debate and emerging stakes over how genetic differences are interpreted by scientists, public advocates, and policy makers and to consider their implications for understandings of responsibility for inequities in health.

The Courage to Race: Rejecting Colorblindness

In 1998 the National Human Genome Research Institute (NHGRI) created a large-scale database of single nucleotide polymorphisms (SNPs) called the DNA Polymorphism Discovery Resource (PDR) and placed it in the public domain. A SNP is a small genetic change or variation that occurs in approximately 3% to 5% of a person's DNA sequence and codes for the production of proteins, although the majority of SNPs are believed to reside in non-coding regions. The purpose in creating the PDR was to provide researchers with samples that reflect the diversity of SNP patterns among the U.S. population. Toward this effort, DNA samples and cell lines from 450 anonymous, unrelated male and female residents of the United States were collected from populations identified as "European-American including non-Hispanic whites, African-American including non-Hispanic blacks, Asian-American including individuals with ancestry from East and South Asia, and Mexican Americans"(Collins, Brooks, & Chakravarti, 1998). Attending to the dual goals of delivering diverse sample sets to researchers while at the same time de-identifying samples of racial and/or ethnic categories, the NHGRI took the historic step of packaging samples from different populations into a "mixed" bundle while at the same time excising racial and ethnic information from individual samples. The goal was to ensure the overall variation of the bundled set and to avoid giving researchers racial or ethnic information on the specific samples they received. To explore the basis for this landmark decision and its implications for the research community, I interviewed policy makers and scientists about the creation of the PDR and its value as a research tool. Even after de-linking this information, program officials at NHGRI said they remained concerned that researchers would attempt to infer the population origins of individual samples, and the institute took the historic step of requiring researchers to sign an agreement that prohibited them from doing so. As one NHGRI program officer suggested, "These policies were put in place to ensure that the repository would be, in a sense, 'colorblind' " (personal interview, April 29, 2003).

The desire for colorblindness at a time when most social scientists and life scientists agreed that race was a product of sociohistorical context and not based on genes appears, on first glance, unnecessary. However, consensus that race is largely socially formed and a poor taxonomy for organizing genetic information did not eliminate, according to program officials, the possibility that providing racial and ethnic information on the DNA samples might result in associations that would, at a minimum, be politically problematic. The decision to go colorblind was influenced in large part by the lingering institutional social memory of protest by indigenous groups against the Human Genome Diversity Project (HGDP) that attempted to sample DNA from global populations (Reardon, 2004). The interviews I conducted with individuals associated with the creation of the PDR revealed overriding concerns over how to "manage" the potential politicization of the repository. The competing concerns for making samples immediately available to the scientific community and the recognition that no palatable policy regarding racial identification was possible resulted in an effort to strip race. In an interview with a member of the working group consulted in the creation of the PDR, the need for expediency also influenced policy:

> The HGDP never had a chance despite what, I believe, were the good intentions of the people who proposed it. . . . In creating the PDR, we knew that there was still a lot of confusion and suspicion around genetic research, and we just didn't have the time to come up with a policy to address all of the possible concerns, at least not in a way that included all the groups we would have wanted to include. You also have to realize that scientists wanted this collection immediately, and we were under a great deal of pressure to move it forward. By taking race off the table, we felt that we would have more time to come up with a sound policy on the "race issues" yet not hold up the science. (personal interview, April 25, 2003)

What became clear from this interview and those with others involved in the project was that the colorblind policy was not borne from a position that racial information on DNA samples was not salient to questions in genetic research. Rather, the PDR policy was a compromise in the face of public scrutiny that many feared might prevent the repository from being developed.

The colorblind PDR however, has become, with some irony, one of the least used of the repositories in the NIH collection. In exploring potential reasons for this, several scientists I interviewed criticized the colorblind policy and suggested that the absence of racial information rendered the samples less useful in conducting genetic association studies. The desire

for racial information was characterized by most interviewees as fundamental and often compared to information on sex. Scientists suggested that
knowing the race of the person who donated the DNA samples would "give
a fuller picture" and "provide better clues as to which populations may
have the variants of interest." Scientists would often cite the importance of
the history of human migration in explaining the value of race for genetic
research. According to one molecular biologist, "What we call races may
change depending on societal norms, but there is no denying that different
populations will have different genetic frequencies that may be extremely
important for medical research. You can't just separate racial information
and pretend, number one, that these do not exist and, number two, pretend
that they are not important" (personal interview, May 15, 2003). The contention that genes are inseparable from race resonated with assertions by
several scientists who suggested that the PDR policy of barring researchers
from determining racial information not only challenges academic freedom,
but is ultimately ineffectual. One researcher chuckled at the idea of sup-
pressing racial and ethnic identifiers, stating that such information was easily determined, claiming that "in some cases you find out that self-reported
identity of the donor is actually wrong; they think they are white, but it
turns out they are actually black or African American" (personal interview,
July 22, 2003). For this scientist, genetic information on allelic frequencies
determines race, even trumping self-reported identity.

Reactions among scientists to the colorblind PDR revealed not only
a generalizable desire for racial information, but also an irritation with a
policy that many perceived as more political than scientific. Expressing dis-
appointment with the NHGRI, one human geneticist and clinician surmised
that the institute "failed the scientific community by caving under political
pressure" (personal interview, June 11, 2003). Another genetic epidemiologist shared his frustration by stating: "The reality is that the population is
segmented and that differences exist. It might be politically correct to avoid
the word 'race,' but at some point we need to tackle the hard issues . . . that
these things matter. In the not so distant future, we are going to see numerous associations and it will not be so easy to take the position that race does
not exist" (personal interview, March 25, 2003). For this scientist and others
who were interviewed on the PDR policy, the challenge of addressing the
meaning of race for human genetics is framed less as a technical problem
for scientists but as a political one. The shift toward searching for difference
presents a dilemma for researchers of human genetic variation who often
dance uncomfortably around the possible implications of their work as
validating a genetic basis for race. The historical polarization of the debate
over race and genes has left little room to interpret emerging human genetic

variation research unself-consciously. The rejection by researchers of a colorblind policy for DNA repositories reveals a growing belief that genetic differences and, in particular, differences among racially identified populations will lead to greater understanding of disease, drug response, and other health-related traits. The calls for open acknowledgment of the importance for racial information are echoed in the sentiments expressed by a molecular biologist who offered, "Everyone is so afraid of saying anything about race because it is not politically appropriate to do so. Political correctness has blinded and gagged science to the point where we no longer have the *courage* to confront the reality of group differences" (personal interview, April 29, 2003, emphasis added).

Courage has become a leitmotiv in the emerging discourse on race and genetics. It is precisely this quality that *New York Times* science writer Nicholas Wade attributes to population geneticist Neil Risch in his reporting of a paper in which Risch and his colleagues suggest that race is not only an appropriate variable in genomic research, but a necessary one (Risch, Burchard, Ziv, & Tang, 2002). Describing Risch as "plunging into an arena where many fear to tread," Wade compares Risch's article to a righteous act against "defenders of political correctness" who, Wade suggests, are "honest and brilliant people" but who, in the end, "are not population geneticists" (Wade, 2002, p. F1). The trope of courage in a context of perceived tyranny of political correctness is one theme that plots the narrative on race in the context of emerging genomic technologies.[3] Perception among many scientists that political correctness has unduly influenced open debate on the validity of race in human genetics research has reconfigured the stakes in the struggle over how to characterize difference. Imbued with courage against self-censorship, heroes of this emerging narrative, such as Risch, are described as taking personal and professional risks in claiming race as "real," or inherent in our genes. In this scenario, social constructionists and others who maintain a framework where race is understood as fluid and contextually determined are seen as policing racial rhetoric. The employment of this emerging discourse on the meaning of race for genes centers on the familiar struggle over objectivity. The allure of "racial realism" in genomic medicine is, in part, its assertion of a politically unmotivated claim on truth that genetic differences among racially identified populations not only exist, but are functionally important, with potentially significant effects on human health. The context of anticipated health benefits is critical in reframing genetic variation as embedded in the history of evolution and global human migration and tethered to conventional ideas around race and racial difference. The promise of improved health advances ethical justification for proffering a basis for linking race to genes.

The Race to Market: Linking Racial Justice to Racial Biology

In an article entitled "Brave New Pharmacy," *Time* magazine (January 15, 2001) heralded an imminent era of genomic medicine that would replace current strategies with "something much more rational and systematic." The public investment, the article suggested, would be returned in improved clinical encounters where "doctors will treat diseases like cancer and diabetes before the symptoms even begin, using medications that boost or counteract the effects of individual proteins with exquisite precision, attacking sick cells while leaving healthy cells alone, and they will know right from the start how to select the best medicine to suit each patient." This now ubiquitous narrative of stealth and precision that has become a hallmark of public understandings of the new genetics is predicated on genomic data that will presumably yield in a straightforward way differences in disease risk and drug response among individuals. Some success has been achieved where new drugs have begun to make dramatic differences in highly circumscribed cases;[4] however, researchers caution that the much anticipated era of individualized genomic medicine is still in its nascency.

Technical decisions made early in the development of a field may create the trajectory on which an area of inquiry unfolds. Which categories scientists use to characterize DNA samples inform the range of questions that may be asked. The search for difference dovetails with practices in clinical medicine where using race as a proxy for biological variables and for disease risk is routine. One cardiologist whom I interviewed put it simply: "Race is a fast and cheap genetic test and it can give you a wealth of information in accessing risk" (personal interview, September 14, 2005). As such, race has become the first filter through which difference is determined. Nowhere is this more apparent than in the emerging landscape of pharmacogenomics (Solus et al., 2004; Yasar et al., 2002; Blaisdell et al., 2004). The salience of race is not limited to assessing risk or identifying genetic variants that are associated with drug response. Race is an important axis of stratifying potential markets and identifying potential consumers. Increasing interest by the pharmaceutical industry and biotech companies in what one corporate executive described as the "as yet untapped racial market niches" indicates a significant shift. More than merely an academic debate fought on the editorial pages of academic journals, the meaning of race and genes is considered big business—one that has coupled itself to the moral mandate of addressing ongoing health disparities.

The case of BiDil, the antihypertensive drug approved by the FDA for use exclusively among African Americans, has become a flashpoint in debates over the relevance of race for genomic medicine. Upon first glance, the prominence of BiDil in this discourse seems misplaced. Genetic data was not presented in the supporting materials to justify the drug's approval by the FDA. Rather, BiDil's maker, Nitromed, Inc., asserted the hypothesis of nitric oxide deficiency as a possible explanation of why the drug may be more efficacious among self-identified African Americans. However, the case of BiDil has become a critical event in the emerging discourse on race, genetic variation, and disparity because it reflects the evolving landscape of governance of personalized medicine. BiDil comes at a critical juncture for the field as it grapples with how to pursue its search for meaningful genetic differences among populations. In the recent FDA "Guidance on the Collection of Race and Ethnicity Data in Clinical Trials" (2005), BiDil is cited as rationale for the adoption of census categories of race and ethnicity as defined by the Office of Management and Budget (OMB). Despite explicit warnings by the OMB that these "represent a social-political construct designed for collecting data on the race and ethnicity of broad population groups in this country, and are not anthropologically or scientifically based" (1997), the FDA followed NIH's well-established policy that requires researchers to sort human subjects into racial and ethnic bins, stating that doing so will "facilitate comparisons across clinical studies analyzed by FDA and data collected by other agencies" (FDA, 2005, p. 6) As a test case, BiDil further institutionalized racial categorization in drug development and solidified institutional support for predominant assumptions of the biological underpinnings of race, emboldening a new trajectory of scientific inquiry that Fullwiley (this volume) describes as the molecularization of race.

Several scholars have discussed the scientific legitimacy of the data presented by Nitromed in their application for FDA approval. Much of this work was conducted in two clinical trials in the early 1980s (the Vasodilator Heart Failure Trials, V-HeFT I & II) and the more recent clinical trial conducted by Nitromed among exclusively self-identified African Americans (the African American Heart Failure Trial, A-HeFT). Careful examination of the data from the two earlier trials has raised questions about the FDA's decision to proceed with the A-HeFT trial (Kahn, 2004, this volume); however, the A-HeFT results appear non-controversial, showing dramatic benefits for those study participants who received the drug (Lapu-Bula et al., 2005; Taylor et al., 2004). My purpose here is not to critique the three clinical trials that led to BiDil's approval by the FDA. Rather, my goal is to examine the framing of BiDil as a matter of social justice by its advocates in the FDA, Nitromed, and African American community. I focus on how BiDil has come to symbolize more than a reduction of hypertension, and is seen as an agent that begins to redress historical inequities in health status. This framing of BiDil is emblematic of the emerging framework of racial realism that supports population-specific medicines and espouses personal responsibility for health.

Responding to those critical of the FDA for setting a precedent for race-based therapeutics, FDA Advisory Committee chairman Stephen Nissen defended the administration's decision to approve BiDil by commending those at Nitromed for courage in addressing the historically underserved African American community: "First of all, I think this was a courageous thing to do, to try to develop a drug for this population which seems to have a disproportionate burden of disease. . . . [Y]es, the road to hell is paved with biological plausability but here is a biological plausible explanation" (transcript from the FDA Hearing on the Approval of BiDil, June 23, 2005). In valorizing Nitromed for developing a drug for the African American population, Nissen alludes to the controversy over the fundamental yet implicit assumption of a biological basis for why the drug may work better in African Americans. Nissen's statement suggests that the disproportionate burden of heart disease suffered by African Americans justifies whatever costs may be associated with accepting putative claims of racial biology.

The persistence of health disparities as moral justification for the approval of BiDil was echoed by several individuals at the FDA hearing speaking on behalf of African American and other organizations that have publicly endorsed FDA approval of the drug. In an impassioned speech, Donna Christensen, chair of the Congressional Black Caucus, stated:

> I want to say to you that today, ladies and gentlemen, you have before you an unprecedented opportunity to significantly reduce one of the major health disparities in the African American community and, in doing so, to begin a process that will bring some degree of equity and justice to the American healthcare system. . . . Addressing these in eliminating the disparities that exist in all aspects of our lives is our highest priority until those gaps are closed. Their continued existence despite our best efforts must not be used to deny treatment to those for whom treatment has been denied and deferred for 400 years. Today this panel is being asked to reverse that history. (Testimony at FDA Hearing on BiDil, June 25, 2005)

The stakes are made plain in Christensen's framing of BiDil as a matter of racial justice in biomedicine. When saddled with the weight of a long history of racism and unequal treatment, the approval of BiDil is recast as an act of moral redemption for a legacy of excluding and exploiting African Americans in biomedical research. However, what has been less amplified in the media and in public statements made by individuals like Christensen are the somewhat paradoxical views within the African American community on BiDil's implication of inherent biological differences among so-called races. Many hold a seemingly incongruous dual position from which they simultaneously maintain that the approval of BiDil represents an important step toward social justice yet, at the same time, vigorously disavow ideas of a distinctive racial biology. A member of the Association of Black Cardiologists, an organization that offered strong public endorsement for the approval of the drug, reveals this paradox. He explained:

> The reality is that the health of blacks in this country has always been a terribly neglected issue. It has and always will be a struggle over resources and, frankly, a result of a long legacy of racism in this country. Do I support BiDil? Sure, I do. Do I think that it is a drug that will only work on blacks? No, I don't. I suspect that race, or at least conventional ideas about race, has little to do with why it works. There are probably a lot of other folks who could benefit from the drug, but am I going to discourage the FDA from approving it? No way. If they want to make drugs that work on black people, then I say that's one win out of a long line of setbacks towards just health care in this country. (personal interview, July 15, 2005)

The hope for a "win" toward social justice obscures the strong rejection of BiDil as a "black drug" by this supporter. There is little doubt that Nitromed needs public support and advocates within the African American community for the company to reap benefits from its development of the drug. However, closer inspection reveals how the dual position of such advocates—that BiDil must be made available to African Americans, but is not evidence of racial biology—diverges with how BiDil is perceived in the wider public imagination. In her essay "Medicine's Race Problem," American Enterprise Institute fellow Sally Satel (2002) cites BiDil as the "smoking gun" to questions over a biological basis for race. Referring to Jay Cohn, the principal investigator whose research resulted in BiDil, Satel suggests that the "ultimate purpose of work like Cohn's and other biological realists is to identify factors that may be genetic in origin"; she criticizes those who warn of the dangers inherent in practicing race-based medicine as "wanting it both ways," arguing that one cannot use race to "root out social injustices in medicine" without accepting that race is "mediated by physiology" (2002).

The link between racial biology and racial justice is an unexpected marriage, though in hindsight it is one that has been developing over the past several decades. The case of BiDil turns on its head the legacy of exploitive recruitment that ultimately served as impetus for the formalization of research ethics guidelines that have led to current protections for research subjects. The pendulum shift from public outcry against the Tuskegee U.S. Public Health Service Study on Syphilis (1932–1972) to the heralding of Nitromed's AHeFT trial reveals a dramatic change from an ethics of protective exclusion to a swing toward inclusion defined by a politics of particularity that insists on an acceptance of bodily difference as a prima facie and necessary step toward achieving justice and reaffirms a conventional logic of race as racial biology.

Language Matters: Reframing Disparity as Difference

To appreciate the significance of the debate over the meaning of race for genomic medicine, it is important to consider this discourse in the context of an ongoing, parallel struggle over defining health disparities. In June 2003 the U.S. Department of Health and Human Services (DHHS) released its National Healthcare Disparities Report. The report was the first created in response to the Healthcare Research and Quality Act of 1999 (Public Law 106–129), which mandates that the department track ongoing disparities among racially identified populations and other "priority populations." This report was largely consistent with the Institute of Medicine's earlier report, *Unequal Treatment: Confronting Racial and Ethnic Disparities in Health Care* (2002). It emphasized that persistent health disparities exist among racially identified populations and stressed the significant personal and "societal price" that are incurred by the national problem.

Six months after submitting the original draft of its report to Congress, the DHHS released the final version on December 23, 2003. A comparison of the two versions reveals several significant changes, including a global replacement of the term "disparity" in the original draft with "difference" in the revised report. This substitution is justified in a section entitled "What Are 'Healthcare Disparities'?" which includes the following statements: "In the absence of consensus on the definition of disparities, this report will focus on presenting the facts. Where we find variation among populations, this variation will simply be described as a 'difference.' *By allowing the data to speak for themselves, there is no implication that these differences result in adverse health outcomes or imply prejudice in any way*" (DHHS, 2003, p. 2, emphasis added). In addition, original findings that "inequality in quality of health-care exists" were revised, and the report asserts that "exceptional quality of health care" is being provided. The revised report deletes language that describes health disparities as significant social and moral problems that require changes in institutional practices and public policy. Instead, the revised report asserts that "priority populations do as well or better than the general population" (DHHS, 2003, p. 6) Other key differences are described in table 18.1.

The revisions to the report did not go unnoticed. On January 13, 2004, the Special Investigations Division of the minority staff of the Government Reform Committee released a DHHS statement from eight members of Congress comparing the two versions of the report. These members wrote DHHS secretary Tommy Thompson to protest what they called "the manipulation

TABLE 18.1

Revisions to the National Health Disparities Report, 2003–2004

Language Describing Key Findings of Original Draft	Language Describing Key Findings of Revised Draft
1. Inequality in quality of health care exists.	1. Americans have exceptional quality of health care; but some socioeconomic, racial, ethnic, and geographic differences exist.
2. Disparities come at a personal and societal price.	
3. Differential access may lead to disparities in quality of health care.	2. Some "priority populations" do as well or better than the general population in some aspects of health care.
4. Opportunities to provide preventive care are frequently missed.	3. Opportunities to provide preventive care are frequently missed.
5. Knowledge of why disparities exist is limited.	4. Management of chronic diseases presents unique challenges.
6. Improvement is possible.	5. There is still a lot to learn.
7. Data limitations hinder targeted improvement efforts.	6. Greater improvement is possible.

of science on healthcare disparities" (letter from U.S. representatives to Thompson, January 13, 2004) and to request copies of all drafts and comments on the disparities report. Following protest by members of Congress and the scientific community, Secretary Thompson admitted that the DHHS was wrong to revise scientific conclusions in the National Healthcare Disparities Report (Bloche, 2004). The Agency for Healthcare Research and Quality—the division of DHHS responsible for drafting the initial report—has since released the original version of the report it had submitted to the department for clearance.

While the DHHS reverted back to its original findings, the revised report, which reframed disparities as differences, reflects an emerging shift in official discourse about why certain racially identified populations experience poorer health status than the general population. The Institute of Medicine's report (2002) concluded that a consistent body of research demonstrates significant variation in medical treatment by race, even when insurance status, income, age, and severity of conditions are comparable (Smedley, Stith, & Nelson, 2002). This research indicates that U.S. racial and ethnic minorities are less likely to receive even routine medical procedures and experience a lower quality of health services. The revised DHHS report challenged a major finding by the Institute of Medicine that such disparities exist due to institutional and attitudinal racism in the health care system. Instead, the revised DHHS report sketched an alternative view of health disparities and refocused the debate onto the role of individual behavior. Arthur J. Lawrence, a deputy assistant secretary of Health and Human Services at the time, defended the revised DHHS report by arguing for more emphasis on "personal responsibility for one's own health status" (*New York Times*, 2004). Echoing this sentiment, then Senate Majority Leader William Frist, in an article entitled "Overcoming Disparities in U.S. Health Care," argues that the best way to eliminate health disparities is through improvement of the care "we deliver to each patient, emphasizing patient dignity and empowerment" (2005, p. 450). Frist writes further, "Patients must be central to our efforts to improve health care. For instance, a person with a chronic illness such as diabetes must essentially 'own' that illness if he or she is to have any hope of effectively managing it. Providers can help with high-quality treatment and the best recommendations, but patients must act on those recommendations. They must stop smoking, eat right, exercise, take their medication, and monitor their blood sugar, based on their own volition and usually outside of the clinical setting. Public policies must encourage patients to embrace *personal responsibility*" (2005, p. 450, emphasis added).

The shift from disparity to difference is a powerful change that reframes questions of heath status inequalities from issues of social justice to problems that are located within individual bodies. The coupling of race with disease and medicine provides fertile ground for assertions of racial biology and the misguided deferral of health disparities to strategies highlighting targeted therapeutics and personal responsibility and discipline. Drawing upon the growing body of genomic data on population differences, the framework of racial realism reassigns who is responsible for health care inequities by emphasizing the notion of a "genetic lottery," while denying systemic problems in health care access and provision. Racial realism shifts the focus from the state to the individual, where social responsibility is reframed as personal responsibility. The struggle over the revisions of the National Healthcare Disparities Report reveals how seemingly subtle shifts in language can change the focus and trajectory of possible solutions to our most serious public health problems.

The Cost of Ambivalence: What Is at Stake?

In December 2003 *Scientific American* featured a cover story with the headline query, "Does Race Exist?" which was followed by a teaser answer: "Science Has the Answer: Genetic Results May Surprise You" (Bamshad & Olson, 2003). The cover image accompanying this headline displayed six young female faces with closely cropped hair arranged in a circle staring out at the reader. All were virtually identical, indistinguishable by age, class, occupation, or education, except for their varying degrees of light to dark skin tone. Without hints of social location or context, this image reflects predominant understandings of race through the conventional prism of skin color: black, white, and variations in between. For readers who venture past the front cover, human geneticist Michael Bamshad and science writer Steve Olson describe the complexity and ongoing ambivalence over how to interpret race and genetic differences. The authors present a nuanced history of human migration and social experience that makes it difficult to equate racial categories with the current global distribution of human genes. They state plaintively that race cannot be "defined as genetically discrete groups" (Bamshad & Olson, 2003, p. 75). Yet the authors emphasize that race remains an important axis of differentiation and that knowing a person's race can provide medically relevant information on the incidence of disease. In an all too familiar ambivalence, they conclude that "human populations are very similar, but they can be distinguished" (Bamshad & Olson, 2003, p. 85).

This contradictory message has become increasingly prevalent in the popular media and scientific literature and signals a pendulum shift in thinking about race and human genetics. Race has become an accepted tool in health care and, as stated by Francis Collins, director of the NHGRI, the focus is now on "genetic variation that is associated with disease risk and how that correlates, in some very imperfect way, with self-identified race, and how we can use that correlation to reduce the risk of people getting sick" (quoted in Henig, 2004, p. 50). Focused scrutiny of genetic difference has fueled a growing infrastructure of biobanks and research initiatives dedicated to finding variants, with much emphasis placed on using race as a tool in identifying variants that may be associated with the onset of disease, differential drug response, and other human traits. Despite predictions that genomics might render racial categories obsolete (Gilroy, 2000) with a new focus on DNA, race has become more institutionalized in the taxonomies used in genetic variation maps, in federal guidelines regulating biomedical research, and in the exploration of new market niches.

Reaction to the three sites of racialization discussed in this chapter—the creation of the colorblind PDR, the approval of BiDil as a "black drug," and the revisions to the DHHS report—provides a glimpse of the contours of the debate over how to interpret difference in genomic medicine. The proliferation of DNA repositories creates a physical and symbolic foundation that is embedded with a priori assumptions about where difference is most likely to be found. The effort to strip race from the PDR makes plain the difficulty of eliminating racial categories in genetic research. The precedent of racially prescribed drugs forces stakeholders to confront the unsettling notion that any achievement of racial justice through targeted therapeutics based on race is impregnated with underlying assumptions of racial biology. The case of BiDil reveals the constraining logic of race that is inextricable from how we think about bodily differences and policy that considers their meaning. The contested revisions of the DHHS report on national health disparities place in stark relief the high stakes in characterizing the relative health of populations in the United States. The seemingly subtle slippage from "disparity" to "difference" endangers the scaffolding of social justice concerns that initially ushered the campaign to end health inequities.

Institutional policies and decisions reveal a struggle over the meaning of race, genes, and health inequities in the United States. This struggle has resulted in a fractured discourse on responsibility. The three events described in this chapter suggest a need for careful scrutiny of the categories we use, the tropes that employ our accounts of race and genetics, and the sociopolitical factors that circumscribe our interpretation of difference. Without such interrogation, we are left with only "mere" difference, which occludes from view what is at stake and threatens our greatest hopes for this genomic age.

NOTES

This research was supported by the National Human Genome Research Institute grant K01HL72465.

1. These included Francis Collins, director of the National Human Genome Research Institute (NHGRI), and Craig Venter, then president and chief scientific officer of Celera Genomics Corporation.

2. See http://clinton3.nara.gov/WH/EOP/OSTP/html/00628_2.html.

3. Hayden White and Paul Ricoeur have been influential in discussing the emplotment of historical writing and the significance of tropes for narratives of events (White, 1973).

4. Identifying the genetic differences among cancerous tumors, for example, has allowed for new therapeutic approaches against breast cancer, such as Herceptin, an antibody that suppresses HER2 protein that is associated with certain tumor growth.

REFERENCES

Bamshad, M. J., & Olson, S. E. (2003) Does race exist? *Scientific American, 289* (6), 23–41.

Blaisdell, J., Jorge-Nebert, L. F., Coulter, S., Ferguson, S. S., Lee, S. J., Chanas, B., et al. (2004). Discovery of new potentially defective alleles of human CYP2C9. *Pharmacogenetics, 14* (8), 527–537.

Bloche, G. (2004). Health care disparities—science, politics, and race. *The New England Journal of Medicine, 350* (15), 1568–1570.

Braun, L. (2002). Race, ethnicity, and health: Can genetics explain disparities? *Perspectives in Biology and Medicine, 45* (2), 159–174.

Brave new pharmacy. (2001, January 15). *Time Magazine,* http://www.time.com/time/magazine/article/0,9171,998963,00.html.

Collins, F. S., Brooks, L. D., & Chakravarti, A. (1998). A DNA polymorphism discovery resource for research on human genetic variation. *Genome Research, 8* (12), 1229–1231.

Department of Health and Human Services (DHHS). (2003, December). National Healthcare Disparities Report. Washington, DC.

Food and Drug Administration (FDA). (2005, September). FDA guidance on the collection of race and ethnicity data in clinical trials. Washington, DC.

Frist, W. (2005). Overcoming disparities in U.S. health care. *Health Affairs, 24* (2), 445–451.

Fullwiley, D. (2008 [this volume]). The molecularization of race: U.S. health institutions, pharmacogenetics practice, and public science after the genome. In B. A. Koenig, S. S.-J. Lee, & S. S. Richardson (Eds.), *Revisiting race in a genomic age* (pp. 149–171). New Brunswick, NJ: Rutgers University Press.

Gilroy, P. (2000). *Against race.* Cambridge: Harvard University Press.

Harding, S. (2002). Representing reality: The Critical Realism Project. *Feminist Economics, 9* (1), 151–159.

Henig, R. M. (2004, October 10). The genome in black and white. *New York Times Magazine,* pp. 47–51.

Institute of Medicine. (2002). *Unequal treatment: Confronting racial and ethnic disparities in health care.* Washington D.C.: National Academies Press. http://www.nap.edu/openbook.php?isbn=030908265X.

Kahn, J. (2004). How a drug becomes "ethnic": Law, commerce, and the production of racial categories in medicine. *Yale Journal of Health Policy, Law, and Ethics, 4* (1), 1–46.

Kahn, J. (2008 [this volume]). Patenting race in a genomic age. In B. A. Koenig, S. S.-J. Lee, & S. S. Richardson (Eds.), *Revisiting race in a genomic age* (pp. 129–148). New Brunswick, NJ: Rutgers University Press.

Lapu-Bula, R., Quarshie, A., Lyn, D., Oduwole, A., Pack, C., Morgan, J., et al. (2005). The 894T allele of endothelial nitric oxide synthase gene is related to left ventricular mass in African Americans with high-normal blood pressure. *Journal of the National Medical Association, 97* (2), 197–205.

Lee, S. S.-J. (2005). "Racializing drug design": Pharmacogenomics and implications for health disparities. *American Journal of Public Health, 95* (12), 2133–2138.

Lewontin, R. C. (1972). The apportionment of human diversity. *Evolutionary Biology, 6,* 381–398.

Office of Management and Budget (OMB). (1997, October 30). Revisions to the standards for the classification of federal data on race and ethnicity. *Federal Register Notice.* http://www.whitehouse.gov/omb/fedreg/1997standards.html.

Omi, M., & Winant, H. (1986/1989). *Racial formation in the United States: From the 1960s to the 1980s.* New York: Routledge.

Ossorio, P., & Duster, T. (2005). Race and genetics: Controversies in biomedical, behavioral, and forensic sciences. *American Psychologist, 60* (1), 115–128.

Pear, R. (2004). Taking spin out of report that made bad into good. *The New York Times.* http://query.nytimes.com/gst/fullpage.html?res=9E0CEFD91E3DF931A15751C0A962 9C8B63&sec=&spon=&pagewanted=al.

Reardon, J. (2005). *Race to the Finish: Identity and governance in an age of genomics.* Princeton: Princeton University Press.

Redon, R., Ishikawa, S., Fitch, K. R., Feuk, L., Perry, G. H., Andrews, T. D., et al. (2006). Global variation in copy number in human genome. *Nature, 444* (7118), 444–454.

Risch, N., Burchard, E., Ziv, E., & Tang, H. (2002). Categorization of humans in biomedical research: Genes, race, and disease. *Genome Biology, 3* (7), comment2007.
1–2007.12.

Rosenberg, N. A., Pritchard, J. K., Weber, J. L., Cann, H. M., Kidd, K. K., Zhivotovsky,
L. A., et al. (2002). Genetic structure of human populations. *Science, 298* (5602),
2381–2385.

Sankar, P., & Cho, M. (2002). Toward a new vocabulary of human genetic variation.
Science, 298 (5597), 1337–1338.

Satel, S. (2002). Medicine's race problem. *Policy Review*, 110, http://www.policyreview.
org/DEC01/satel.html.

Smedley, B., Stith, A., & Nelson, A. (2002). *Unequal treatment: Confronting racial and
ethnic disparities in health care.* Institute of Medicine of the National Academies
Report. Washington, DC: National Academies Press.

Solus, J. F., Arietta, B. J., Harris, J. R., Sexton, D. P., Steward, J. Q., McMunn, C., et al.
(2004). Genetic variation in eleven Phase I drug metabolism genes in an ethnically
diverse population. *Pharmacogenomics, 5* (7), 895–931.

Taylor, A. L., Ziesche, S., Yancy, C., Carson, P., D'Agostino, R., Jr., Ferdinand, K., et al.
(2004). Combination of isosorbide dinitrate and hydralazine in blacks with heart
failure. *New England Journal of Medicine, 351* (20), 2049–2057.

Wade, N. (2002, July 30). Race is seen as real guide to track roots to disease. *The New
York Times*, p. F1.

Waxman, H., et al. (2004, January 13). Letter from U.S. Representatives to DDHS Secretary Tommy Thompson on the National Healthcare Disparities Report.

White, H. (1973). *Metahistory. The historical imagination in nineteenth-century Europe.* Baltimore and London: Johns Hopkins University Press.

Yasar, U., Aklillu, E., Canaparo, R., Sandberg, M., Sayi, J., Roh, H. K., et al. (2002, November 1). Analysis of CYP2C9*5 in Caucasian, Oriental, and black-African populations.
European Journal of Clinical Pharmacology, 58 (8), 555–558.

CONTRIBUTORS

LAWRENCE D. BOBO is the Martin Luther King Jr. Centennial Professor at Stanford University. He also serves as director of Stanford's Center for Comparative Study in Race and Ethnicity and of the Program in African and African American Studies. He was formerly the Norman Tishman and Charles M. Diker Professor of Sociology and of African and African American Studies at Harvard University. His research interests include racial attitudes and relations, social psychology, public opinion, and political behavior.

DEBORAH A. BOLNICK is assistant professor of anthropology at the University of Texas at Austin. She received her PhD in anthropology from the University of California at Davis and is interested in contemporary American understandings of race, ethnicity, and genetics. Her research focuses on the patterns of human genetic variation and how they are shaped by culture, language, history, and geography.

MOLLY J. DINGEL is assistant professor of sociology at North Dakota State University in Fargo. From 2005 until 2007, she was a post-doctoral research fellow at the Mayo Clinic in Rochester, Minnesota. Her research interests center on the social, legal, and ethical dimensions of new medical technologies and research, with a special focus on behavioral genetics. Her current work investigates the legal and social issues surrounding addiction genomics research, including how that research intersects with racial categories, policy decisions, clinical applications, and our own sense of identity and disease.

JOHN DUPRÉ is professor of philosophy of science and director of the Economic and Social Research Council Centre for Genomics in Society at the University of Exeter, United Kingdom. He previously taught for 14 years at Stanford University and also at Birkbeck College, London. His most recent book is *Darwin's Legacy: What Evolution Means Today* (Oxford University Press, 2003), which has been translated into German and Spanish.

MARCUS W. FELDMAN is professor of biological sciences at Stanford University. He uses applied mathematics and computer modeling to simulate and analyze the process of evolution. He also studies the evolution of modern humans using models for the dynamics of molecular polymorphisms, especially DNA variants. He is a fellow of the American Academy of Arts and Sciences and of the California Academy of Science and is the author of more than 350 scientific papers and five books on evolution, ecology, demography, and mathematical biology.

DUANA FULLWILEY is an anthropologist of science and medicine and is assistant professor in the anthropology and African and African American studies departments at Harvard University. Her work explores intersections of identity and molecular genetics in the United States as well as in France and in Senegal, West Africa. She is currently completing a book entitled *The Enculturated Gene: Making Sense of Sickle Cell Difference in Modern Africa* (Princeton University Press, forthcoming).

DAVID B. GOLDSTEIN is director of the Center for Population Genomics and Pharmacogenetics and professor of molecular genetics and microbiology at Duke University. His research focuses on the clinical implications of human genetic population structure and methodological questions in quantitative genetics. He is a member of the editorial boards of *Current Biology, Molecular Biology and Evolution, Human Genomics*, and the *Annals of Human Genetics*. He received his PhD in population genetics from Stanford University in 1994.

HENRY T. GREELY is the Deane F. and Kate Edelman Johnson Professor of Law and professor, by courtesy, of genetics at Stanford University. He specializes in legal and social issues arising from advances in the biosciences and in health law and policy and chairs the California Advisory Committee on Human Embryonic Stem Cell Research and the steering committee of the Stanford University Center for Biomedical Ethics. He also directs the Stanford Center for Law and the Biosciences and the Stanford Program in Neuroethics.

SALLY HASLANGER is professor of philosophy in the department of linguistics and philosophy at the Massachusetts Institute of Technology; she also teaches regularly in the MIT women's studies program. Her publications have addressed topics in metaphysics, epistemology, and feminist theory, with a recent emphasis on feminist epistemology and topics in philosophy of social science. She is co-editor (with Charlotte Witt) of *Adoption Matters: Philosophical and Feminist Essays* (Cornell University Press, 2005), (with Elizabeth Hackett) of *Theorizing Feminisms* (Oxford University Press, 2005), and (with Roxanne Marie Kurtz) of *Persistence* (MIT Press, 2006).

JONATHAN KAHN is associate professor of law at Hamline University School of Law in St. Paul, Minnesota. His scholarship focuses on the history and politics of federal classificatory schemes, with particular attention to issues of race and biotechnology. He is currently working on a two-year project entitled "The Place of Race in Patenting and Drug Development." He holds a JD and a PhD in U.S. history.

RICK A. KITTLES is associate professor of genetic medicine at the University of Chicago. His research interests are in genetic ancestry and genetic and environmental effects on complex diseases and traits. His work on tracing the genetic ancestry of African Americans has brought his research into contact with issues of race, ancestry, identity, and group membership.

BARBARA A. KOENIG, an anthropologist who studies contemporary biomedicine, is professor of medicine at the Mayo Clinic College of Medicine in Rochester, Minnesota, and faculty associate at the Center for Bioethics, University of Minnesota. Koenig is one of a small number of anthropologists who works within the interdisciplinary field of bioethics. For 10 years she served as executive director of the Stanford University Center for Biomedical Ethics. Her research focuses on two areas: end-of-life care and the ethical, social, and political implications of new biomedical technologies, particularly those within the genomic sciences. Koenig's National Institutes of Health–funded research examines the ethical and policy implications of emerging knowledge in the genetics and neurobiology of addiction.

SANDRA SOO-JIN LEE is a senior research scholar at the Stanford Center for Biomedical Ethics. She received her undergraduate degree in human biology from Stanford University and her PhD in medical anthropology from the joint program at the University of California, Berkeley and San Francisco. Dr. Lee is the recipient of a Rockefeller Foundation Humanities Fellowship and a National Research Service Award and a Research Scientist Development Award from the National Human Genome Research Institute. Her forthcoming book focuses on the interpretation of difference in genomic sciences and its implications for understandings of race and justice.

SALLY LEHRMAN is an award-winning reporter who specializes in medicine and science policy reporting, with an emphasis on genetics, race, and sexuality. She has written for *Scientific American*, *Nature*, *Health*, the *Washington Post*, *Salon.com*, *The DNA Files*, served as science content editor for public radio documentary series, and is the author of *News in a New America*. She is also a recipient of a John S. Knight Fellowship, a shared Peabody/Robert Wood Johnson Award for excellence in health and medical programming, and the Columbia/Du Pont Silver Baton (both for *The DNA Files*).

RICHARD C. LEWONTIN is Alexander Agassiz Research Professor at Harvard University. His research has concentrated on both experimental and theoretical studies of genetic variation within and between populations, including humans. He is a member of the American Academy of Arts and Sciences. His book *The Genetic Basis of Evolutionary Change* is a classic in the field of population genetics (Columbia University Press, 1974).

JONATHAN MARKS is professor of anthropology at the University of North Carolina, Charlotte. His primary area of research is molecular anthropology—the application of genetic data to illuminate our place in the natural order—or more broadly, the area of overlap between (scientific) genetic data and (humanistic) self-comprehension. He is the author of *Human Biodiversity* (Aldine de Gruyter, 1995) and *What It Means to Be 98% Chimpanzee* (University of California Press, 2002).

ALONDRA NELSON is assistant professor of sociology, African American studies, and American studies at Yale University. Her research areas include the sociocultural implications of genetics, African American health advocacy, and racial formation processes in biomedicine and technoculture. She is the co-editor (with Thuy Linh N. Tu) of *Technicolor: Race, Technology, and Everyday Life* (New York University Press, 2001) and author of *Body and Soul: The Black Panther Party and the Politics of Health and Race* (University of California Press, forthcoming).

JENNY REARDON is assistant professor of sociology at the University of California, Santa Cruz, and adjunct research professor of women's studies at Duke University. She is the author of *Race to the Finish: Identity and Governance in an Age of Genomics* (Princeton University Press, 2005).

SARAH S. RICHARDSON is a doctoral candidate in the Program in Modern Thought and Literature at Stanford University. Her research focuses on philosophical approaches to modeling the social dimensions of scientific knowledge, theories of science and democracy, and race and gender in the contemporary biosciences.

PAMELA SANKAR is a medical anthropologist and associate professor of bioethics in the department of medical ethics at the University of Pennsylvania. Her current work examines how medical and forensic genetics researchers use and define race.

MARK D. SHRIVER is associate professor of anthropology and genetics at Penn State University, where he works on the genetics of normal and disease variation. He works on admixture and admixture mapping and is a leader in the development of this technology for identifying genes for complex traits. His recent work is on the genetics of skin pigmentation.

JACQUELINE STEVENS is associate professor at the University of California at Santa Barbara. Her recent work focuses on the political-theoretical dimensions of state health policies in an era of molecular medicine. She is the author of *Reproducing the State* (Princeton, 1999).

KIMBERLY TALLBEAR is assistant professor of science, technology, and environmental policy in the department of environmental science, policy and management (ESPM) at the University of California, Berkeley. TallBear is a member of the Sisseton-Wahpeton Oyate in South Dakota and is also descended from the Cheyenne and Arapaho Tribes of Oklahoma.

SARAH K. TATE is a postdoctoral researcher in the department of clinical and experimental epilepsy at University College London. She received her PhD in pharmacogenetics from University College London in 2005. Her research focuses on genetic predisposition to epilepsy and the pharmacogenetics of anti-epileptic drugs.

INDEX

Note: Page numbers followed by a *t* indicate tables;
page numbers followed by an *f* indicate figures.